Ultra Precision Bearings

Related titles:

Eleventh European fluid machinery congress
(ISBN 978 0 85709 091 1)

Solving tribology problems in rotating machines
(ISBN 978 1 84569 110 3)

Tribology and dynamics of engine and powertrain
(ISBN 978 1 84569 361 9)

Ultra Precision Bearings

Frank Wardle

ELSEVIER

AMSTERDAM • BOSTON • CAMBRIDGE • HEIDELBERG • LONDON
NEW YORK • OXFORD • PARIS • SAN DIEGO
SAN FRANCISCO • SINGAPORE • SYDNEY • TOKYO
Woodhead Publishing is an imprint of Elsevier

WP
WOODHEAD PUBLISHING

Woodhead Publishing is an imprint of Elsevier
80 High Street, Sawston, Cambridge, CB22 3HJ, UK
225 Wyman Street, Waltham, MA 02451, USA
Langford Lane, Kidlington, OX5 1GB, UK

British Library Cataloguing-in-Publication Data
A catalogue record for this book is available from the British Library

Library of Congress Control Number: 2014949584

ISBN 978-0-85709-162-8 (print)
ISBN 978-0-85709-218-2 (online)

For information on all Woodhead Publishing publications
visit our website at http://store.elsevier.com/

Typeset by RefineCatch Limited, Bungay, Suffolk

'Picoace is a grinding machine featuring an ultra precision air spindle, hydrostatic rotary table, slides and lead screws to achieve optical quality finishes and flatness's on silicon wafers and other brittle materials used in the electronics industry.'
Courtesy of LoadPoint

Contents

List of figures and tables

Figures

Tables

Preface

The material presented in this book is based on the author's past experience, initially in the ball bearing industry and later in the air bearing and machine tool industries. In developing high- and ultra-precision bearings and machinery it became clear there was very little in the way of guidance for machine designers needing to achieve extreme precision or low motion errors. Yet over the last four or five decades this has become an increasingly important aspect of performance relevant to a growing number of new bearing applications in the computer, electronics and optical industries.

Traditionally, books on bearings have tended to emphasise the load carrying capacity and/or life aspect of performance where the failure criteria are generally related to wear, fatigue or seizure mechanisms. In ultra precision machinery, running accuracy, stiffness, damping and heat generation are more relevant bearing performance parameters, and bearing life may be limited by deterioration in any of these. As a result, this book primarily aims to provide a basic understanding of the mechanisms upon which these parameters depend for a range of ultra precision bearing types. The mechanisms identify most, if not all, of the influencing variables and at worst, enable the machine designer to make an informed judgement about a particular design scenario. But in many cases the mechanisms are used to develop reasonably accurate analytical formulae that enable the bearing or machine performance to be determined and optimised at the design stage.

In real machinery the freedom to select or design bearings is usually restricted by practical constraints such as, for example, space, aspect ratio, ease of assembly and access. In addition, the bearing design process is often a compromise requiring a trade off between dissimilar parameters such as stiffness and heat generation. Thus, in many applications, the challenge is to find the most suitable bearing rather than a mathematically optimised design. For this reason the preferred approach, adopted in this book, is to provide the designer with a calculation that they can use for their situation rather than standardised design charts.

A running accuracy of 0.1 μm peak-peak is used as a criterion to distinguish ultra precision bearings from other types. Effectively this limits the book to the rolling bearings – single-row radial and angular contact ball bearings; hydrostatic and aerostatic externally pressurised fluid film bearings; hydrodynamic and aerodynamic herringbone and spiral groove fluid film bearings. For the fluid film bearings, the analysis is focused on the simple geometries, circular or annular pad and cylindrical journal that are most easily manufactured to high precision and are hence most widely used. But the basic principles apply equally to conical and spherical bearing forms.

Encouragement to write the book has come from colleagues in the precision machining industry, friends and family, particular thanks go to Professor Eric Marsh of Penn State University for many helpful comments and suggestions. Acknowledgements also go to my son David, who has programmed many of the bearing calculations and my daughter Amanda who has helped with illustrations and diagrams. Last but not least is my wife Vivien, without whose patience and understanding, the book would not have been completed.

About the author

Frank has 40 years experience in an industrial research environment, gained in the ball bearing, air bearing and precision machine tool industries.

Twenty years was spent as a research manager in the ball bearing industry, focusing on the development of special purpose bearing products, such as silent running bearings for military applications, high speed bearings for aircraft engines and high precision bearings for machine tools. Reduction in bearing noise and motion errors resulted from wide ranging improvements to manufacturing processes and control over application conditions, where he gained first-hand experience of manufacturing machinery and test equipment alike. He was given a best paper award by the I. Mech. E in 1988, for his work on quiet running bearings and another best paper award by the I. Mech. E's Tribology Group in 1989, for pioneering the use of ceramic balls in machine tool bearings.

Twelve years was spent as Technical Director with Loadpoint Ltd, developing air bearing products, manufacturing processes and precision machinery. Air bearing developments covered both aerostatic and aerodynamic bearing types. The former was related to diamond turning, grinding and disc verification applications, where low motion error was an important requirement, whilst the latter was aimed at optical scanning applications. He developed a novel electrochemical micro-machining process to machine the helical and spiral grooves in aerodynamic bearings and in the mid-1990s this was used to support hydrodynamic bearing developments in the computer hard disc drive industry. The process is now widely used by the hard disc drive and medical industries for fluid film bearing manufacture. Machine tool developments included a diamond turning lathe and a silicon wafer grinding machine.

Frank started Ultraprecision Motion Ltd in 2006, continuing industrial research and development in machine tool systems. Developments to date include Ultramill, an ultra precision five-axis micro-milling machine, micro-recessed air bearings for improving the stiffness and accuracy of ultra precision machine tool slideways and rotary tables, and Tempconet, an intelligent and adaptable machine tool control system.

1

Introduction

DOI: 10.1533/9780857092182.1

Abstract: A motion error criterion is used to distinguish ultra precision bearings from standard precision bearing types. In this chapter, bearing motion error is defined and methods of measurement explained. The bearing types that can meet a low motion error criterion are given and other performance parameters of importance to ultra precision machinery are outlined.

Key words: motion error, bearing vibration, static stiffness, damping, bearing friction, bearing power loss.

1.1 Benefits of ultra precision bearings

The key feature that distinguishes ultra precision bearings from other types of bearing is accuracy of rotation or in the case of linear bearings, straightness of motion. In a large majority of bearing applications this aspect of performance is unimportant and is usually neglected, but over the last four or five decades there have been a growing number of new and emerging applications where accurate bearing motion is critical to product performance. These are generally associated with the electronics, computer, optical, instrumentation and medical industries, where advances in micro-electronics have led to the development of new and innovative products or miniaturisation has demanded more accurate manufacturing processes and measuring equipment. In addition, research into traditional machining processes such as turning and grinding have not only led to a better understanding of the factors that influence surface integrity, form and finish but also the true importance of these parameters on surface functionality. Friction levels and wear rates in sliding contacts, peak stress levels in load bearing contacts, distortion and reflectance of optical surfaces are, for example, all critically dependent on the geometry and finish of component surfaces.

Advances in the micro-electronics industry over the last four decades have been considerable. Of central importance has been the miniaturisation and integration of electronic circuits and the development of the manufacturing processes that have enabled this to happen. Circuits containing literally millions of transistors have been reduced in area to a few square centimetres and are built up or etched into silicon wafers. Wafers, 300 or 400 mm diameter, are cut from single crystal billets and ground down to as little as 50 μm in thickness. As silicon is a brittle material and prone to generating micro-cracks during machining, wafer grinding machines, referred to as back grinders, rely upon ultra precision bearings to rotate the grinding wheel and the wafer. Within this industry there is a clear correlation between risk of micro-cracking and bearing precision and there has been a trend for machine manufacturers to move away from precision ball bearings on to air bearings, achieving the highest possible motion accuracy of all bearing types.

A single wafer may contain thousands of circuits and these are separated out by dicing. Dicing machines are essentially high precision saws where a diamond grinding wheel some tens of microns wide is rotated at high speed on an ultra precision spindle and used to cut through the wafer between the rows of circuits. Again there is a clear correlation between the incidence and size of edge chipping and spindle precision. Almost all spindles used for dicing silicon now employ ultra precision air bearings. Other materials used by the electronics industry, such as germanium and sapphire for LED arrays, are at least as brittle as silicon and require similar manufacturing processes and precision.

The computer industry has also seen huge advances since the 1970s. Along with massive increases in computer speed and memory size, storage media have increased in capacity from just a few tens of kilobytes to a terabyte or more. On hard disc drives, data is written to and read from rotating discs via electronically controlled heads. The data is stored in a series of closely spaced 'circular' tracks easily accessed by the heads and track spacing is a critical parameter directly determining the storage capacity of the disc. As disc rotation relies upon bearings, their motion accuracy largely determines the track spacing. Clearly tracks cannot be allowed to cross, so the spacing must accommodate the largest expected motion errors. Driven by the need to increase storage capacity the disc drive industry has made large advances in bearing precision, initially with ball bearings and more recently with fluid film bearings. Ball bearing motion errors as low as 50 nm have been reported in the technical press but even so the switch to fluid film bearings towards the end of the 1990s led to a four-fold increase in disc capacity.

The optical industry has also experienced major advances in manufacturing technology. Whereas lapping and polishing operations were used on a vast majority of lenses and mirrors four decades ago, now this is limited to a relatively low number of specialised components. Instead lenses may be turned directly or moulded from turned dies, requiring no further finishing and therefore giving substantial benefits in manufacturing cost, precision and flexibility. This has spawned a complete industry of ultra precision machine tool manufacturers and is still driving the standards for motion error of ultra precision bearings. Modern-day ultra precision lathes and milling machines frequently achieve motion errors of 10 nm or less using high-quality aerostatic or hydrostatic spindles.

With the advances described in the preceding paragraphs and with modern ultra precision bearings achieving extremely low motion errors, it is clear that there is a demand for metrology equipment such as roundness testers and coordinate measuring machines that can provide an ultra precision measurement capability. As these machines require moving axes to perform their measurements, so too do they require ultra precision bearings. Aerostatic bearings have always been the preferred choice for the linear axes on the most accurate coordinate measuring machines and are now becoming widely used on roundness testers gaining preference over hydrodynamic bearings. Other types of instrumentation requiring ultra precision motion include goniometers used for scientific experiments with X-rays or light.

Micro-machining is a relatively new field of application for ultra precision bearings. It is emerging as a sizeable industry in its own right and embraces the requirements of the electronic component, medical and transducer industries, where there is widespread demand for small, complex three-dimensional (3-D) components and for components with microscopic surface features. Figure 1.1 is an example of micro-milling performed on an ultra precision micro-milling machine using a high-speed air spindle fitted with a diamond tool. This was manufactured as part of a research project investigating the effect of using onboard liquid cooling in integrated circuits – an example of a possible future application for ultra precision micro-machining. Many other types of miniature components are still under development and may be expected to increase the size of the micro-machining market still further in the future.

The above is not intended to be a comprehensive list of ultra precision bearing applications but merely to outline some of the main applications

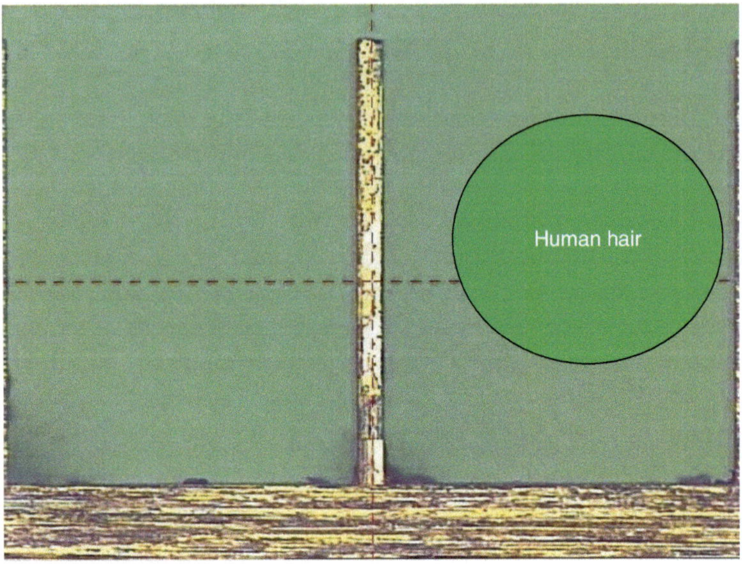

Figure 1.1 Micro-milled upstand – 15 μm wide (courtesy of Brunel University)

and some of the benefits they can offer. It is an increasingly important aspect of bearing performance and one that is not adequately covered by presently available technical publications. Thus, this book is focused on ultra precision bearings and aims to give the reader a basic insight into types of ultra precision bearings currently available, the factors that influence their running accuracy and other aspects of their performance that are important in ultra precision machinery; thereby enabling the machine designer to make the most suitable bearing selection and define the optimum set-up in any given application. Not all ultra precision bearing types are covered, as the focus is on rolling element and fluid film bearings that have so far found widespread use, namely:

- single row radial ball and angular contact bearings;
- herringbone and spiral groove hydrodynamic bearings;
- herringbone and spiral groove aerodynamic bearings;
- hydrostatic bearings;
- aerostatic bearings.

Bearings are required to enable both rotary and linear motion. In this text only hydrostatic and aerostatic bearings are considered to meet the standards of linear motion error normally required of ultra precision machinery. Another notable absence from the book is magnetic and other types of active bearings that to date have found very limited application in the ultra precision field. In the research laboratory these have shown encouraging results in terms of low motion errors and dynamic stability, but applications have so far been highly specialised.

1.2 A systems approach

Ultra precision bearings usually form part of a system and it is the overall system performance that is important. There is little point in using ultra precision bearings if other parts of the system are not of a matching precision. For example, the low motion errors of an ultra precision bearing are unlikely to be achieved if the shaft it supports is driven via noisy machine elements such as belts and pulleys or gears. For this reason ultra precision bearings tend to be integrated into ultra precision sub-systems such as spindles and rotary tables or in the case of linear bearings, slide units. This also has the added advantage of reducing the amount of ultra precision engineering needed by the end user. A large majority of modern sub-systems will include some form of direct drive, usually a precision electric motor in addition to other components specific to function. For example, the machining spindle shown in Figure 1.2 comprises an aerostatic bearing arrangement, in this case two journal and one double-sided thrust bearings, a DC brushless electric motor, a tool and tool holding arrangement. From an ultra precision view point, the performance of the spindle is specified in terms of:

- the motion error at the tool tip over a range of speeds;
- by how much the tool deflects under a given cutting load–static stiffness;
- by how much the tool tip changes in position with time–thermal growth;
- dynamic stiffness or damping. Specifying motion error over a range of speeds encompasses the dynamic characteristics of the spindle – how well it is balanced, whether there are critical speeds (resonances) in the speed range and if so, how well they are damped;
- bearing load capacity does not have a direct influence on precision but must be taken into account in selecting bearing size, particularly for aerostatic, aerodynamic and hydrodynamic bearing types for which load capacities are limited.

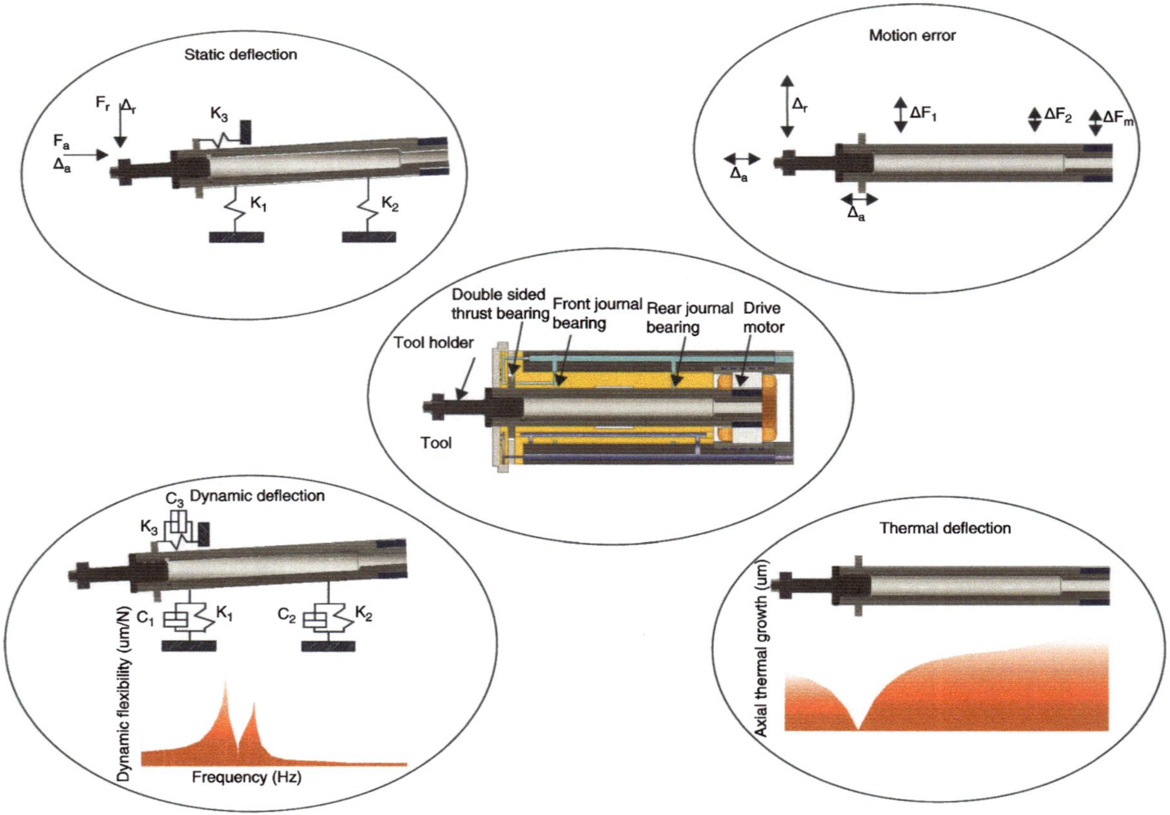

Figure 1.2 Ultra precision bearing system

For the spindle of Figure 1.2, motion error measured at the tool tip is still mostly due to motion error of the bearings, but it may also include some contribution from the drive motor. In addition to generating a steady drive torque, even the most precise electric motors still generate small dynamic magnetic forces, the frequency of which depends on motor speed and the number of poles. The forces are minimised by the manufacturer's attention to the motor's internal design and by the use of a high stiffness bearing arrangement.

The system design and manufacturing precision also influences spindle motion errors. Whilst imperfections on the bearing surfaces may be expected to be a primary source of motion error, factors such as bearing arrangement, alignment and fits and tolerances may induce bearing motion error or at the very least influence the levels of motion error generated by given bearing surface imperfections. For example, increasing the overhang of the tool from the front bearing or reducing bearing spacing increases the system leverage, in turn amplifying the motion error at the tool position for given bearing motion errors. Similarly, misalignment or distortion of bearing housing components may increase the motion errors inherent in the bearings. It follows that it is not only the manufacturing quality of the bearings that is important but also the system design and bearing application details.

1.3 Bearing motion error

1.3.1 Causes

Measured in terms of displacement ultra precision bearings have peak–peak motion errors typically in the range 10 to 100 nm and can be influenced by many subtle effects. In all of the bearing types included in this book the causes of motion error are attributed to one of three sources:

1. *Manufacturing quality*: namely the size and distribution of imperfections on the critical load bearing surfaces;
2. *Bearing design*: how effective the bearing is in attenuating or amplifying the motion error generated by a given size of surface imperfection;
3. *External influences*: how system and application factors such as alignments and applied loads affect bearing motion error.

No surface is perfectly flat or perfectly round. Bearing surfaces are produced by any of several conventional manufacturing processes – turning, milling, grinding, honing or lapping – and as a result contain the surface geometry and roughness errors characteristic of these processes. To put this in perspective, an average surface roughness (R_a) of 10 nm from lapping or honing is generally considered excellent within the bearing industry; however, an average roughness of 10 nm corresponds to a peak to valley height (R_t) of typically 100 nm, around the upper limit of the motion error expected of ultra precision bearings. Interestingly enough the surface geometry characteristic produced by turning, milling and grinding is highly dependent on the bearings used in these machines. Thus the very best bearings are required to produce the very best bearings – a situation that almost all bearing manufacturers are aware of but not all put into practice.

All bearing designs are prone or sensitive to particular mechanisms that produce motion errors. For example, ball bearings have a discrete number of balls and it is easy to envisage that in a bearing subject to radial load, the approach of the inner ring with respect to the outer ring will vary cyclically as balls pass through the load vector. Similarly, hydrodynamic and aerodynamic bearings are prone to an instability known as 'half speed whirl', whilst aerostatic bearings may suffer from another type of instability frequently referred to as 'pneumatic hammer'. These mechanisms are generally well understood and in almost all cases can be avoided or minimised through fine tuning of bearing design and matching of the design to application requirements.

There are a wide range of external effects that can influence bearing motion errors. Factors such as contamination, misalignment and distortion of bearing surfaces have an intuitively obvious effect but there are many application related variables that have a significant but not so obvious effect on motion error. A basic awareness of these is useful in optimising the bearing set-up and keeping motion errors to a minimum.

1.3.2 Types of motion error

Motion error parameters usually measured on ultra precision bearings and sub-assemblies relate to the applications for which the bearings are used. For rotational motion, a fundamental distinction is made between motion errors that are harmonic of the basic rotor speed–synchronous motion error and those that are not - asynchronous motion error.

An example of the significance of these two types of motion error is shown in Figure 1.3, where the outside diameter of a cylindrical work piece is being turned on a lathe. In Figure 1.3(b) the spindle supporting the work piece is assumed to have only asynchronous motion error of a frequency much higher than the spindle's speed of rotation. The error becomes machined into the work piece surface in the form of waviness, but as it is asynchronous, with spindle speed the waviness generated in successive revolutions is shifted in phase leaving a surface with a 'rough' appearance. Figure 1.3(a) shows the surface generated when the spindle motion error is synchronous with spindle speed, and as the waviness is now in phase for successive revolutions, the surface takes on a smoother, lobed appearance. The difference between the two types of surface is highly significant for many precision components. The rougher surface has less ability to generate lubricant films in tribological components and gives poorer light transmission and reflectance in optical components. Furthermore, if the cylindrical component was a grinding wheel being subjected to single point dressing, the lobing produced by synchronous motion error of the spindle would compensate for the motion error so the surface of the wheel would appear to be of constant radius when viewed at any one angular position. This cannot occur with asynchronous motion error, which is dressed into the wheel, increasing its 'roughness' and reducing its ability to generate a good finish in subsequent grinding operations.

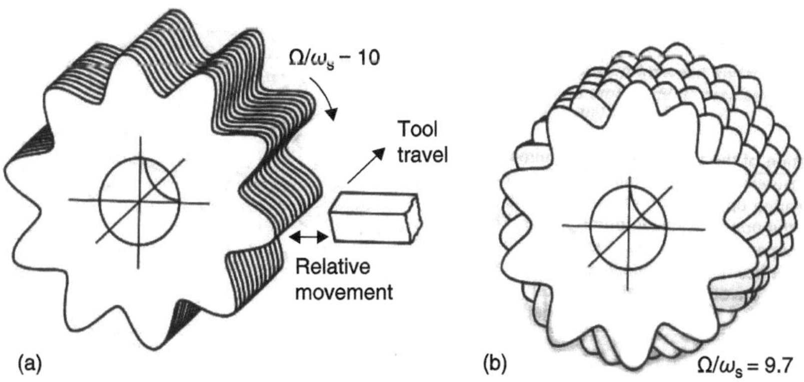

Figure 1.3 **Effect of (a) synchronous, and (b) asynchronous motion error on surface form**

1.3.3 Measurement of motion error

Throughout the bearing industry there are a number of tests and specifications that relate to motion error. In this text the parameters and notation defined in the specification ASME B89.3.4 [1] is used for bearings that undergo rotational motion. Although this specification is aimed at spindles, it defines motion error parameters that are relevant to many ultra precision bearing applications, which have absolute values and may therefore be independently verified; and are amenable to practical measurement. For many types of fluid film bearings, motion error measurements are in any case only practical on the assembled system of which they are part. However, ball bearings can be tested individually, but the parameters and measurement method defined in ASME B89.3.4 are still relevant and used within the ball bearing industry as a quality control check on some types of critical use bearings.

Motion error is the relative movement between the stationary and rotating parts of a bearing and in general has five components. In Figure 1.4, a bearing is shown relative to a fixed Cartesian Coordinate

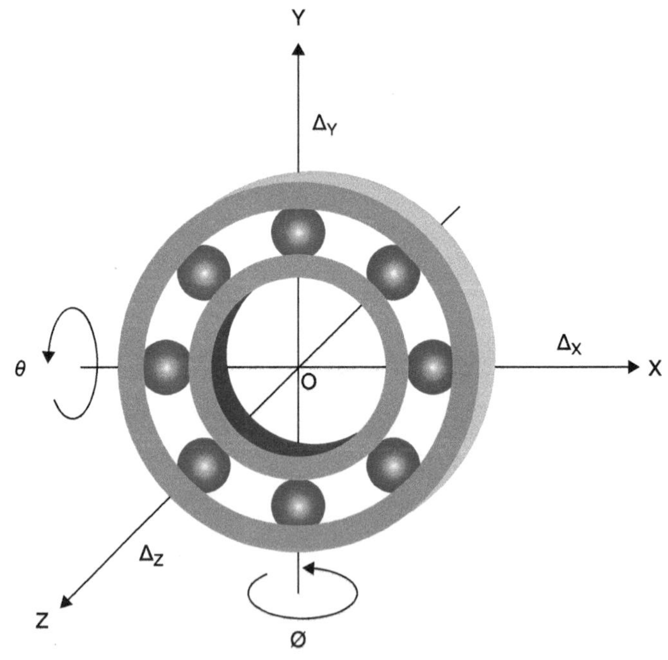

Figure 1.4 Notation and components of motion error produced by rotating bearings

Frame OXYZ. One ring of the bearing is stationary and the other rotates at constant speed about the axis OZ. Either the outer ring or the inner ring may rotate but as a result a small amplitude relative motion between the rings is generated that has three linear components – one axial is Δ_Z, two radial is Δ_X and Δ_Y and the two tilt components are θ, about the OX-axis and ϕ about the OY-axis.

1.3.4 Motion error parameters

A comprehensive classification of motion errors, the methods of measurement and data analysis is given in Marsh [2]. Here, the motion error parameters most commonly found in practice are described.

Figure 1.5 shows the basic measurement concept used to determine the motion error of a ball bearing. Its inner ring is mounted on a mandrel supported by an accurate spindle running at constant speed. The bearing is subject to a light axial load and a measurement sensor, normally a

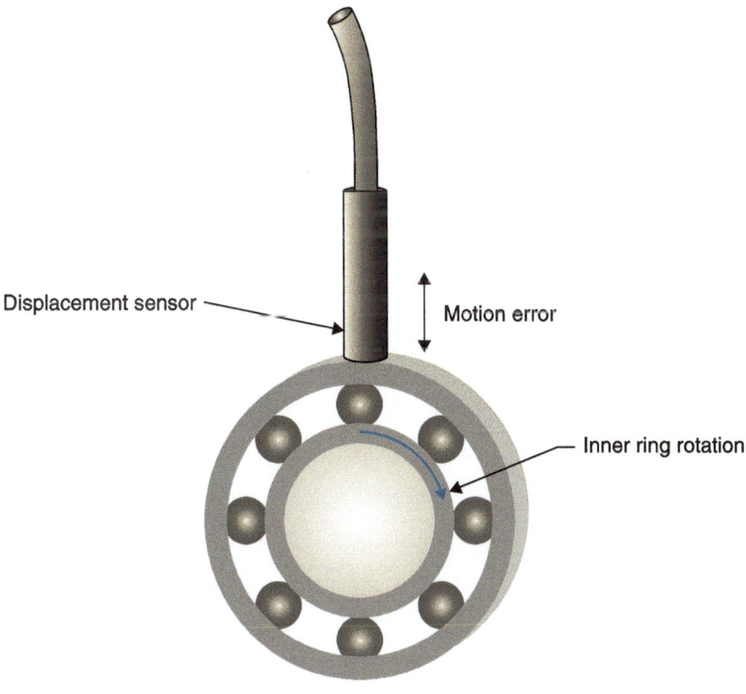

Figure 1.5 Measurement of ball bearing motion error

displacement sensor such as a capacitance gauge is either positioned with its axis perpendicular to the bearing's axis of rotation to pick up radial motion error or parallel to the axis of rotation if axial or tilt motion errors are required. The measurement is performed over a time period corresponding to several, for example, five revolutions of the bearing and is subsequently displayed as a polar plot of motion error amplitude versus angle of rotation.

Figure 1.6 shows an example of such a measurement performed in a radial direction. This plot shows the raw data recorded by the sensor and is referred to as the 'Total Motion Error'. It contains all of the parameters of interest to ultra precision bearing applications, each of which can be extracted from Figure 1.6 and quantified by applying suitable signal processing techniques. Figures 1.7(a) and (b) show the synchronous and asynchronous components of Figure 1.6 respectively. That is to say, Figure 1.7(a) contains all motion error frequencies that are harmonic of the rotational speed of the tested bearing, whilst Figure 1.7(b) contains

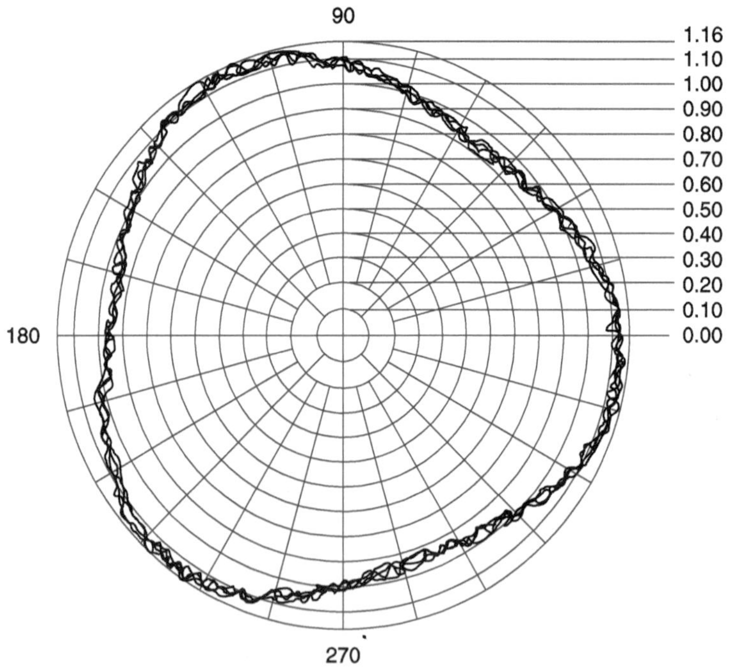

Figure 1.6 Total motion error

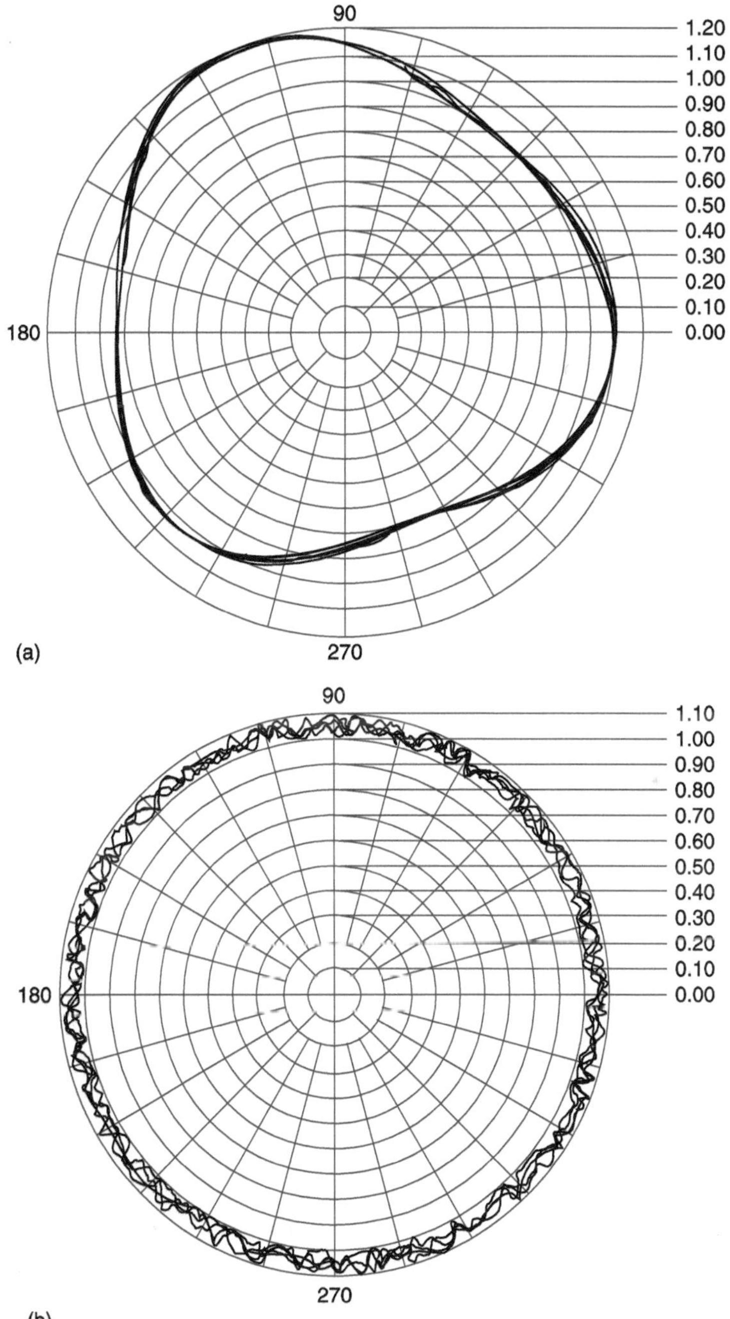

Figure 1.7 (a) Synchronous motion error; and (b) asynchronous motion error

all other motion error frequencies. Appropriately these parameters are known as 'Synchronous Error Motion' and 'Asynchronous Error Motion'.

Particularly for radial motion error measurements, the synchronous error motion shown in Figure 1.7(a) is commonly separated into two components, a 'Fundamental Error Motion' and a 'Residual Error Motion' (Figures 1.8(a and b)).

The Fundamental Error Motion contains only one frequency corresponding to the speed of rotation of the bearing. For radial measurements it may, for example, represent a basic eccentricity somewhere in the bearing or test set-up and is not always significant. In Figure 1.5, the test arbour is designed to have all types of motion error to be negligible compared to typical measured values. However, basic eccentricity between the bearing's inner ring and the test arbour is likely to generate a significant Fundamental Error Motion, so in this case it is not a good indication of bearing quality and may be discarded. On the other hand, the Residual Motion Error shown in Figure 1.8(b) is a far better indicator of bearing running accuracy.

The measurements in Figures 1.6 to 1.8 may be quantified by a number of different parameters but it is commonplace to find the 'Least Squares' best fit centre and then to place Maximum Inscribed and Minimum Circumscribed circles on the measurement. The radial separation of the centres of the circles then represents a 'Peak to Peak' value of the error motion.

Fluid film bearings are usually measured when assembled as part of a system such as a spindle. In this case the displacement sensor is mounted over a rotating target surface attached to the shaft supported by the bearings. However, the displacement sensor now measures not only the motion error of the shaft but also any geometrical errors present in the target surface. For a radial error motion measurement, out of roundness of the target surface is recorded along with the shaft's motion error. As the motion error of ultra precision fluid film bearings may be comparable in magnitude to the geometrical errors in the most accurately manufactured target surfaces then a correction must be made. A measurement procedure was proposed that involved two measurements, one with the target surface fixed at some angular position relative to the shaft and the second with it moved through 180 degrees. The procedure is attributed to Donaldson and described in Marsh [2]. By adding or subtracting the two measurements, geometrical errors on the target surface can be separated from shaft motion errors.

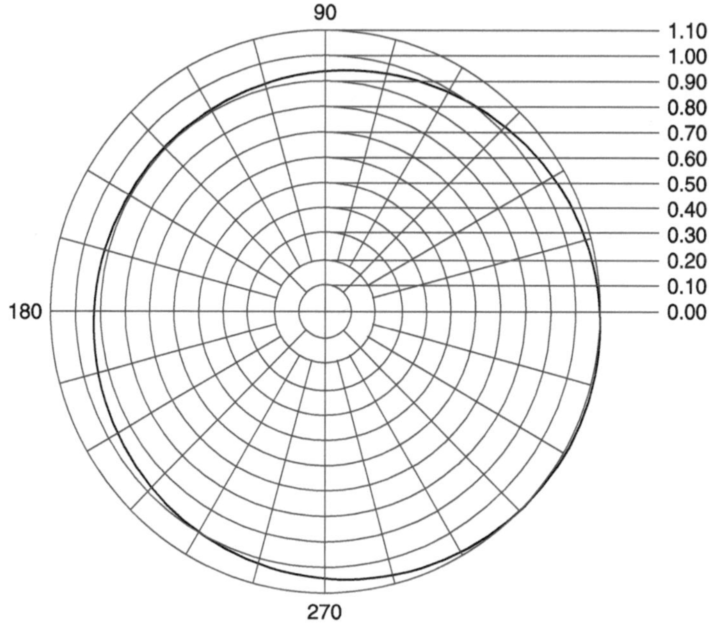

(a)

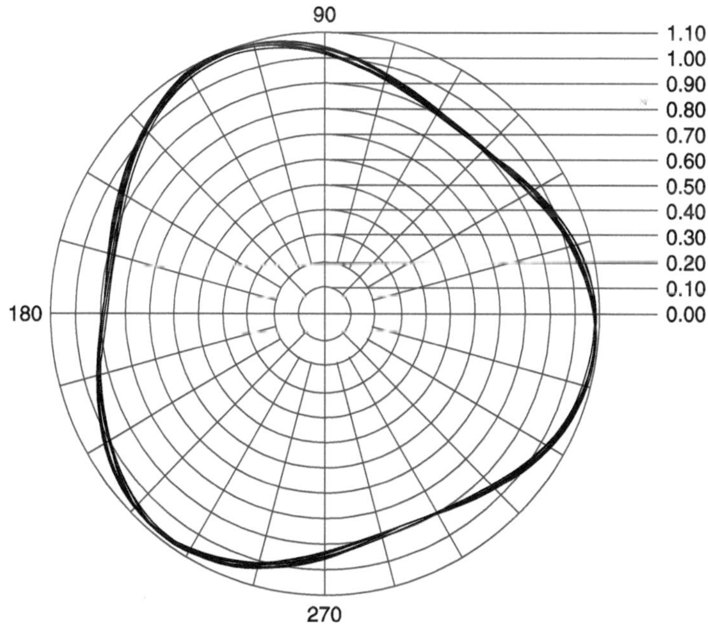

(b)

Figure 1.8 (a) Fundamental error motion; and (b) residual synchronous error motion

1.3.5 Frequency analysis

Frequency analysis provides a useful way of breaking down arbitrary motion error signals into a series of simple harmonic terms. It is based upon Fourier analysis and is very efficiently implemented in digital signal processing systems as a Fast Fourier Transform (FFT) [3]. Figure 1.9 shows an example of a displacement–time signal recorded on a rotating system and the corresponding frequency analysis. In this case the complex time signal is seen to contain several series of frequencies. In this example, a rotating system with a shaft speed of 12.5 Hz was monitored. This frequency and higher-order harmonics at 37.5, 62.5 and 75 Hz are clearly visible. Many other frequencies relating to ball bearings and drive motor are also present. In general, rotating systems will exhibit motion errors containing several series of harmonics, each of which relate to different aspects or components of the system. The main benefit of frequency analysis is therefore to obtain diagnostic information with which to identify the likely sources of motion error and to help reduce their amplitude should they be unacceptable.

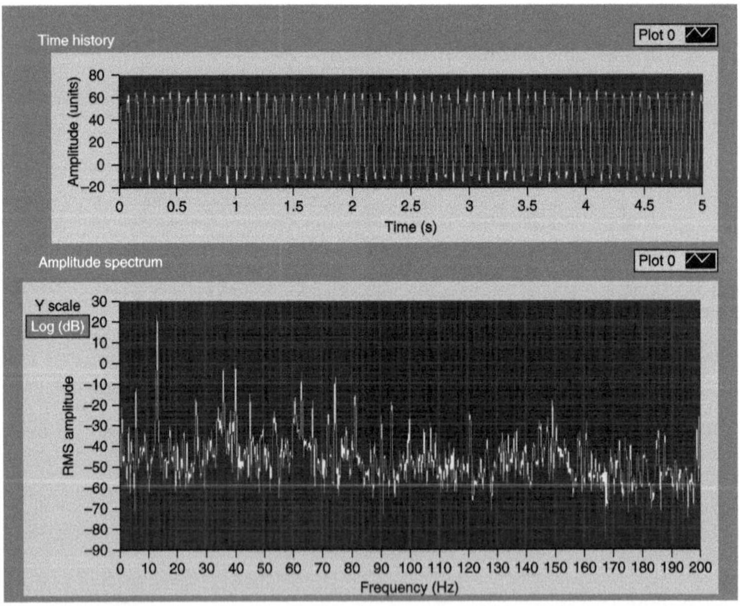

Figure 1.9 Displacement time history and frequency analysis

One of the purposes of this book is to define transfer functions between particular surface imperfections on bearing surfaces and the resulting motion error signatures. To consider both arbitrary surface imperfections and arbitrary motion error signatures to be made up of a series of simple harmonic terms in line with Fourier analysis provide a convenient way of deriving the required relationships. Furthermore, once the transfer functions are defined, they can be used to aid interpretation of motion error signals in the frequency domain.

1.3.6 Vibration measurement in the ball bearing industry

Within the ball bearing industry, the motion error parameters described earlier are confined to a few high precision bearing types. However, almost every ball bearing, regardless of its precision or application, is subjected to a simple vibration test as part of quality control procedures. This test was originally developed in the 1950s in collaboration with the US Military in order to define low vibration bearings suitable for use in life support systems on nuclear submarines. The test was found to be very sensitive to a wide range of bearing faults and has since become universally adopted within the ball bearing industry. It is also sensitive to the features that create motion error and for this reason is briefly described here. A full specification of the test can be found in [4].

The inner ring of the test bearing is mounted on a mandrel attached to a precision spindle and the outer ring is subjected to an axial load, as for the bearing shown in Figure 1.5. The spindle has fluid film bearings designed to have low vibration limits compared to those being measured and rotates at a constant speed of 1800 rpm. The sensor is a velocity sensor and is mounted with its axis in a radial plane with the stylus contacting the outer ring of the bearing. The sensor signal is conditioned to display RMS velocity in three frequency bands, 50–300, 300–1800 and 1800–10 000 Hz. Permissible vibration limits are defined in [4]. The permissable limits were developed in the 1960s and continuous developments within the ball bearing industry now mean that these are easily surpassed.

The test provides a relative measure of vibration and any correlation with absolute vibration or motion error in any given application is empirical. Nonetheless, the test is effective in picking out any departure from normality and the low and medium frequency bands in particular are sensitive to the parameters that dominate motion error.

1.4 Static stiffness

1.4.1 Rotating bearings

The static stiffness characteristics of ultra precision bearings are critically important in machine design. For example, machine tools are required to meet a loop stiffness criterion such that under cutting conditions, the cutting force does not produce a relative deflection between tool and work piece that exceeds the machine's specified accuracy. There are many components and sub-systems in the loop and all will contribute more or less to the total deflection between tool and work. In a majority of cases it is the work or tool holding spindle that is the weak link and produces the largest single component of deflection. Furthermore, spindle deflection is largely determined by the choice of bearings and their arrangement.

Bearing stiffness also has a critical influence on a machine's dynamic performance. Resonant frequencies occurring in machine sub-systems such as slides and spindles may result in relatively high levels of vibration that in the case of a machine tool will create poor work geometry or surface finish. Similarly, in measuring equipment resonances may be excited by extraneous vibration limiting their resolution and measuring accuracy. Common practice is to tune resonances to as high a frequency as possible in an attempt to minimise their effect on machine performance. Since resonances in slide systems and critical speeds of spindles are dependent upon bearings, this means maximising the bearings' static stiffness.

Static stiffness relates the static loads applied to a bearing to the resulting deflections. Bearing load–deflection characteristics can vary widely depending on bearing type and the conditions under which it is used. Figure 1.10 shows some typical relationships. Aerostatic bearings can be near linear over most of their permissible load range but hydrodynamic bearings and ball bearings exhibit non-linear load–deflection relationships with stiffness generally increasing with load.

In a large majority of ultra precision bearing applications, it is the small amplitude stiffness that is of interest and this is defined to be the slope of the load–deflection curve at a specified level of load or deflection. For example, in Figure 1.10, stiffness is defined at a non-dimensional load, L, and is therefore given by the gradient dP/dX at L. In general, a detailed knowledge of the bearing's static load–deflection characteristics is required in order to determine its small amplitude stiffness at the designed operating condition.

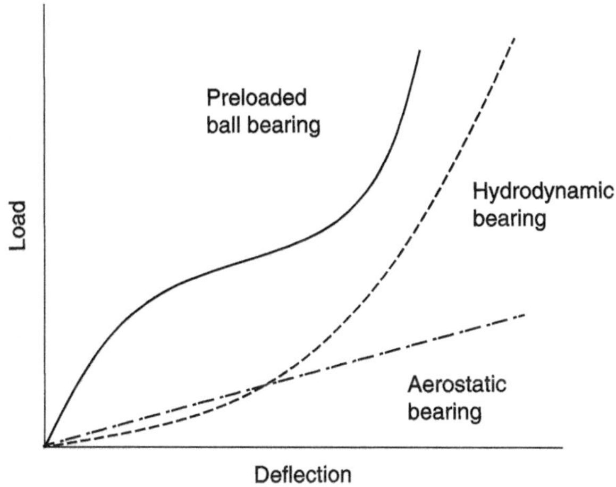

Figure 1.10 Hypothetical bearing load-deflection characteristics

The notation used to define bearing stiffness is shown in Figure 1.11; for convenience a ball bearing is shown but the notation applies equally well to fluid film bearings. OXYZ is a fixed Cartesian Coordinates Frame of reference and OZ is the axis about which the bearing rotates. The axes OX and OY are mutually orthogonal and contained in a radial plane normal to OZ. Forces F_x, F_y and F_z are applied along the OX, OY and OZ axes respectively while moments M_θ and M_ϕ are applied about the OX and OY axes respectively. The forces and moments are small values that add to or subtract from the applied bearing loads and they result in linear deflections x, y and z and angular deflections θ and ϕ that also increment to the deflections produced by the applied bearing loads.

The static stiffness of the bearing is defined to relate the applied forces and moments to the deflections and tilts such that:

$$\begin{bmatrix} F_x \\ F_y \\ F_z \\ F_\theta \\ F_\phi \end{bmatrix} = \begin{bmatrix} K_{xx} & K_{xy} & K_{xz} & K_{x\theta} & K_{x\phi} \\ K_{yx} & K_{yy} & K_{yz} & K_{y\theta} & K_{y\phi} \\ K_{zx} & K_{zy} & K_{zz} & K_{z\theta} & K_{z\phi} \\ K_{\theta x} & K_{\theta y} & K_{\theta z} & K_{\theta\theta} & K_{\theta\phi} \\ K_{\phi x} & K_{\phi y} & K_{\phi z} & K_{\phi\theta} & K_{\phi\phi} \end{bmatrix} \begin{bmatrix} \Delta_x \\ \Delta_y \\ \Delta_z \\ \Delta_\theta \\ \Delta_\phi \end{bmatrix} \quad [1.1]$$

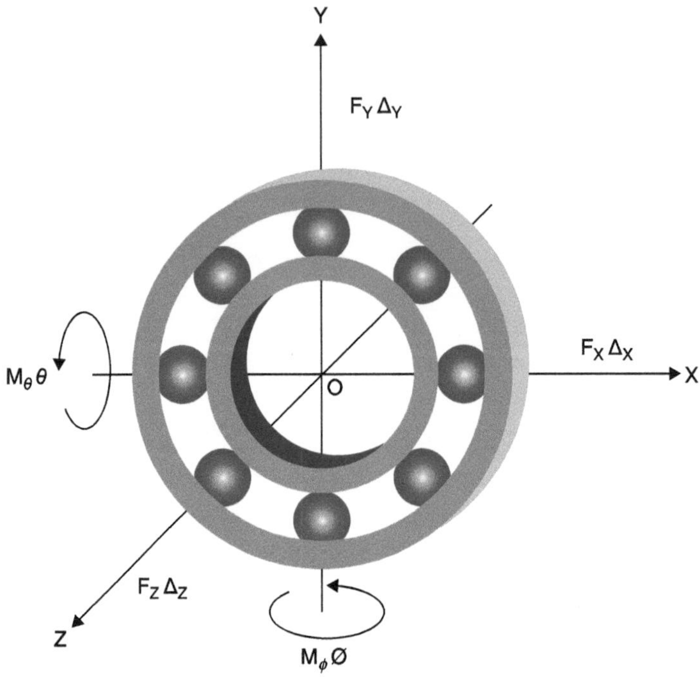

Figure 1.11 **Notation and coordinate system used to define static stiffness**

The stiffness matrix is a generalised stiffness matrix that contains the principal terms K_{xx}, K_{yy}, K_{zz}, $K_{\theta\theta}$ and $K_{\phi\phi}$, as well as cross-coupling terms. Note that for consistency of units:

$$F_\theta = M_\theta / D_p \quad \text{and} \quad \Delta_\theta = \theta . D_p$$

$$F_\phi = M_\phi / D_p \quad \text{and} \quad \Delta_\phi = \phi . D_p$$

where M_θ and M_ϕ are moments, θ and ϕ are angular deflections and D_p is the pitch circle diameter of the bearing. Not all cross-coupling terms are significant and different bearing types will be characterised by different terms. For example, an axially preloaded ball bearing subjected to a radial load will deflect in radial, axial and tilt directions (Figure 1.12). The reason is due to the internal geometry of the bearing that results in small elastic deformations and movements of the balls that experience a change in load. For this specific example the relevant terms in the stiffness matrix are K_{yy}, K_{yz} and $K_{y\theta}$.

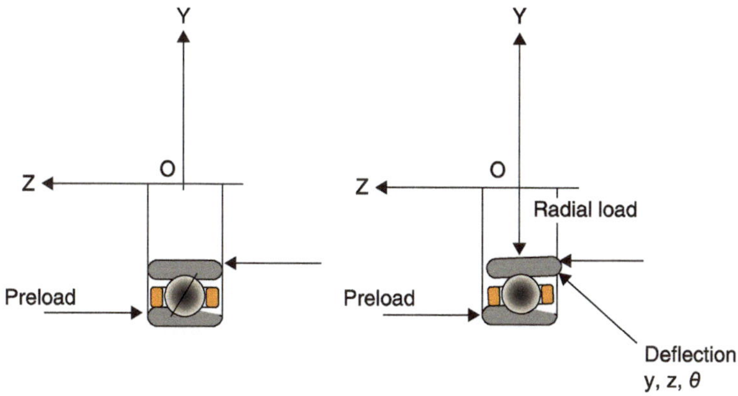

Figure 1.12 Deflections produced by radial load applied to preloaded ball bearing

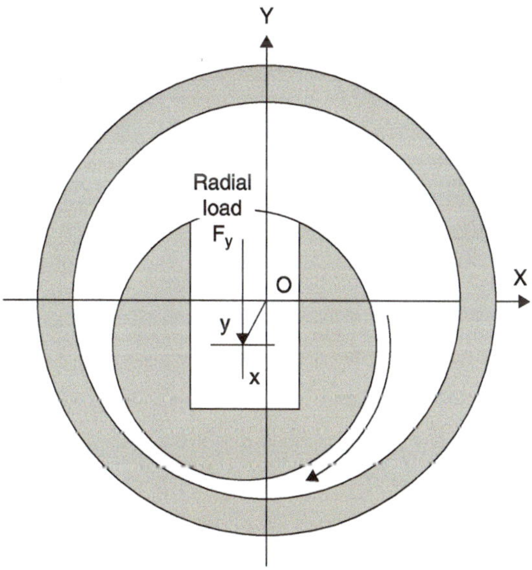

Figure 1.13 Hydrodynamic bearing subject to radial load

Another example for which cross-coupling terms are clearly significant is the hydrodynamic journal bearing (Figure 1.13), where the application of radial load produces a variation in bearing gap with angular position. The viscous drag in the bearing is no longer uniformly distributed around the bearing, but is a maximum where the gap is a minimum. This imbalance creates a component of deflection normal to the direction of the applied load so that the position of the minimum gap is not in line with the applied load. In this example the relevant stiffness terms are K_{yy} and K_{yx}.

1.4.2 Linear bearings

The notation and coordinate system used to define linear bearing stiffness is shown in Figure 1.14. The bearing comprises a moving carriage and a fixed base. A Cartesian Reference Frame OXYZ is chosen to have its origin O fixed in the moving carriage and its axis OZ being parallel to the axis along which linear motion takes place. Axes OX, OY and OZ remain fixed in that direction but are free to move in translation in the directions OX, OY and OZ. The carriage is subject to incremental forces F_x and F_y that act along the OX and OY axes respectively and incremental moments M_θ, M_ϕ, and M_ψ that act about the OX, OY and OZ axes. In response to the forces F_x and F_y, the carriage and hence the origin O of the frame OXYZ is displaced by small amounts x and y, whilst the moments M_θ, M_ϕ, and M_ψ produce angular deflections of the carriage with respect to OXYZ of θ, ϕ and ψ respectively. Bearing stiffness relates the applied incremental forces and moments to the linear and angular deflections.

As for rotating bearings, stiffness is represented by a 5×5 matrix:

$$\begin{bmatrix} F_x \\ F_y \\ F_\theta \\ F_\phi \\ F_\psi \end{bmatrix} = \begin{bmatrix} K_{xx} & K_{xy} & K_{x\theta} & K_{x\phi} & K_{x\psi} \\ K_{yx} & K_{yy} & K_{y\theta} & K_{y\phi} & K_{y\psi} \\ K_{\theta x} & K_{\theta y} & K_{\theta\theta} & K_{\theta\phi} & K_{\theta\psi} \\ K_{\phi x} & K_{\phi y} & K_{\phi\theta} & K_{\phi\phi} & K_{\phi\psi} \\ K_{\psi x} & K_{\psi y} & K_{\psi\theta} & K_{\psi\varphi} & K_{\psi\psi} \end{bmatrix} \begin{bmatrix} \Delta_x \\ \Delta_y \\ \Delta_\theta \\ \Delta_\phi \\ \Delta_\psi \end{bmatrix} \qquad [1.2]$$

where now

$F_\theta = M_\theta / L_\theta$ and $\Delta_\theta = \theta . L_\theta$

$F_\phi = M_\phi / L_\phi$ and $\Delta_\phi = \phi . L_\phi$

$F_\psi = M_\psi / L_\psi$ and $\Delta_\psi = \psi . L_\psi$

and L_θ, L_ϕ and L_ψ are characteristic lengths of the carriage.

For linear bearings, the angular components of stiffness are particularly important as applied loads are usually offset from the geometric centre of

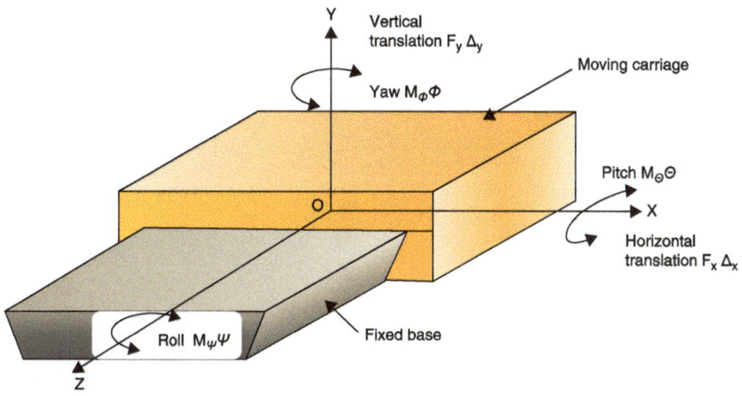

Figure 1.14 Notation and coordinate system for linear bearings

the carriage. The offset distances can be substantial so that even small angular tilts result in significant movements at the positions at which the loads are applied.

1.5 Damping

1.5.1 Types of damping

Bearings make a vital contribution to machine or system damping levels and in fact are often the dominant source of damping. For example, in machine tools, cutting forces may have a significant dynamic component arising from intermittent cutting or local variations in cut depth or work piece material properties. As spindle bearings are located close to the cutting point and must transmit the cutting forces, their damping levels have a critical effect on the amplitude of any vibration generated by dynamic cutting force components. Similarly, disc drives may experience externally applied disturbing forces for which bearing damping levels help to reduce the incidence of read/write errors and in the extreme to reduce the risk of drive damage. Damping also limits the amplitudes of vibration generated by rotor resonances. In ultra precision machines, this helps to control motion amplitudes generated by errors that occur at multiples of the basic rotor speed. In high speed machinery, bearing damping will limit the amplitudes of vibration reached as the rotor is accelerated through its resonances.

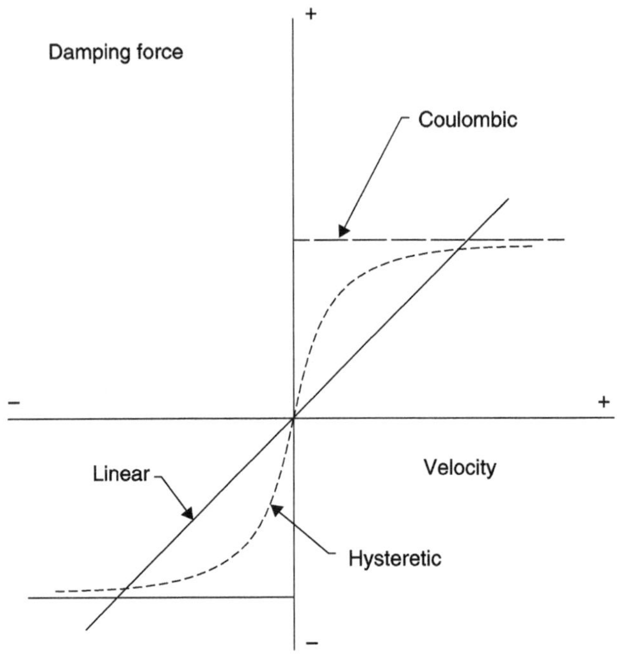

Figure 1.15 Types of damping associated with bearings

In practice, the damping generated by bearings may arise from several sources. For example, a ball bearing may generate squeeze film damping at the lubricated ball–raceway contacts. However, interfaces at the inner ring–shaft and in particular the outer ring–housing joint may, in some situations, provide additional sources of damping that cannot be ignored.

Damping forces can therefore exhibit different characteristics or combinations of characteristics (Figure 1.15). Damping arising from sliding motion at an interface tends to be Coulombic in nature, exhibiting a stick–slip character; damping due to the normal approach of two surfaces forming an interface tends to be hysteretic, whereas damping due to squeezing an oil film is generally near linear. That is to say, that the damping force is proportional to the velocity of approach of the surfaces that the oil film separates. In the case of bearings it is convenient to consider all damping to be linear and represented by a damping coefficient C such that:

$$F_d = C.dx/dt$$

where F_d is dynamic force applied to the bearing, x is the resultant deflection across the bearing and dx/dt is the velocity associated with the deflection, x. Thus, damping forces for rotational bearings are described by a generalised damping matrix for which there are three linear and two angular components of deflection:

$$\begin{bmatrix} F_x \\ F_y \\ F_z \\ F_\theta \\ F_\phi \end{bmatrix} = -i\omega \begin{bmatrix} C_{xx} & C_{xy} & C_{xz} & C_{x\theta} & C_{x\phi} \\ C_{yx} & C_{yy} & C_{yz} & C_{y\theta} & C_{y\phi} \\ C_{zx} & C_{zy} & C_{zz} & C_{z\theta} & C_{z\phi} \\ C_{\theta x} & C_{\theta y} & C_{\theta z} & C_{\theta\theta} & C_{\theta\phi} \\ C_{\phi x} & C_{\phi y} & C_{\phi z} & C_{\phi\theta} & C_{\phi\phi} \end{bmatrix} \begin{bmatrix} \Delta_x \\ \Delta_y \\ \Delta_z \\ \Delta_\theta \\ \Delta_\phi \end{bmatrix} \qquad [1.3]$$

where Δ_x, Δ_y, Δ_z, Δ_θ and Δ_ϕ are small amplitude, simple harmonic, time dependent movements between the inner and outer parts of the bearing and are given by $\Delta_x = \Delta_x.\cos(\omega.t + \phi)$, etc., i is the square root of −1 and ω is the frequency of the movement. For consistency of units:

$$F_\theta = M_\theta / D_p \quad \text{and} \quad \Delta_\theta = \theta.D_p$$

$$F_\phi = M_\phi / D_p \quad \text{and} \quad \Delta_\phi = \phi.D_p$$

The damping forces oppose the velocity of approach of the inner and outer bearing parts and are 90 degrees out of phase with the relative displacement. For linear bearings, damping forces are given by:

$$\begin{bmatrix} F_x \\ F_y \\ F_\theta \\ F_\phi \\ F_\psi \end{bmatrix} = -i\omega \begin{bmatrix} C_{xx} & C_{xy} & C_{x\theta} & C_{x\phi} & C_{x\psi} \\ C_{yx} & C_{yy} & C_{y\theta} & C_{y\phi} & C_{y\psi} \\ C_{\theta x} & C_{\theta y} & C_{\theta\theta} & C_{\theta\phi} & C_{z\psi} \\ C_{\phi x} & C_{\phi y} & C_{\phi\theta} & C_{\phi\phi} & C_{\phi\psi} \\ C_{\psi x} & C_{\psi y} & C_{\psi\theta} & C_{\psi\varphi} & C_{\psi\psi} \end{bmatrix} \begin{bmatrix} \Delta_x \\ \Delta_y \\ \Delta_\theta \\ \Delta_\phi \\ \Delta_\phi \end{bmatrix} \qquad [1.4]$$

where now there are three components of angular motion – pitch, yaw and roll.

As for stiffness, not all of the cross-coupling terms will be significant and those that are will depend on the bearing type and the application conditions.

1.5.2 Measurement of damping

In practice, bearing damping levels are notoriously variable. Apparently identical systems may exhibit very different damping factors, and similarly dismantling and rebuilding a system with the same components can significantly change the level of damping. Since the accurate measurement of bearing damping also requires a sophisticated test rig in which the bearing is isolated from its surroundings yet subject to the loads, speeds and temperatures representative of real applications, approximate measurements on real systems are frequently made. A common method involves measuring the dynamic response of a system and interpreting the magnitude and shape of the response curve local to resonances.

For example, in Figure 1.16, a simple bearing–shaft system has bearings represented by linear springs in parallel with viscous dampers. In this case the bearings are the only significant source of damping. Applying a dynamic force to some point on the system and measuring the deflection at that point enables the dynamic flexibility of the system to be determined. In general, the system will exhibit several resonances. For measurements

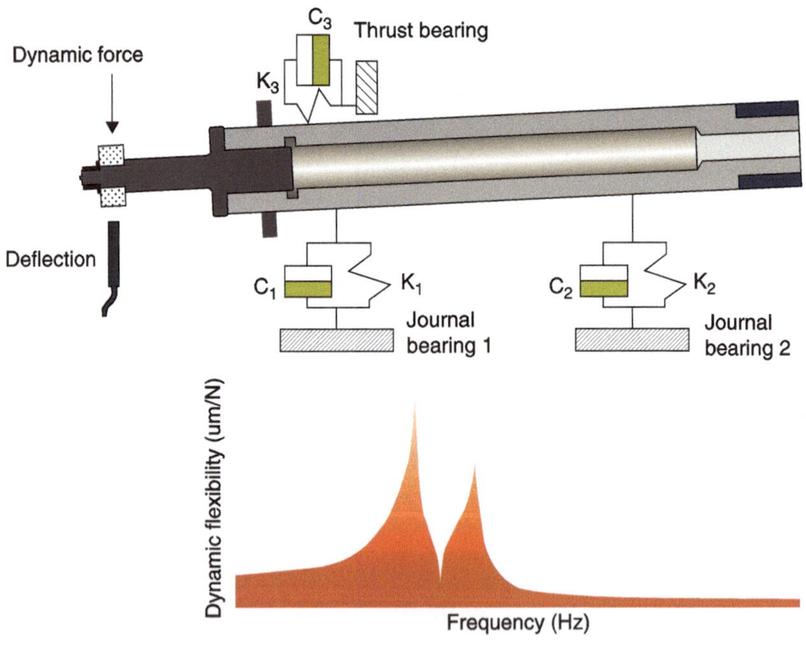

Figure 1.16 Dynamic response of a bearing-shaft system

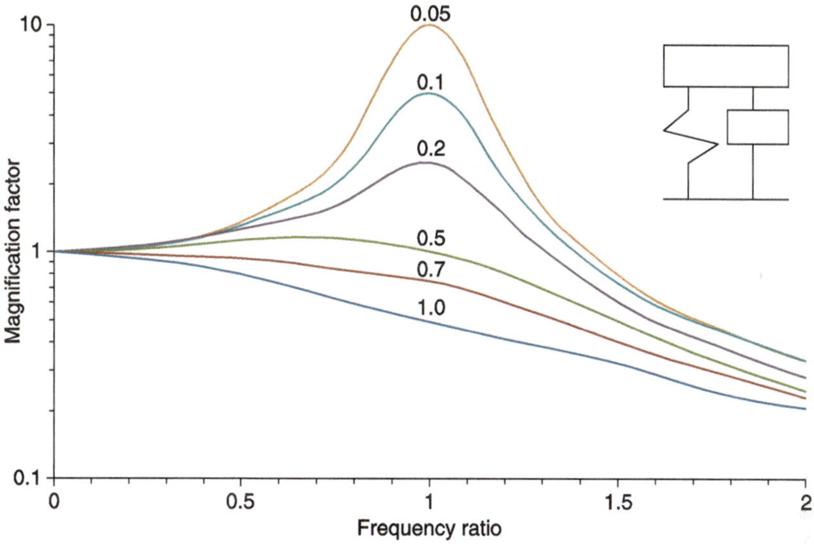

Figure 1.17 Effect of damping ratio on magnification factor

in a radial direction, there will be at least two resonances, the frequencies of which are dependent on the stiffness of the journal bearings. As bearing damping levels are the main factors limiting the amplitudes of the response of these resonances, then inspection of the response curve enables damping levels to be estimated.

The methods described here are based on a comparison with an idealised response curve for a single degree of freedom spring–mass system with a linear viscous damper. Such a system will have a single resonant frequency [5], dependent on the static stiffness, K, of the spring and the magnitude, m, of the mass:

$$\omega = (K/m)^{0.5}$$

At resonance, the amplitude of the response, i.e. the deflection per unit force, is given by:

$$\Delta/F = 1/(\omega C)$$

where C is the damping coefficient. The response curves for the single degree of freedom system, expressed in non-dimensional form, are shown in Figure 1.17. Frequency Ratio is the ratio of the frequency of interest and ω to the resonant frequency, p:

$$\omega_r = \omega/p$$

The Damping Ratio, γ, is the ratio of the damping coefficient, C, to a critical damping coefficient, C_{cr}:

$$\gamma = C/C_{cr}$$

and the magnification factor is:

$$B = 1/[(1 - \omega^2/p^2)^2 + (2\gamma\omega/p)^2]^{0.5} \qquad [1.5]$$

The critical damping coefficient is defined as:

$$C_{cr} = K/p$$

In other words, the critical damping coefficient is defined by the static stiffness and resonant frequency of the system.

A simple way of determining bearing damping coefficients from a measured system's dynamic response curve is to use the magnification factor at resonance to determine the damping ratio, γ, based on Figure 1.7. As the bearing's static stiffness, K, and the resonant frequency, p, are readily determined, the bearing's damping factor can be found from:

$$C = \gamma.C_{cr} = \gamma K/p$$

The above provides an approximate estimate of damping but often suffices as an accurate measurement of bearing damping levels usually require a sophisticated test set-up. Limitations to the approximate measurement are that it can be influenced by the structure of which it is part and that it is assumed that both bearings of the system of Figure 1.16 have the same damping coefficients. When a system's response curve has several resonances, each with different magnification factors, the resonance with the highest magnification factor provides a lower bound estimate of damping.

Heavily damped systems that can occur with oil lubricated bearings may not exhibit clearly defined bearing dependent resonances, in which case the phase of the response curve may be used to identify the resonant frequency. It should be noted that a phase lag of 90 degrees occurs at resonance [5].

1.6 The bearing as a dynamic machine element

The equations for stiffness and damping (Eqs 1.1 and 1.3, 1.2 and 1.4) can be combined to completely describe the bearing's dynamic characteristics. For rotating bearings:

$$\begin{bmatrix} F_x \\ F_y \\ F_z \\ F_\theta \\ F_\phi \end{bmatrix} = \begin{bmatrix} K_{xx} - i\omega C_{xx} & K_{xy} - i\omega C_{xy} & K_{xz} - i\omega C_{xz} & K_{x\theta} - i\omega C_{x\theta} & K_{x\phi} - i\omega C_{x\phi} \\ K_{yx} - i\omega C_{yx} & K_{yy} - i\omega C_{yy} & K_{yz} - i\omega C_{yz} & K_{y\theta} - i\omega C_{y\theta} & K_{y\phi} - i\omega C_{y\phi} \\ K_{zx} - i\omega C_{zx} & K_{zy} - i\omega C_{zy} & K_{zz} - i\omega C_{zz} & K_{z\theta} - i\omega C_{z\theta} & K_{z\phi} - i\omega C_{z\phi} \\ K_{\theta x} - i\omega C_{\theta x} & K_{\theta y} - i\omega C_{\theta y} & K_{\theta} - i\omega C_{\theta z} & K_{\theta} - i\omega C_{\theta\theta} & K_{\theta\phi} - i\omega C_{\theta\phi} \\ K_{\phi x} - i\omega C_{\phi x} & K_{\phi y} - i\omega C_{\phi y} & K_{\phi z} - i\omega C_{\phi z} & K_{\phi\theta} - i\omega C_{\phi\theta} & K_{\phi\phi} - i\omega C_{\phi\phi} \end{bmatrix} \begin{bmatrix} \Delta_x \\ \Delta_y \\ \Delta_z \\ \Delta_\theta \\ \Delta_\phi \end{bmatrix} \qquad [1.6]$$

and for linear bearings:

$$\begin{bmatrix} F_x \\ F_y \\ F_\theta \\ F_\phi \\ F_\psi \end{bmatrix} = \begin{bmatrix} K_{xx} - i\omega C_{xx} & K_{xy} - i\omega C_{xy} & K_{xz} - i\omega C_{xz} & K_{x\theta} - i\omega C_{x\theta} & K_{x\phi} - i\omega C_{x\phi} \\ K_{yx} - i\omega C_{yx} & K_{yy} - i\omega C_{yy} & K_{yz} - i\omega C_{yz} & K_{y\theta} - i\omega C_{y\theta} & K_{y\phi} - i\omega C_{y\phi} \\ K_{\theta x} - i\omega C_{\theta x} & K_{\theta y} - i\omega C_{\theta y} & K_{\theta z} - i\omega C_{\theta z} & K_{\theta\theta} - i\omega C_{\theta\theta} & K_{\theta\phi} - i\omega C_{\theta\phi} \\ K_{\phi x} - i\omega C_{\phi x} & K_{\phi y} - i\omega C_{\phi y} & K_{\phi z} - i\omega C_{\phi z} & K_{\phi\theta} - i\omega C_{\phi\theta} & K_{\phi\phi} - i\omega C_{\phi\phi} \\ K_{\psi x} - i\omega C_{\psi x} & K_{\psi y} - i\omega C_{\psi y} & K_{\psi\theta} - i\omega C_{\psi\theta} & K_{\psi\varphi} - i\omega C_{\psi\varphi} & K_{\psi\psi} - i\omega C_{\psi\psi} \end{bmatrix} \begin{bmatrix} \Delta_x \\ \Delta_y \\ \Delta_\theta \\ \Delta_\phi \\ \Delta_\psi \end{bmatrix} \qquad [1.7]$$

Individual stiffness and damping elements in Eqs 1.6 and 1.7 are determined from established bearing analyses, covered in the following chapters of this book or from experimental data.

Equations 1.6 and 1.7 enable bearings to be integrated into general machine design procedures, such as those based on Finite Element Analysis (FEA). Such analyses are well suited to the modelling of machine structures where flexibility and inertia is distributed. Treating the bearings as discrete components with stiffness and damping coupling static and rotating parts of the machine is an efficient way to model the overall system. The analysis enables important machine design parameters such as static stiffness, dynamic stiffness and resonant frequencies to be determined. An example is shown in Figure 1.18. FEA is used to model the distributed mass and flexibility of a machine tool structure containing a spindle, rotary table and linear slide. All bearings are defined as point elements having stiffness and damping.

The example shows two modes of vibration of the structure that depend upon the bearing characteristics; a spindle bending mode at 142 Hz and a slide 'roll' mode of vibration at 164.5 Hz. Performing this type of analysis at the design stage of the machine enables its configuration to be adjusted and evaluated, avoiding obvious design errors and speeding up the development cycle.

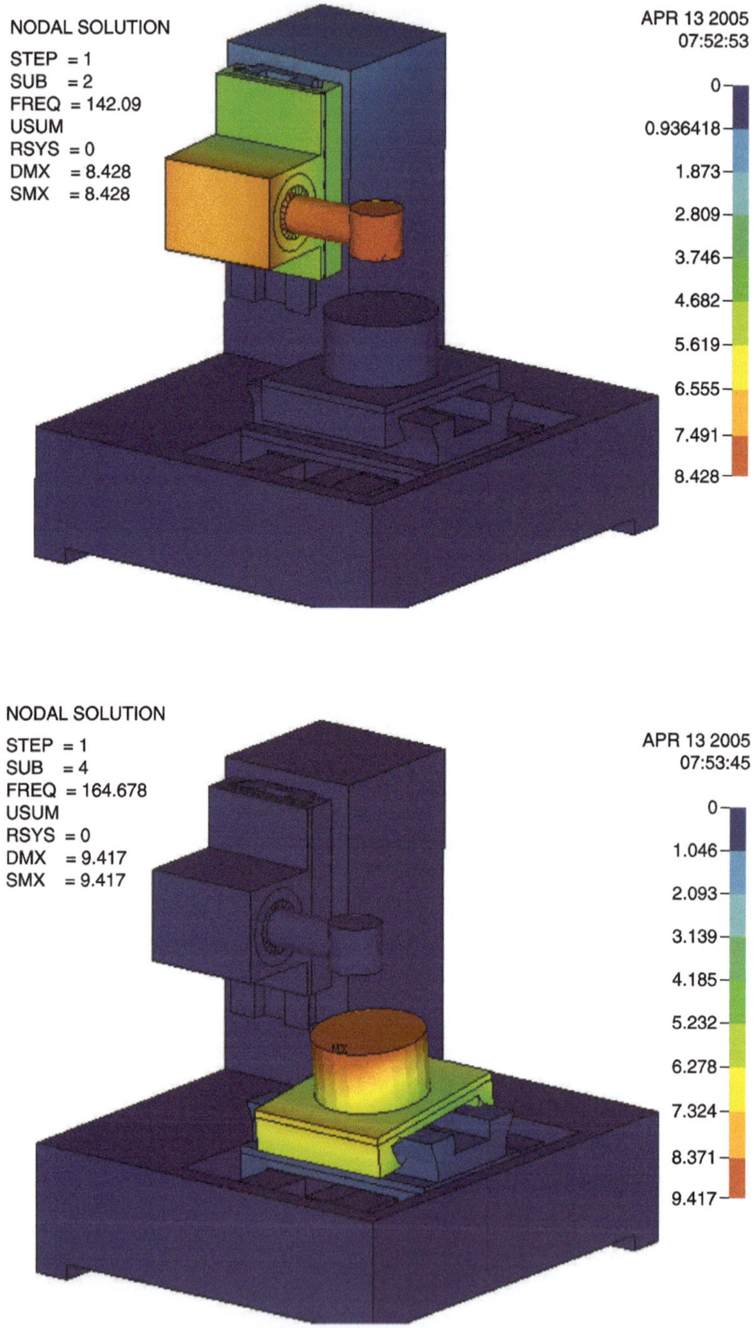

Figure 1.18 FEA of machine tool modes of vibration (courtesy of Brunel University)

1.7 Thermal characteristics

No bearing is completely frictionless, as even air bearings can consume significant amounts of power that ends up as heat in the adjacent machine structure, creating unwanted thermal expansions. Apart from being sources of heat generation, bearings are also poor conductors of heat, creating temperature gradients between moving and stationary parts of the machine that add to the system's overall thermal distortion. Lubrication systems may also have a significant influence on the thermal characteristics of bearings in as much as high lubricant flow rates have the capacity to remove bearing generated heat, thus providing some degree of control over temperature rise and thermal distortion. However, the beneficial cooling effect must be balanced against additional power losses, for ball bearings friction losses within the bearing generally increases with lubricant flow rate. Similarly, for hydrostatic bearings, high oil flow rates supplied at high pressure increase the pumping losses.

To put bearing power consumption into perspective, even a 1 °C rise or change in temperature occurring on ultra precision machines can produce thermal distortions that lead to an unacceptable loss of accuracy and this is why they are commonly placed in temperature controlled environments. Bearing heat generation is therefore completely unwanted and must be controlled. As the lubrication system alone is seldom sufficient, then the addition of independent cooling systems is often required. Although bearings may not necessarily be the largest source of heat generation on a machine, they are often in critical positions where relatively small amounts of heat generation can have a disproportionately large effect on machine performance. Furthermore, since the bearings support moving components of the machine, removing the heat uniformly in such a way as not to create temperature gradients within the machine structure can provide challenges to cooling system design.

Bearing power consumption is speed dependent, increasing with speed raised to some power depending on bearing type. Bearings that are required to operate at a sustained high speed are therefore the most critical. Examples are machine tool spindle bearings of any type and fluid film disc drive bearings. The machine tool spindle supports a tool overhung at one end (Figure 1.19). The spindle shown is an air spindle using two cylindrical journal bearings to support radial loads and a double sided thrust bearing to take axial load in two directions. In a spindle of this type, the thrust bearing also locates the axial position of the shaft with respect to the housing and is normally positioned close to the spindle nose in order to minimise the length of shaft and hence the thermal expansion between the

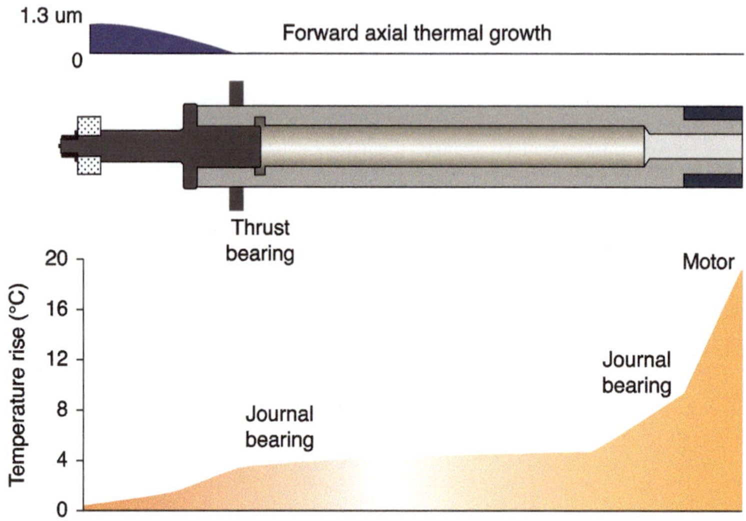

Figure 1.19 **Thermal distortion of machining spindle**

location plane and the work piece, in this case a distance of 100 mm. In this example the journal bearing diameter is 40 mm and the shaft rotates at 32 000 rpm. Each journal consumes 85 Watts and the double sided thrust bearing consumes 110 Watts. Since external cooling is only introduced to the bearing housing, temperature gradients are set up between the bearings and the cooling jackets. Even though the maximum temperature rise that occurs at the thrust bearing is only 3 °C, the forward axial thermal growth of the shaft is 1.3 μm. The highest temperatures are generated by the motor located at the rear end of the spindle to minimise the effect on tool position.

Hydrodynamic bearings used in hard disc drives present a different challenge. Here the main issue is simply to minimise bearing power loss so that in portable equipment such as laptops, camcorders and cameras, battery life is maximised and there is no need for parasitic equipment such as cooling fans. But the bearings must meet stiffness and load capacity requirements, so the main issue facing the designer is to achieve these requirements with minimum power loss.

1.7.1 Thermal model

For the purpose of estimating the temperature rise and thermal distortion in machine structures, the bearing is assumed to generate a power loss,

P_B, which is uniformly distributed around its circumference (Figure 1.20). In general, an amount P_L is carried away by the lubricant so that the power that transmits to the structure is $P_B - P_L$. For simplicity, half of this is assumed to transmit directly to the housing and half to the shaft. In addition to generating a power loss, the bearing also acts as a thermal conductor with a conductance C_B that again is uniformly distributed around its circumference. In most cases conductance may be neglected and bearing power loss assumed to transmit directly to housing and shaft in equal amounts. This is a simplification that provides an upper bound estimate of the temperature rise and hence the thermal distortion in these components. In ball bearings, for example, determining exactly where the heat is generated can require sophisticated computer models such as that defined in Gupta [6]. These models are primarily intended to aid understanding and optimisation of bearing design rather than aid the estimation of thermal distortion in adjacent structures.

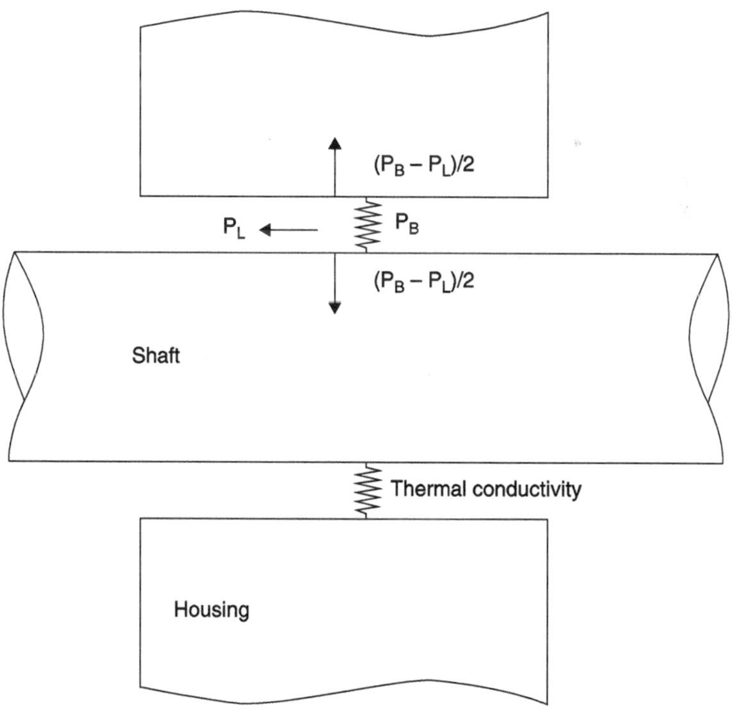

Figure 1.20 Thermal model of bearing

1.8 Closure

In this chapter we have defined the bearing parameters of primary importance in designing ultra precision machines, namely motion error, static stiffness, damping and power consumption. In the next five chapters the influence of bearing design and bearing mounting variables upon these parameters are explained for the principle ultra precision bearing types – ball bearings, hydrodynamic bearings, aerodynamic bearings, hydrostatic and aerostatic bearings.

1.9 References

1. ASME B89.3.4 (2010), *Axes of Rotation: Methods for Specifying and Testing*, USA: ASME.
2. Marsh, E.R. (2008), *Precision Spindle Metrology*, Pennsylvania: DEStech Publications Inc.
3. Newland, D.E. (1975), *An Introduction to Random Vibrations and Spectral Analysis*, 2nd edition, Harlow, UK: Longman.
4. AFBMA Standard (1970), *Rolling Bearing Vibration and Noise: Method of Measuring*, American National Standards Institute Inc.
5. Timoshenko, S., Young, D.H. and Weaver, W. (1974), *Vibration Problems in Engineering*, 4th edition, New York: John Wiley & Sons Inc.
6. Gupta, P.K. (1979), 'Dynamics of rolling element bearings, Part III: Ball bearing analysis', *Transactions of the ASME*, **101**: 312–18.

1.10 Notation

C – Damping coefficient
C_{cr} – Critical damping coefficient
D_p – Pitch circle diameter
F – Force
i – $\sqrt{-1}$
K – Static stiffness
L – Characteristic length
M – Moment
M – mass
P – Power
p – Resonant Frequency
t – time

X, Y, Z – Linear coordinates
OXYZ – Coordinate system

β – Magnification factor
θ, ϕ, ψ – Angular coordinates
γ – Damping ratio

Subscripts

X, Y, Z, θ, ϕ, ψ – Coordinate directions
B – Bearing
L – Lubricant

2

Ball bearings

DOI: 10.1533/9780857092182.37

Abstract: An important aspect of ultra precision applications is to select the most appropriate bearing type and arrangement. Ball-bearing manufacturers supply a wealth of information and design data but not much of it relates to the criteria of particular interest to ultra precision machine designers, namely motion error, static stiffness, damping and heat generation. Whilst these factors are dependent on the internal design of the bearing and under the control of bearing manufacturers they are also substantially influenced by application variables associated with housing and shaft design. This chapter provides a means of estimating the static stiffness of ball bearings subject to different mounting arrangements and types of load; a basic understanding of the mechanisms that influence damping and to provide guidance on the selection of damping coefficient; a basic understanding of the mechanisms that create motion error and how they are influenced by bearing application variables; and a means of estimating bearing torque and power loss.

Key words: rolling bearings, ball bearings, angular contact bearings, single row radial ball bearings, hybrid ball bearings.

2.1 Scope

Of all bearing industries, the rolling bearing industry has been the most successful in standardising its products but taking into account the number of parameters necessary to select a particular bearing design diameter, section, internal clearance, cage type, lubricant type and whether or not seals or shields are required, means that leading manufacturers may each have more than 50 000 standard designs to select from. However, very few of these bearings meet the criterion used in this book to define ultra precision bearings, namely a motion error of less than 100 nm peak to peak.

There is a vast selection of rolling bearing types: single-row radial ball bearings, angular contact bearings, cylindrical roller bearings, tapered roller bearings, needle roller bearings, spherical roller bearings, double-row and self-aligning ball bearings are just some of the principal types. Furthermore, within each of these categories there is a wide range of variants; single-row radial ball bearings, for example, are available in different grades of precision, with different ball sizes and internal clearances, with or without seals or shields and with or without a pre-charge of grease as a lubricant. In this text only two types of rolling bearing are singled out as being suitable for ultra precision engineering applications and these are the single-row radial ball bearing and the angular contact ball bearing (Figures 2.1(a and b)). Even then, not all of the variants available within these types may generate the low motion error that is a basic requirement for ultra precision applications. Generally only the precision grades or low noise grades of ball bearing are likely to produce low motion errors. These types of ball bearing are widely used in high precision machine tools, quiet running electric motors, computer disc drives and instrumentation, where they provide good but not exceptional running accuracy at a competitive price.

Apart from good running accuracy, ball bearings as distinct from other types of rolling bearings, have a number of other attributes that are useful in ultra precision applications. They have low friction that minimises thermal distortion and enables them to achieve high rotational speeds.

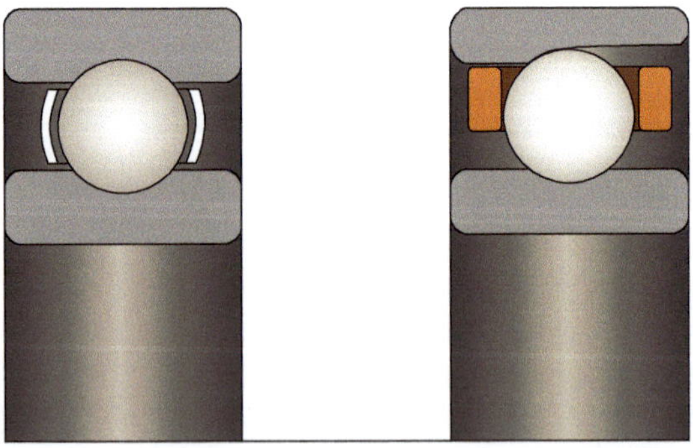
(a) (b)

Figure 2.1 (a) Single-row radial ball bearing; and (b) angular contact ball bearing

They are versatile in terms of the types of load they can take and the types of lubrication they may be used with. They are easy to mount and preload in order to generate good static stiffness, eliminate free play and hence centre the rings. Collectively these factors have made ball bearings a popular choice in most types of precision equipment and as a result the manufacturing processes used to make ball bearings have been continuously improved to further reduce motion errors now, making them a clear choice over other types of rolling bearing in most high precision applications.

The difference between single-row radial and angular contact ball bearings is their construction and hence the type of loads they can support. Single-row radial ball bearings have inner and outer rings with shoulders on both sides of the ball complement (Figure 2.1(a)). With the balls being trapped by the ring geometry, assembly generally involves placing the inner ring inside the outer ring, fully off centre, filling the gap between the two raceways with balls, snapping the inner ring into a central position and then spacing out the balls ready for cage insertion. The assembly process thus limits the number of balls in the bearing and hence its load carrying capacity, but it can support axial loads in two directions as well as radial loads. An angular contact bearing has one shoulder relieved on one of the rings, usually the outer ring (Figure 2.1(b)). The diameter of the relief is carefully controlled for the purpose of assembly, which comprises of putting the cage and ball complement in position around the inner ring, heating the outer ring and then moving it axially into position over the balls. When the outer ring cools the relief diameter is just sufficient to prevent the bearing from being separated. Thus the angular contact bearing can have a greater number of balls and hence a higher load carrying capacity and static stiffness than a single-row radial ball bearing, but it can only support axial load in one direction and radial load.

Single-row radial ball bearings are favoured in precision engineering applications such as computer disc drives and precision electric motors, where low motion errors or low noise are a primary requirement. Angular contact bearings, on the other hand, are widely used in precision applications such as machine tool spindles and rotary tables where static stiffness is also important.

In this chapter, the factors that influence the motion error, determine static stiffness, damping and the heat generation associated with single-row radial and angular contact ball bearings are of primary interest. A simple theoretical model is developed that first enables ball bearing stiffness to be estimated and second provides a framework for interpreting the effect of different parameters upon motion error. The model is used to assess the effect of both internal bearing parameters and external or application variables on the type and level of motion error. Damping and heat generation are treated more empirically in this text, as in practice they can both vary significantly for apparently identical bearings in identical applications. Such variation is not well explained, even by the complex computer models used in the rolling bearing industry [1], and it is therefore considered that such a detailed treatment is not justified here. It is stressed that the following is by no means a comprehensive coverage of ball bearing technology, the aim being to focus on aspects of the technology especially relevant to ultra precision machinery. For a more comprehensive coverage of rolling bearing technology, the reader is referred to [2].

2.2 Bearing sections

For the same nominal diameter, ball bearings are available in a range of sections (Figure 2.2); heavier sections, i.e. 7307 and 7407 are also available but are generally confined to heavy load applications. The larger the section, the larger the ball diameter that can be used, and consequently the higher the load carrying capacity of the bearing. However, compared

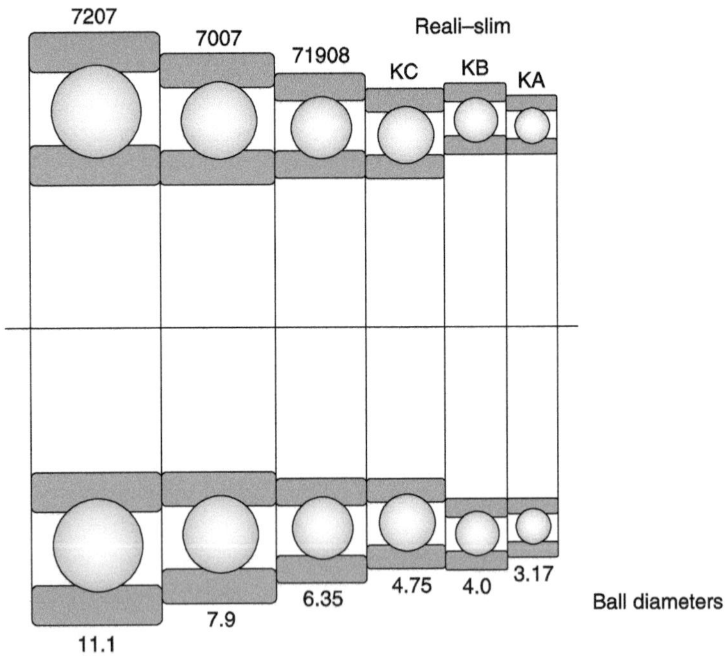

Figure 2.2 Sections available on precision angular contact bearings of nominally 50 mm pitch circle diameter

on the basis of the same applied load, bearings with a smaller section and smaller balls provide greater static stiffness. At first sight this may seem surprising but it is a fundamental property of ball–race contacts that under the same load the elastic deflections at the contact reduce with ball size. The most widely used in precision applications are the '19', '0' and '2' series bearings and some ball bearing manufacturers only offer these bearing sections as standard in precision grades.

2.3 Bearing arrangements

Single-row radial bearings are not matched in any way and are mounted independently, normally one at each end of a rotor (Figure 2.3). One bearing, in this case, the right-hand bearing shown in Figure 2.3, has both inner and outer rings axially located while the left-hand bearing has one ring, the outer ring, free to slide to take up any differential thermal expansion between rotor and stator. In lightly loaded and in most precision bearing applications, an axial spring load is applied to the sliding ring to ensure all balls within both bearings are in firm rolling contact with both raceways. This has several beneficial effects, in that it eliminates free play within the bearing and the possibility of whirl or axial instability of the rotor; it centres the rings, eliminates skidding of the balls ensuring low motion error and it maximises the static stiffness of the bearing. In the arrangement shown in Figure 2.3, the two bearings may be of a different size and have a different section to suit the specific requirements of an application. For example, in Figure 2.3, the outside diameter of the left-hand bearing is made larger than the diameter of the rotor, just to simplify assembly.

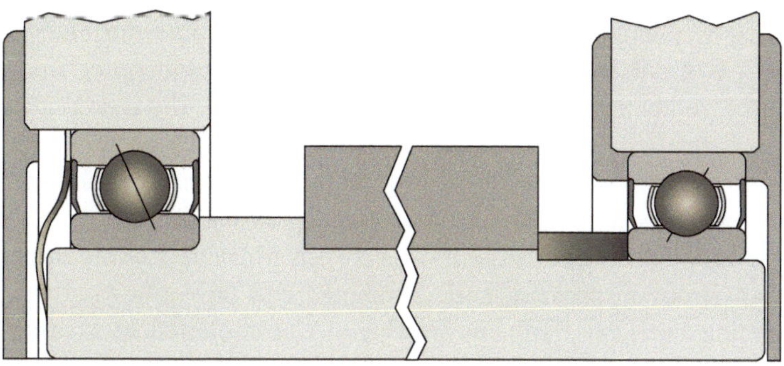

Figure 2.3 Example of single-row radial ball bearing arrangements

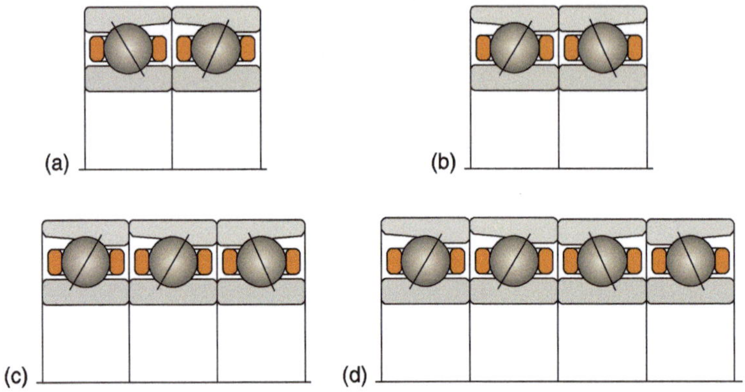

Figure 2.4 **Popular angular contact ball bearing arrangements: (a) face–face; (b) back–back; (c) triple; and (d) quadruple bearing sets**

The idea of subjecting ball bearings to a parasitic axial load, as described above, is all the more important when using angular contact bearings. The term 'axial preload' is used to describe the axial load applied to angular contact bearings. It has a large effect on bearing performance and in some applications much effort is put into finding the optimum preload. But first and foremost angular contact bearings must be set up with an axial preload or a controlled amount of axial clearance, simply to prevent the balls from running over the edge formed by the relief on one of the raceways. If this happens there is a risk that the bearing may separate, but even if it does not, this will certainly damage the ball surfaces to an extent where the bearing cannot be used to meet low motion error requirements.

To simplify preloading, angular contact bearings are available in matched sets and with inner and outer ring end faces machined flush or with a small step between them. When inner and outer rings are mounted and clamped in position by an axial force, the correct preload is automatically set. Bearings are available with different 'face steps' to generate high or low preloads as required by the application. This method of preloading is termed fixed displacement and it should be noted that any differential thermal expansion between inner and outer rings will have an effect on preload level. Machined end faces not only simplify preloading but also allow the bearings to be mounted in a variety of ways. Figure 2.4 shows some common angular contact bearing arrangements. The bearings may be mounted in a face–face arrangement (Figure 2.4(a)) or back-to-back (Figure 2.4(b)) or they may be stacked in arrangements of three, four or even more bearings (Figures 2.4(c and d)). These arrangements are normally used on machining spindles and positioned close to the spindle nose. The tail end of the spindle may be supported by a single, spring loaded ball bearing or a back–back pair of angular contact bearings with inner rings axially clamped together and the outer rings free to slide to take up differential thermal expansion.

For axially loaded angular contact bearings, the point at which the line through inner and outer ball–raceway contacts intersects the bearing axis is the effective position of the bearing. In face–face arrangements, the effective centres of the bearings are close together somewhere inboard of the physical bearing positions, so the overall tilt stiffness of the set is low. This also means that the arrangement is insensitive to misalignment and it is therefore favoured in applications where high misaligning loads are present. The effective centres of bearings in a back-to-back arrangement lie outboard of the actual bearing positions and the span between the centres means that the arrangement can have a high tilt stiffness. Back–back bearing arrangements are therefore commonly used on machine tool spindles where stiffness is of paramount importance.

A disadvantage of fixed displacement preload is its sensitivity to thermal expansion and in practice this limits the speed of the bearing arrangement. Spring preloading is therefore used on high speed bearing arrangements. The highest speeds are generally obtained with two bearings (Figure 2.5(a)), but bearings can be stacked, arranged in tandem (Figure 2.5(b)) to support heavier loads or provide higher stiffness.

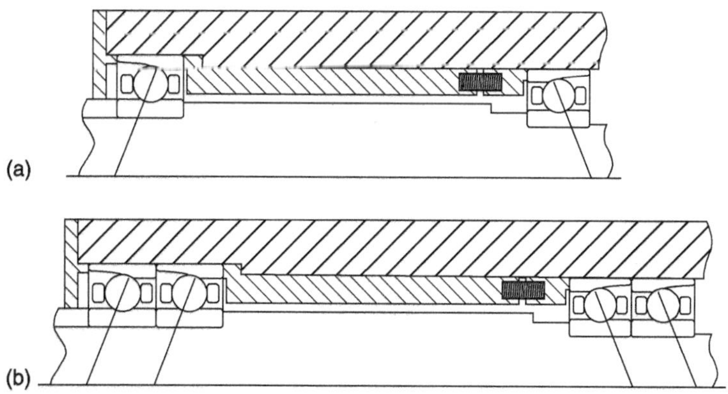

Figure 2.5 Spring preloaded angular contact bearing arrangements

2.4 Bearing tolerances

Bearing tolerances are defined in several international standards: ISO 492:2002, ISO 199:1997, DIN 620–2:1998, DIN 620–3:1982 and ABMA 20–1987. Bearings are usually manufactured to conform to one of five precision grades – normal or classes 6, 5, 4 or 2, where class 2 is the highest precision. Tolerances apply to the bearings external surfaces – bore, outside diameter and end faces, and to the radial and side face runouts of the assembled bearing – eccentricity and face wobble. Tables 2.1 and 2.2 summarise the tolerances for class 2 precision (the highest precision grade) bearing rings. Tolerances for all bearing classes are available from most manufacturer's catalogues.

All tolerances in Tables 2.1 and 2.2 are given in µm. Radial and face runouts represent eccentricity and misalignment between the ring's external surfaces and the raceway and for the rotating ring contribute directly to radial and axial fundamental motion errors. In many precision bearing applications, radial runout may be compensated for by balancing or by machining any critical locating surfaces after the system has been assembled. However, axial runout cannot be compensated for and is a critically important parameter for ball bearings.

Whilst the basic radial and face runouts are specified in international standards, the bearing's total motion error is not. The rolling element and raceway surface geometry and finish have a large influence on motion error and specification of these surfaces falls outside the scope of international standards. It follows that even the high precision grades of ball bearings do not necessarily have low total motion errors. Motion error standards are entirely the responsibility of the bearing manufacturer and are not usually published. Furthermore, the standards may vary from one manufacturer to another. The main benefits of using higher precision bearings are generally to achieve lower fundamental components of motion error; more rigid machine assemblies through better control over ring to housing fits; to allow closer tolerances between rotating and stationary parts or to reduce the extent of balancing required on high speed rotors.

2.5 Ring fits and mounting parameters

The importance of ring fits, shaft and housing design in ultra precision bearing applications cannot be overstated. These factors control the alignment and distortion of the bearing rings and determine whether

Table 2.1 Class 2 inner ring tolerances

d(mm)	Δdmp		Δds		Vdp	Vdmp	Kia	Sd	Sia	ΔBs*		VBs
Over Incl.	High	Low	High	Low	Max.	Max.	Max.	Max.	Max.	High	Low	Max.
2.5 10	0	–2.5	0	–2.5	2.5	1.5	1.5	1.5	1.5	0	–40/–250	1.5
10 18	0	–2.5	0	–2.5	2.5	1.5	1.5	1.5	1.5	0	–80/–250	1.5
18 30	0	–2.5	0	–2.5	2.5	1.5	2.5	1.5	2.5	0	–120/–250	1.5
30 50	0	–2.5	0	–2.5	2.5	1.5	2.5	1.5	2.5	0	–120/–250	1.5
50 80	0	–4	0	–4	4	2	2.5	1.5	2.5	0	–150/–250	1.5
80 120	0	–5	0	–5	5	2.5	2.5	2.5	2.5	0	–200/–380	2.5
120 150	0	–7	0	–7	7	3.5	5	2.5	2.5	0	–250/–380	2.5
150 180	0	–7	0	–7	7	3.5	5	4	5	0	–250/–380	4

d – Bore diameter
Δdmp – Single plane mean bore diameter deviation
Δds – Deviation of single bore diameter
Vdp – Bore diameter variation in a single radial plare
Vdmp – Mean bore diameter variation
Kia – Radial runout of inner ring
Sd – Variation of inner surface generatrix inclination with face
Sia – Outer ring face runout with raceway
ΔBs – Deviation of single inner ring width
VBs – Inner ring width variation
ΔBs – High limit for single and combined bearings
Low limit for single/combined bearings

Table 2.2 Class 2 outer ring tolerances

D(mm)	ΔDmp		ΔDs		VDp	VDmp	Kea	SD	Sea	ΔCs	Vcs
Over Incl.	High	Low	High	Low	Max.	Max.	Max.	Max.	Max.	As	1.5
6 18	0	−2.5	0	−2.5	2.5	1.5	1.5	1.5	1.5	for	1.5
18 30	0	−4	0	−4	4	2	2.5	1.5	2.5	inner	1.5
30 50	0	−4	0	−4	4	2	2.5	1.5	2.5	ring	1.5
50 80	0	−4	0	−4	4	2	4	1.5	4	of	2.5
80 120	0	−5	0	−5	5	2.5	5	2.5	5	the	2.5
120 150	0	−5	0	−5	5	2.5	5	2.5	5	same	2.5
150 180	0	−7	0	−7	7	3.5	5	2.5	5	bearing	4
180 250	0	−8	0	−8	8	4	7	4	7		5

D – Outside diameter
ΔDmp – Single plane mean outside diameter deviation
ΔDs – Deviation of single outside diameter
VDp – Outside diameter variation in a single radial plane
VDmp – Mean outside diameter variation
Kea – Radial runout of outer ring
SD – Variation of outside surface generatrix inclination with face
Sea – Outer ring face runout with raceway
ΔCs – Deviation of single outer ring width
Vcs – Outer ring width variation

there is any radial or axial free play in the bearing assembly. Shaft and housing diameter tolerances as well as geometrical accuracy and roughness specifications are defined in most ball bearing manufacturer's catalogues in relation to their own products. Shaft and housing tolerances are also defined in ISO 1101 and surface roughness classes in ISO 1302.

2.5.1 Ring fits

In ultra precision bearing applications, these specifications are often used as a starting point for developing the most appropriate housing and shaft seating specifications. For example, the static stiffness of a rolling bearing is influenced by the clearance fit of the outer ring in its housing. Where stiffness is critically important and the bearing does not have to slide to take up thermal expansion, the ISO recommended clearance may be reduced or even changed to a light interference. Understanding the influence of ring fits and mounting parameters on bearing performance is therefore essential in getting the best performance out of a particular arrangement.

Table 2.3 is an example of ring-shaft tolerances and fits for class 2 precision inner rings used with Js3 shaft tolerances taken from ISO 1101. For rolling bearings it is essential that any ring experiencing a relative rotating load is made an interference fit in order to prevent fretting corrosion at the ring–shaft or ring–housing interface. The inner ring is usually the rotating ring and in the presence of static loads should therefore be an interference fit on the shaft. If the loads are light, then a low interference is sufficient to prevent fretting and has minimal effect on ring diameter and distortion levels. Table 2.3 is for the highest grade of precision bearing and illustrates just how accurately shaft seatings must be manufactured to achieve a light interference fit.

The mean fit gives an interference of between 1.25 and 3.5 μm, depending on ring diameter, but even for these high precision tolerances, the fits may vary from 5 μm clearance to 12 μm interference for the larger diameter rings. This level of clearance risks fretting corrosion, whilst the high interference reduces the internal clearance of the bearing, affecting its static stiffness. Also, the fits do not take into account the effect of thermal expansion. Heat generated by ball bearings generally results in strong temperature gradients local to the rings, so even for modest temperature rises, differential expansion between the ring and shaft can significantly change the fit, reducing the designed level of interference or even creating clearance.

Table 2.3 Example of inner ring–shaft tolerances and fits

d (mm)	Class 2 ring tol.		Js3 shaft tol.		Ring–shaft fit		
Over Incl	High	Low	High	Low	High	Low	Mean
2.5 10	0	2.5	+1.25	−1.25	+3.75	−1.25	+1.25
10 18	0	−2.5	+1.5	−1.5	+4	−1.5	+1.25
18 30	0	−2.5	+2	−2	+4.5	−2	+1.25
30 50	0	−2.5	+2	−2	+4.5	−2	+1.25
50 80	0	−4	+2.5	−2.5	+6.5	−2.5	+2
80 120	0	−5	+3	−3	+7	−3	+2
120 150	0	−7	+4	−4	+11	−4	+3.5
150 180	0	−7	+5	−5	+12	−5	+3.5

In high precision bearing applications, the effect of manufacturing tolerances on ring fits are often minimised by matching seating diameters to ring diameters with the aim of achieving a particular level of fit regardless of the actual ring diameter. The target fit may also allow for a level of differential thermal expansion expected in the application. However, the benefit in doing this may be limited in applications where the bearing must operate over a wide range of speeds, therefore experiencing a wide range of operating temperatures.

Outer rings are normally stationary and do not usually experience a significant rotating load. As a result they are usually made a clearance fit in housings. Outer rings may either be located in order to control the axial position of the shaft and to support axial loads in two directions or they may be free to slide to take up any axial thermal expansions. The two situations require different fits. For located outer rings a low level of clearance allows for easy assembly yet minimises loss in stiffness and radial free play at the ring–housing interface. For sliding rings the clearance is somewhat larger to prevent radial thermal expansion of the ring at its normal operating temperature from reducing the clearance to zero. Table 2.4 is an example of class 2 precision outer rings used with H5 and JS4 precision housing

Table 2.4 Example of outer ring–housing tolerances and fits

D (mm)		Class 2 ring tol.		H5 housing tol.		Housing–ring fit (sliding fit)			JS4 housing tol.		Housing–ring fit (transition fit)		
Over	Incl.	High	Low	High	Low	High	Low	Mean	High	Low	High	Low	Mean
6	18	0	−2.5	+7	0	+9.5	0	+4.75	+2.5	−2.5	+5	−2.5	+1.75
18	30	0	−4	+9	0	+13	0	+6.5	+3	−3	+7	−3	+2
30	50	0	−4	+11	0	+15	0	+7.5	+3.5	−3.5	+7.5	−3.5	+2
50	80	0	−4	+13	0	+17	0	+8.5	+4	−4	+8	−4	+2
80	120	0	−5	+15	0	+20	0	+10	+5	−5	+10	−5	+2.5
120	150	0	−5	+18	0	+23	0	+11.5	+6	−6	+11	−6	+2.5
150	180	0	−7	+18	0	+25	0	+12.5	+6	−6	+13	−6	+2.5
180	250	0	−8	+20	0	+28	0	+14	+7	−7	+15	−7	+4

tolerances defined in ISO1101. H5 tolerances generate a relatively high mean clearance, enabling outer rings to slide freely, whereas JS4 tolerances generate a relatively low mean clearance, more suited to locating rings.

The difficulties with manufacturing tolerances discussed in the preceding section in relation to shaft seatings also apply to housing seatings. Housing diameters may be matched to individual bearing diameters to gain better control over the level of clearance in a particular application. The target clearance may also take into account differential thermal expansion between the ring and housing.

2.5.2 Seating geometry and roughness

Shaft and housing geometry and surface roughness specifications are defined in terms of practical measurement parameters and can be found in most manufacturer's catalogues. The specifications are designed to limit the misalignment and distortion of bearing rings to a level that does not create significant parasitic loads, obvious increases in motion error or bearing running temperature. From a motion error point of view they are not comprehensive but represent a good starting point for developing application specific specifications.

Figure 2.6 is an example of the parameters normally used to specify the geometry and roughness of bearing seatings on shafts, and Table 2.5 gives the permissible limits for class 2 precision bearings. Class 2 is the highest level of bearing precision and the limits for the shaft parameters given in Table 2.5 are matched to the bearing precision requiring the shaft manufacturing process to be precise and well controlled.

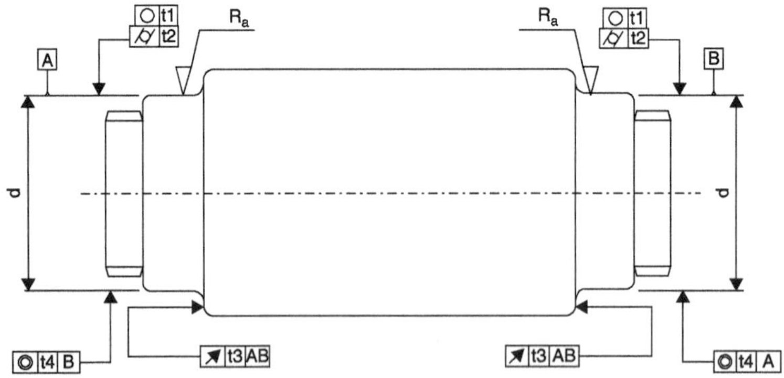

Figure 2.6 Example of shaft geometry and roughness parameters

Table 2.5 **Class 2 shaft geometry and roughness specification**

d (mm)	Class 2 Limit (μm)				
Over Incl.	Circularity *t1*	Cylindricity *t2*	Runout *t3*	Coaxiality *t4*	Roughness *Ra*
2.5 10	0.5	0.5	1.2	2.5	0.1
10 18	0.6	0.6	1.5	3	0.1
18 30	0.7	0.7	2	4	0.1
30 50	0.7	0.7	2	4	0.1
50 80	1	1	2.5	5	0.1
80 120	1.2	1.2	3	6	0.2
120 180	1.7	1.7	4	8	0.2

The simple practical parameters specified have an empirical correlation with bearing ring distortion and hence its levels of motion error. In other words, the same parameter levels may not necessarily generate the same motion errors in apparently identical situations. For example, circularity does not take into account whether the shaft seating has 2-point, 3-point or 10-point lobing, all of which may have a different effect on bearing motion error. Similarly, runout is primarily intended to define the misalignment of abutments with respect to the axis of rotation and does not give a good indication of abutment flatness that may contain long wavelength undulations.

In most precision bearing applications, the shaft seating is not the only component influencing the bearing's inner ring distortion. For example, inner rings are usually clamped by lock nuts, or some other form of clamping mechanism so that they can take axial load in two directions. The integrity of the clamping mechanism is also important. Practical experience shows that non-uniform clamping pressures will distort inner rings even if the shaft abutment it is pressed against is perfectly flat and square to the axis of rotation. Lock nuts are notorious for generating uneven clamping forces. Clearance between the shaft and lock nut threads

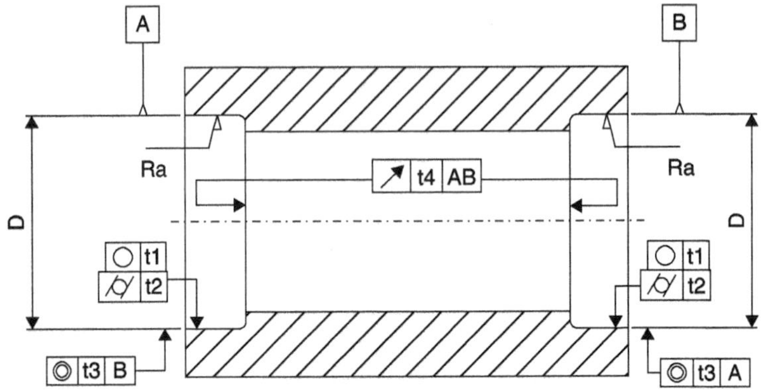

Figure 2.7 Example of housing geometry and roughness parameters

allows the nut to move off centre generating misalignment of its end faces, pushing the ring on one side.

Figure 2.7 and Table 2.6 show examples of the parameters normally used to specify bearing seating geometry and roughness in housings. As for shafts, a full set of specifications can be found in bearing manufacturer's catalogues. Because outer rings are generally a clearance fit in housings, ring distortion levels for the same level of seating error are lower than for shafts and this is reflected in slightly higher permissible parameter limits.

Outer rings that provide axial location are usually clamped and as for shafts, the clamping mechanism must generate a uniform clamping pressure around the ring. Similarly, on spring preloaded outer rings that must be free to take up axial thermal expansion, the spring force should be evenly distributed around the ring.

Table 2.6 Class 2 housing geometry and roughness specification

D (mm)	Class 2 Limit (μm)				
Over Incl.	Circularity *t1*	Cylindricity *t2*	Runout *t3*	Coaxiality *t4*	Roughness *Ra*
10 18	0.6	0.6	1.5	3	0.2
10 30	0.7	0.7	2.0	4	0.2
30 50	0.7	0.7	2.0	4	0.2

50 80	1.0	1.0	2.5	5	0.2
80 120	1.2	1.2	3.0	6	0.4
120 180	1.2	1.2	4.0	8	0.4
180 250	2.2	2.2	5.0	10	0.4

2.6 Ball bearing load–deflection characteristics

Ball bearings can support axial, radial and moment loads or any combination of these. All externally applied loads are reacted by internal loads generated at the ball–raceway contacts for which the ball load distribution is a function of the applied load vector, the internal geometry of the bearing and elastic deformation at the contacts. Load–deflection analysis is of central importance to ball bearing design. It is an essential step in determining ball-raceway contact stresses, static load capacity and fatigue life. It also enables the bearing's static stiffness characteristics to be determined, which is of most interest here.

Analytical equations describing the relationship between the external loads applied to a bearing and the approach of its inner and outer rings are readily derived but they are non-linear and generally require a numerical solution. In the following sub-sections, an iterative solution scheme is presented. An initial guess at the inner to outer ring approach is made and the elastic deformation at each ball–raceway contact is determined from a consideration of the internal geometry of the bearing. The load associated with the elastic deformation at each contact is then derived using Hertzian contact mechanics and finally the ball loads are summed to find the principal components of the total bearing load. These are compared with the applied loads and if the two are significantly different, the ring approach is adjusted and the calculation process is repeated.

2.6.1 *Internal geometry*

The first step in the analysis is to define the key parameters describing the internal geometry of the bearing. Not all of these parameters are catalogued and available to the bearing end user, so some guidance to selecting approximate values is also provided.

Parameters

For given applied loads, the internal geometry of a ball bearing determines the loads, stresses and elastic deformation associated with individual ball-raceway contacts and hence the overall life and stiffness of the bearing. Sub-sections 2.6.2 and 2.6.3 show how these loads and stresses may be calculated and hence how bearing stiffness may be determined. Some internal parameters necessary to determine bearing stiffness such as radial internal clearance and contact angle are available from manufacturer's catalogues; others such as raceway radius remain proprietary information. However, the parameters tend to follow well established design rules and can generally be estimated sufficiently accurately for the purpose of determining static stiffness.

Figure 2.8 shows a cross-section through a single-row radial ball bearing; ball centres lie on a diameter known as the pitch circle diameter, D_p. A good approximation to D_p can be obtained by taking the average of the bearing bore and outside diameter. The bearing contains N equi-

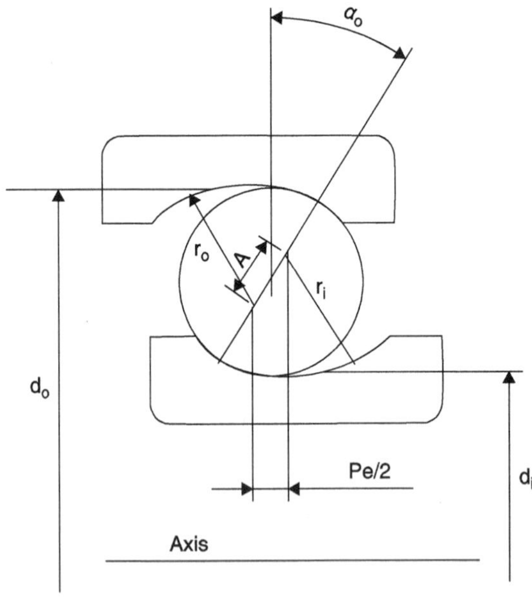

Figure 2.8 Geometrical parameters

spaced balls all with the same diameter, D, and both parameters can be estimated using the approximate calculations given in the following sub-section. The inner and outer raceways have cross-sectional radii r_i and r_o respectively. These radii must be slightly larger than the ball radius in order for the ball to contact each raceway at a single point. Their values are a compromise, where too high a radius means only a small contact area and high stresses at the ball–raceway contact under load, whereas too low a value generates a long contact area under load that extends across the raceway generating high friction when the bearing is in motion. Typically, r_i and r_o have values in the range:

$$1.025D < r_i < 1.035D \quad \text{and} \quad 1.04D < r_o < 1.05D$$

For analytical purposes, it is convenient to define the parameters:

$$\phi_i = D/2r_i \quad \text{and} \quad \phi_o = D/2r_o$$

known as the osculations of the inner and outer ball–raceway contacts.

The bearing is also generally arranged to have a small, positive radial internal clearance, P_d, which is normally tabulated in manufacturer's catalogues. The clearance is given by:

$$P_d = d_o - d_i - 2D \qquad [2.1]$$

The ratio $s = D/D_p$ defines the bearing section. Heavier section bearings have higher values of s and the advantage of being able to carry higher loads without incurring loss of life.

In high precision applications, ball bearings are almost always used with an axial load, the effects of which are to centralise inner and outer rings and eliminate any radial free play (Figure 2.8). All balls contact both inner and outer raceways and share the axial load equally. The angle subtended by a line passing through the centres of inner and outer contacts and a radial plane is known as the contact angle, α. α is load dependent and as the axial load approaches zero it approaches a limiting value known as the free contact angle α_o. α_o then is simply dependent on bearing geometry and can be determined by inspection of Figure 2.8:

$$\text{Cos } \alpha_o = (r_i + r_o - D - P_d/2)/(r_i + r_o - D) \qquad [2.2]$$

In the case of an angular contact ball bearing, α_o is an important design parameter that determines its ratio of axial to radial stiffness and load carrying capacity. Angular contact bearings are available in a range of standard contact angles varying from 12 degrees to as much as 60 degrees. The contact angle is not normally specified as a design parameter on single-row radial ball bearings and is usually less than 12 degrees. However, such bearings are available with different standard internal clearances, P_d, primarily to allow for the effects of interference fitting of inner rings on shafts and differential thermal expansion. Associated with a single-row radial ball bearing is the endplay:

$$P_e = 2(r_i + r_o - D)\ \sin\alpha_o \qquad [2.3]$$

that determines the maximum axial movement between inner and outer rings under no load.

Approximate parameter calculations

The parameters described in the preceding sub-section are necessary for calculation of bearing stiffness. Some are not readily available from manufacturers but may be estimated sufficiently accurately for the purpose of bearing selection using the following formulae:

$$D_p = (D_o + D_i)/2$$

$$D = 0.3(D_o - D_i)$$

$$r_o = 0.525D$$

$$r_i = 0.515D$$

For angular contact bearings:

$$N = \mathrm{Int}(0.8\,\pi D_p/D)$$

and for single-row radial ball bearings:

$$N = \mathrm{Int}(0.6\,\pi D_p/D)$$

where D_o and D_i are the outside diameter and bore diameter of the bearing respectively and N is the number of balls.

Kinematic relationships

The orbital and spinning speeds of the rolling elements in a precision bearing determines many of the vibration and error motion frequencies that it will produce. These speeds are not harmonic of shaft speed and result in the bearing producing relatively high levels of asynchronous motion error when compared to fluid film bearing types. A completely general analysis of rolling element motion is complex, as it needs to take into

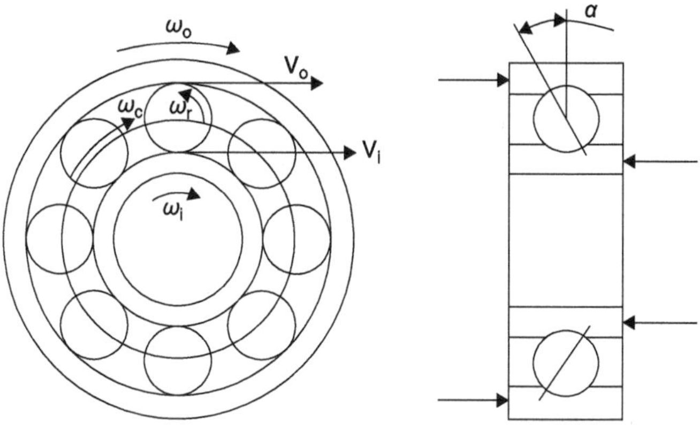

Figure 2.9 **Rolling speeds**

account spinning and sliding motions occurring at rolling element–raceway contacts in addition to rolling motion. Sliding and spinning motions are particularly important in high speed, lightly loaded bearings, where the effects of lubricant drag acting on the cage and ball surfaces can overcome traction forces at the raceway contacts. This is not the case in the vast majority of high precision bearing applications for which a good approximation to rolling element orbital and spinning speeds can found from a consideration of bearing geometry and ring speeds.

The case of a slow speed ball bearing subject to axial load and with both inner and outer rings rotating is first considered. Referring to Figure 2.9, the velocities of the raceway contacts at inner and outer contact diameters may be written as:

$$V_i - \frac{1}{2}\omega_i(D_p - D\cos\alpha)$$

And:

$$V_o = \frac{1}{2}\omega_o(D_p + D\cos\alpha)$$

Neglecting any gross slip at the rolling element–raceway contacts, then the orbital speed of the ball centres is the average of inner and outer contact speeds:

$$V_m = \frac{1}{2}(V_i + V_o)$$

The orbital speed of the balls is:

$$\omega_c = 2V_m/D_p$$

$$\omega_c = \tfrac{1}{2}\,\omega_i(1-\gamma) + \tfrac{1}{2}\,\omega_o(1+\gamma) \qquad [2.4]$$

where:

$$\gamma = D\cos\alpha / D_p \qquad [2.5]$$

Ball spinning speed is determined from the difference in speeds between ball centre and raceway contact speeds. The difference between ball centre and inner raceway angular speeds is:

$$\omega_{ci} = \omega_c - \omega_i$$

Again, assuming no gross slip at the raceway contact, surface velocities of ball and raceway are equal:

$$\tfrac{1}{2}\,\omega_{ci}\,(D_p - D\cos\alpha) = \tfrac{1}{2}\,\omega_r D$$

substituting for ω_{ci}:

$$\omega_r = (\omega_c - \omega_i)\,(1-\gamma)\,D_p/D$$

substituting for ω_c and simplifying gives:

$$\omega_r = \tfrac{1}{2}\,(\omega_o - \omega_i)\,(1-\gamma^2)\,D_p/D \qquad [2.6]$$

For the usual situation of inner ring rotation, Eqs 2.4 and 2.6 simplify to:

$$\omega_c = \tfrac{1}{2}\,\omega_i\,(1-\gamma)$$

$$\omega_r = -\tfrac{1}{2}\,\omega_i\,(1-\gamma^2)\,D_p/D \qquad [2.7]$$

respectively. The negative sign in Eq. 2.7 simply denotes that the inner ring and balls rotate in opposite directions. In a vast majority of bearing applications, Eqs 2.4 and 2.7 enable ball orbital and spinning speeds to be determined to within an accuracy of ±2%. To cater for extreme situations of high speed or light load a full dynamic analysis is required. Research on this subject can be found in [1–7].

2.6.2 Contact mechanics

In normal designs of ball bearings the curvature of the ball is higher than that of the raceway, so that under no load the contact is theoretically a single point. However, even under the lightest of loads, elastic deformation occurs and a small contact area develops. In typical ball–raceway contact geometries, the contact areas are small in size compared to the radii of curvature of the surfaces and the normal contact stresses are high, being more than 1 GN/m^2 under steady loading conditions and as high as 3 GN/m^2 under transient loading conditions. The relationships between normal load and the parameters, contact stress, contact area and normal approach are of fundamental importance to bearing design. The first satisfactory analysis of two elastic bodies pressed into contact is attributed to Hertz, who in 1882 published a classical paper [8], analysing the elastic deformation of two glass lenses in contact. Within the bearing industry the analysis is so widely accepted that the stresses developed in ball–raceway contacts are referred to as Hertzian stresses. A full treatment of contact mechanics is given in [9], and here only a brief summary of the essential steps in the analysis necessary to define the relationship between normal load and normal approach of a ball and a raceway is made.

Contact geometry

A general case is considered in which the surface geometry of both bodies may have different principal radii of curvature about orthogonal axes (Figure 2.10). Both surfaces are assumed to be perfectly smooth, i.e. free from asperities, waviness and discontinuities that may affect stress distribution, locally within the contact. For such surfaces the undeformed profiles in the vicinity of the contact may be approximated to quadratic functions:

$$z_1 = x_1^2/2R'_1 + y_1^2/2R''_1 \qquad [2.8]$$

and:

$$z_2 = -(x_2^2/2R'_2 + y_2^2/2R''_2) \qquad [2.9]$$

where the x-y plane is tangential to the contact and z is the normal distance of the surface from the plane (Figure 2.10). The principal radii of curvature of a surface at a point are the maximum and minimum radii of all possible cross-sections passing through that point. In the most general case, the principal axes of each body may be inclined at some angle to each other. However, in the case of a ball contact, the principal radii of curvature are the same and can be chosen to align with the principal radii of curvature of the raceway. The separation of the surfaces can then be expressed as:

$$h = Ax^2 + By^2 = x^2/2R' + y^2/2R'' \qquad [2.10]$$

where R' and R'' are the principal relative radii of curvature:

$$R' = R_1' + R_2' \qquad \text{and} \qquad R'' = R_1'' + R_2''$$

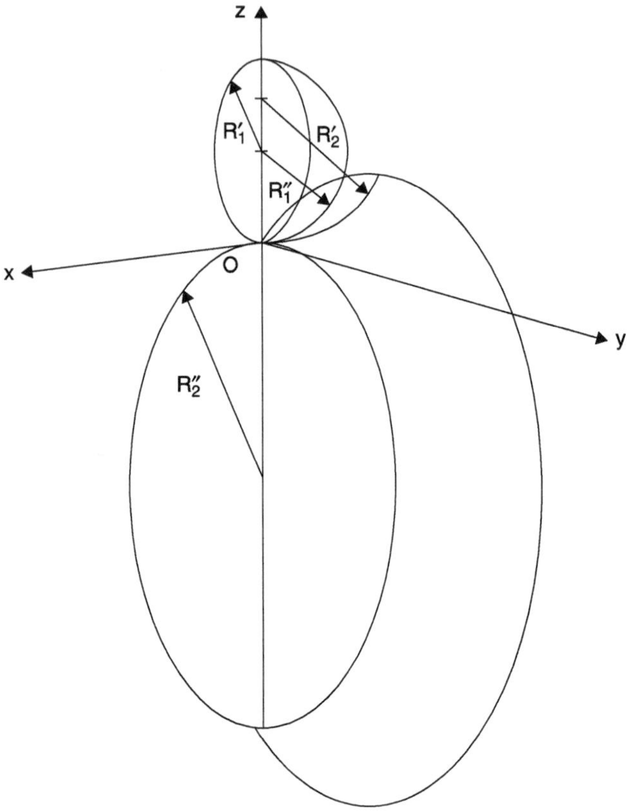

Figure 2.10 Contact geometry

It is convenient to define curvature sum and curvature difference for the contact:

$$(A+B)=0.5(1/R_1'+1/R_1''+1/R_2'+1/R_2'') \quad [2.11]$$

and:

$$(A-B)=0.5((1/R_1'-1/R_1'')+(1/R_2'-1/R_2'')) \quad [2.12]$$

It should be noted that whilst radius of curvature is always positive, curvature may be either negative or positive. Convex surfaces are defined to have positive curvature and concave surfaces negative curvature.

It is clear from Eq. 2.10 that contours of constant separation of the undeformed surface outside of the contact are elliptical with an axis length ratio of $(A/B)^{1/2}$. From observation of the interference fringe patterns formed by two lenses pressed in contact, Hertz hypothesized that the shape of the contact was also elliptical.

Elastic deformation

In considering lens deformation, Hertz was concerned with elastic non-conforming surfaces and his analysis was based on a number of assumptions:

1. The proportional limit of the material is not exceeded so that all deformations are purely elastic.
2. The contact dimensions are small in comparison to the radii of curvature of the surfaces at the contact.
3. The radii of curvature of the contacting surfaces are very large in comparison to the dimensions of these areas.
4. Loading is normal to the contact interface and that surface tractions are negligible.

In the general case where surface separation is given by Eq. 2.10, he further assumed that the contact is elliptical in shape with semi major and minor axes of lengths a and b. For the above assumptions, Hertz found that a parabolic pressure distribution:

$$p = p_0 \{1 - (x/a)^2 - (y/b)^2\}^{1/2} \qquad [2.13]$$

produces normal displacements within the contact ellipse of the form:

$$u = (1 - v^2)(L_c - M_c x^2 - N_c y^2)/\pi E$$

so that the combined normal displacement is:

$$u_{z1} + u_{z2} = (L_c - M_c x^2 - N_c y^2)/\pi E' \qquad [2.14]$$

where:

$$\frac{1}{E'} = \frac{1 - v_1^2}{E_1} + \frac{1 - v_2^2}{E_2}$$

$$A = M_c/\pi E' = (p_0/E')(b/e^2a^2)\{\mathbf{K}(e) - \mathbf{E}(e)\} \qquad [2.15]$$

$$B = N_c/\pi E' = (p_0/E')(b/e^2a^2)\{(a^2/b^2)\mathbf{E}(e) - \mathbf{K}(e)\} \qquad [2.16]$$

$$\delta = L_c/\pi E' = (p_0/E')b\mathbf{K}(e) \qquad [2.17]$$

where:

$$E^* = (1 - \nu^2)/\pi E \qquad \text{and} \qquad \mathbf{E}(e) \text{ and } \mathbf{K}(e)$$

are complete elliptic integrals of argument:

$$e = (1 - b^2/a^2)^{1/2}$$

$$E = \int_0^{\pi/2} [1 - e^2 \sin^2 \phi]^{\frac{1}{2}} d\phi \qquad [2.18]$$

$$K = \int_0^{\pi/2} [1 - e^2 \sin^2 \phi]^{-\frac{1}{2}} d\phi \qquad [2.19]$$

The pressure p_0 is the maximum pressure that occurs at the centre of the contact and because the pressure distribution is ellipsoidal is related to total contact load by:

$$P = (2/3)\, p_0 \pi ab \qquad [2.20]$$

To find the shape and size of the contact ellipse, put:

$$B/A = (R'/R'') = [(a/b)^2\mathbf{E}(e) - \mathbf{K}(e)]/[\mathbf{K}(e) - \mathbf{E}(e)] \qquad [2.21]$$

and:

$$\begin{aligned}(AB)^{1/2} &= \tfrac{1}{2}(1/R'R'') \\ &= p_0 b/(E'a^2e^2)[\{(a/b)^2\mathbf{E}(e) - \mathbf{K}(e)\}\{\mathbf{K}(e) - \mathbf{E}(e)\}]^{1/2}\end{aligned} \qquad [2.22]$$

Equations 2.20 to 2.22 can be rearranged to give expressions for the semi major and semi minor axes of the contact ellipse and the normal approach of the surfaces:

$$a = [3P\{\mathbf{K}(e) - \mathbf{E}(e)\}R'/(\pi E'e^2)]^{1/3} \qquad [2.23]$$

$$b = [3P\{\mathbf{K}(e) - \mathbf{E}(e)\}R'/(\pi E'e^2)]^{1/3}[\mathbf{E}(e)/\{R'(\mathbf{K}(e) - \mathbf{E}(e)\} + R''\mathbf{K}(e)]^{1/2} \qquad [2.24]$$

$$\delta = \mathbf{K}(e)[9P^2e^2/(8\pi^2E'^2R'\{\mathbf{K}(e) - \mathbf{E}(e)\}]^{1/3} \qquad [2.25]$$

Equations 2.23 and 2.24 may be solved numerically by iterative techniques. Once a, b and hence e have been determined, substitution into Eqs 2.20 and 2.25 enables the maximum contact pressure and the normal approach to be determined. Equation 2.25 may be rewritten as:

$$P = K'\delta^{3/2} \qquad [2.26]$$

where:

$$K' = \pi E'[8R'(\mathbf{K}(e) - \mathbf{E}(e)]^{1/2}/(3e\mathbf{K}(e)^{3/2}) \qquad [2.27]$$

K′ may be regarded as a compliance factor that is completely dependent on the material properties and geometry of the surfaces in contact. Equation 2.26 is a simple relationship relating normal load and elastic deflection that can be used to determine the ball load distribution in a complete bearing. The contact stiffness is found by differentiating Eq. 2.26:

$$K_c = \partial P / \partial \delta = 3/2 K' \delta^{1/2} \quad [2.28]$$

It should be noted that load–deflection relationships are non-linear for ball–raceway contacts with stiffness increasing with both load and deflection.

2.6.3 Bearing loads and deflections

The load–deflection characteristics of ball bearings are highly non-linear and generally require a numerical solution scheme. The method of solution outlined here consists of imposing a known approach of the inner and outer rings, calculating the elastic deformations at each ball–raceway contact to determine ball loads and then to compare the sum of the ball loads with the known loads applied to the bearing. If the loads do not balance, the ring approach is adjusted and the calculation repeated until convergence is achieved. Ball bearings may be subject to misalignment, radial, axial and moment loads either with or without fixed displacement preload. The ring approach is therefore, in general, a three-dimensional vector. The analysis is subject to the usual assumptions that ring and ball materials are perfectly elastic and isotropic, the rings are rigid and balls and rings have perfect geometry.

A first step in the analysis is to consider a single ball subject to a local ring approach. In Figure 2.11, we first consider the situation where the rings are concentric and coplanar. The distance between raceway centres is $r_o + r_i - D - P_d/2$ and the centres lie in the same radial plane. The inner raceway centre i is then displaced radially by Δ_r and axially by Δ_a to i'. The distances between ball and outer and ball and inner raceway centres is, in general, the difference in radius of curvature plus any elastic deformation occurring at the contacts:

$$l_o = r_o - D/2 + \delta_o \quad [2.29]$$

and:

$$l_i = r_i - D/2 + \delta_i \quad [2.30]$$

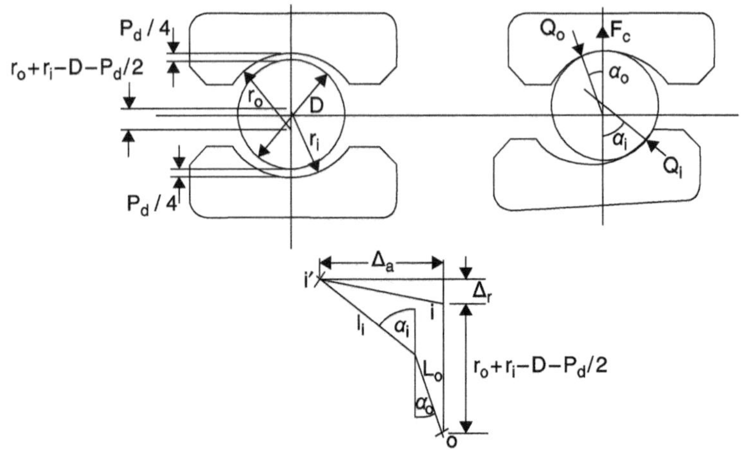

Figure 2.11 **Loads and deflections for a single ball**

The distances l_i and l_o are related to the local ring approaches by:

$$l_i \cos\alpha_i + l_o \cos\alpha_o = \Delta_r + r_o + r_i - D - P_d/2 \qquad [2.31]$$

and:

$$l_i \sin\alpha_i + l_o \sin\alpha_o = \Delta_a \qquad [2.32]$$

In Figure 2.11, it is also assumed that the contact angles at inner and outer raceways may be unequal due to the effect of centrifugal force acting upon the balls. This is significant in high speed bearings and has the effect of increasing the outer raceway contact load, reducing the outer contact angle and increasing the inner contact angle. Equilibrium of ball loads is given by:

$$Q_o \cos\alpha_o = F_c + Q_i \cos\alpha_I \qquad [2.33]$$

$$Q_o \sin\alpha_o = Q_i \sin\alpha_I \qquad [2.34]$$

where $F_c = ½ M d_m \omega_c^2$ and M is the mass of the ball. The contact loads are related to contact deflections by:

$$Q_o = K_o' \, \delta_o^{3/2} \qquad [2.35]$$

and:

$$Q_i = K_i' \, \delta_i^{3/2} \qquad [2.36]$$

Equations 2.33 to 2.36 enable the contact loads Q_o and Q_i to be determined in terms of the local ring approaches Δ_a and Δ_r. A straightforward method of solution is to first neglect ball centrifugal force, i.e., put $F_c = 0$ for which Eqs 2.33 and 2.34 give $Q_o = Q_i = Q$ and $\alpha_o = \alpha_i = \alpha$. Equations 2.35 and 2.36 may then be rearranged to give:

$$\delta_o + \delta_i = [(\Delta_r + r_o + r_i - D - P_d/2)^2 + \Delta_a^2]^{1/2} - (r_o + r_i - D) \quad [2.37]$$

and Eqs 2.35 and 2.36 rearranged to give:

$$K_o'^{2/3} \delta_o = K_i'^{2/3}\ \delta_I \quad [2.38]$$

so that δ_o and δ_i may be found directly and α and Q found by back substitution.

It should be noted that a positive contact load will only occur provided:

$$[(\Delta_r + r_o + r_i - D - P_d/2)^2 + \Delta_a^2]^{1/2} > [r_o + r_i - D] \quad [2.39]$$

The effect of centrifugal force at the outer contact is then found by adding it to the static contact force. Resolving forces in axial and radial directions:

$$Q_a = Q \sin \alpha \quad [2.40]$$

and:

$$Q_r = Q \cos \alpha + F_c \quad [2.41]$$

where:

$$Q_o = [Q_a + Q_r]^{1/2} \text{ and } \alpha_o = \text{Tan}^{-1} (Q_a/Q_r) \quad [2.42]$$

The parameters δ_i, α_i and Q_i can now be found by back substitution into Eqs 2.40, 2.41 and 2.42, respectively.

The local ring approaches at individual ball positions may be expressed in terms of the global ring approach. Referring to Figure 2.12, the bearing is assumed to have a Cartesian coordinate frame OXYZ fixed at its geometric centre. With respect to this frame, the inner ring is displaced a distance of z in the axial direction and distances x and y in orthogonal radial directions. In addition, the ring is tilted by an angle θ about the OX-axis and an angle γ about the OY-axis. The bearing has N balls equi-spaced at an angle of $2\pi/N$ radians. Ball number j is situated at an azimuth angle ϕ_j to the OX axis where $\phi_j = 2\pi j/N$. From inspection of Figure 2.12, the local axial and radial ring approaches at ball j may be written as:

$$\Delta_{aj} = z + D_p/2 \sin\theta \sin\phi_j + D_p/2 \cos\gamma \sin\phi_j \quad [2.43]$$

and:

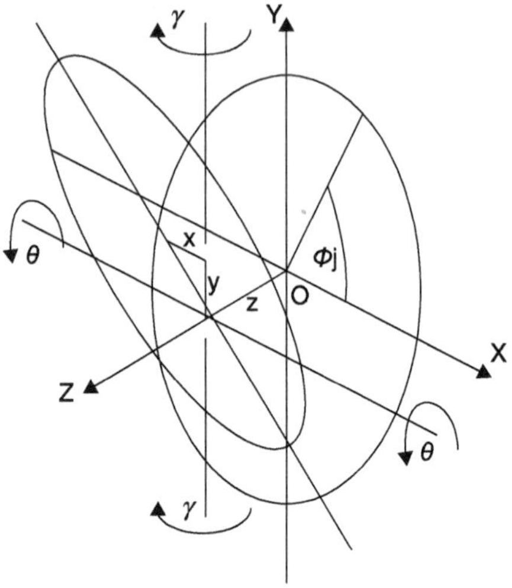

Figure 2.12 Relative approach of inner and outer rings

$$\Delta_{rj} = x\cos\phi_j + y\sin\phi_j \quad [2.44]$$

Finally, the total load supported by the bearing is found by summing ball loads:

$$F_z = \sum_{j=1}^{N} Q_{ij}\sin\alpha_{ij} \quad [2.45]$$

$$F_x = \sum_{j=1}^{N} Q_{ij}\cos\alpha_{ij}\cos\phi_j \quad [2.46]$$

$$F_y = \sum_{j=1}^{N} Q_{ij}\cos\alpha_{ij}\sin\phi_j \quad [2.47]$$

$$M_{xx} = \sum_{j=1}^{N} Q_{ij} D_p/2\sin\alpha_{ij}\sin\phi_j \quad [2.48]$$

$$M_{yy} = \sum_{j=1}^{N} Q_{ij} D_p/2\sin\alpha_{ij}\sin\phi_j \quad [2.49]$$

2.6.4 Method of solution

In the general case, where the bearing is subjected to known loads and moments or tilts, an iterative solution consists of guessing initial values of the ring approach vector x, y, z, θ, γ; calculating all ball loads and hence the net internal loads; finding the difference between net internal and applied loads and if this exceeds a tolerance adjusting the ring approach vector and repeating the calculation. The following steps define a simple solution algorithm:

1. Input ball and ring material properties – E, ρ, v
2. Input internal geometry – N, D, D_p, r_i, r_o, P_d
3. Input bearing ring speeds – ω_i, ω_o
4. Input applied loads – F_x, F_y, F_z
5. Input applied moments – M_{xx}, M_{yy} or tilts θ, γ
6. Calculate ball centrifugal force
7. Initial guess at inner-outer ring position – x, y, z, θ, γ
8. Calculate total elastic approach and contact angle at all ball–raceway contacts – δ_{oj}, δ_{ij}, α_{oj}, α_{oj} $1<j<N$
9. Calculate all ball loads – Q_{oj} , Q_{oj} $1<j<N$
10. Calculate net internal loads and moments – $\Sigma Q_{ij}\sin\alpha_j$, $\Sigma Q_{ij}\cos\alpha_j\sin\phi_j$, $\Sigma Q_{ij}\cos\alpha_j\cos\phi_j$, $\Sigma 0.5 D_p Q_{ij}\sin\alpha_j\sin\phi_j$, $\Sigma 0.5 D_p Q_{ij}\sin\alpha_j\cos\phi_j$
11. Check difference between net internal loads and applied loads is less than a tolerance
12. If not – adjust x, y, z, θ, γ and go to 6 or if so – end.

2.7 Static stiffness characteristics of ball bearings

The static stiffness of ball bearings used in high precision machinery is an important aspect of performance, but for a given bearing type and size it can vary widely depending on application conditions. Mounting arrangement, speed, applied loads, in particular axial preload, can all have a significant effect on the static stiffness of a given bearing design. Being able to estimate bearing stiffness at the machine design stage is a useful activity, enabling the designer to avoid excessively flexible structures or machine designs that run for prolonged periods at critical speeds. Simplified ball bearing stiffness calculations are available in, for example [2]; furthermore, precision bearing manufacturer's catalogues frequently contain stiffness design charts or graphs. In both cases the calculation or the design data is only valid for a limited range of conditions and the limitations are not always made clear. In contrast, the load–deflection model described in Section 2.6 provides a generalised calculation enabling ball bearing stiffness to be determined for all application conditions, provided some basic internal geometry parameters of the bearing are known or can be estimated using the formulae given in Sub-section 2.6.1. For this reason, it is used in this section, to determine the effect of application variables upon bearing stiffness for examples of practical bearing designs and mounting arrangements. One of the aims of this section is to illustrate just how much the stiffness of a given bearing design can vary, depending on how it is set up.

Because the load-deflection relationship for ball bearings is highly non-linear, stiffness is most reliably obtained by using the analysis of Section 2.6 to first determine the load–deflection relationship for the bearing and then to evaluate the effect of an increment in load or deflection. Stiffness determined in this way is the small amplitude stiffness that, for example, determines the vibration behaviour of a structure or the critical speeds of a rotor but in extracting it from the load–deflection relationship, an appreciation of its sensitivity to load is gained. The five generalised components of stiffness given by Eq. 1.1 and illustrated by Figure 2.11 can be determined using this approach. However, it first necessary to define three different ball bearing mounting situations common to precision bearing applications:

a. spring preload with controlled inner and outer ring alignment;
b. spring preload with uncontrolled inner or outer ring alignment;
c. face step preload.

The right-hand bearing shown in Figure 2.3 is an example of 'a'. It is a single bearing used to provide axial location of the shaft and the alignment of both inner and outer rings are controlled by the squareness of shaft or housing abutments. Preload is applied to this bearing by the spring acting on the left-hand bearing being transmitted through the shaft. Spring preload may be assumed to be constant in magnitude.

The left-hand bearing in Figure 2.3 is an example of 'b'. The inner ring alignment is controlled by an abutment on the shaft but the outer ring is deliberately made a clearance fit in the housing in order to allow it to slide in the event of axial thermal expansion. Spring preload is applied directly to the outer ring. The alignment of a ring that is free to slide is not controlled and under the action of radial load it will tilt to maintain zero moment loading.

Face step preload applies to Figures 2.4(a to c). In all cases, both inner and outer ring alignments are controlled and maintained perpendicular to the bearing axis. Face step preload controls the axial position of inner rings with respect to outer rings and an applied radial load has the effect of inducing an axial preload, increasing the value set by the face step.

The stiffness characteristics of the above three mounting situations are very different and are examined in more detail in the following sections.

2.7.1 Spring preloaded bearing with controlled inner and outer ring alignment

A starting point is to consider the load–deflection characteristics of a single ball bearing mounted with the alignment of both its inner and outer rings controlled. This is the case of a bearing providing axial location of the rotor where both inner and outer rings are clamped. Ideally, the tilt components of shaft and housing deflections are small so that the inner and outer rings can be considered to remain square to the bearing's axis. For this situation the relative approach of the bearing rings only has three translational components acting in axial (z) and radial (x and y) directions. This is a theoretically ideal situation as in practice shafts and sometimes housings bend to allow bearing rings to tilt. Figure 2.13 illustrates the arrangement and Figure 2.14 presents the axial load–deflection relationship. The bearing considered is a 7006 angular contact bearing running at low speed:

- number of balls 14
- ball diameter (mm) 7.14
- pitch circle diameter (mm) 42.5
- inner track radius (mm) 3.68
- outer track radius (mm) 3.75
- free contact angle (°) 15.0

The axial–load deflection curve illustrates an increase in axial stiffness with load that is typical of ball bearings. This increase is due to two factors, a non-linear elastic deformation at the ball–raceway contacts and the fact that bearing contact angle increases with load.

Figure 2.15 shows how the static stiffness of the 7006 angular contact ball bearing used in the previous example depends on preload for radial loads small in magnitude. Over a practical range of preload, both radial and axial components of stiffness vary by more than a factor of two. The fact that radial stiffness is three to four times higher than axial stiffness throughout the preload range is due to the bearing's relatively low contact

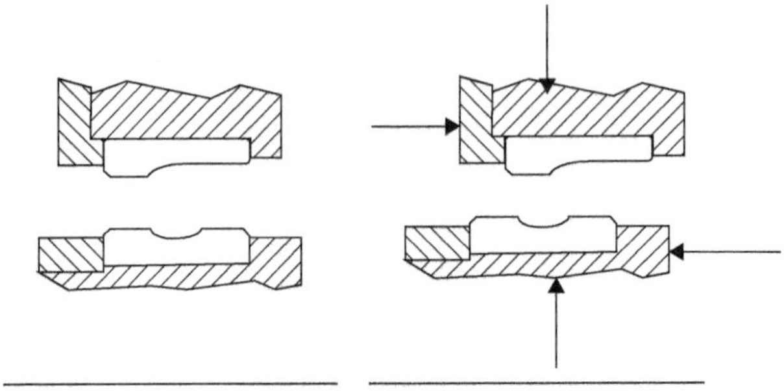

Figure 2.13 Relative ring movements for spring loaded controlled alignment bearing subject to axial and radial loads

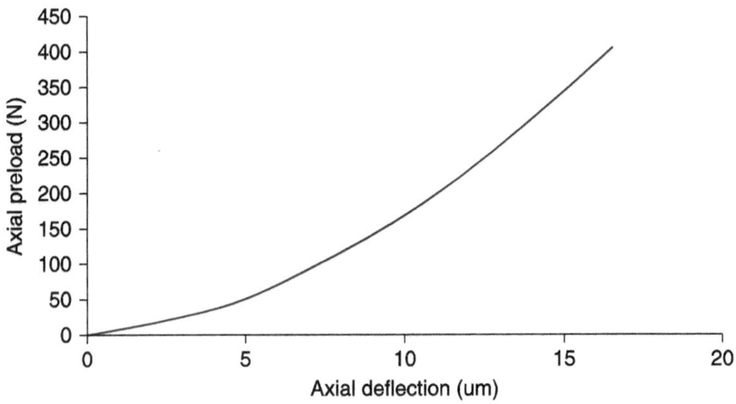

Figure 2.14 Axial load–deflection characteristics of 7006 angular contact ball bearing

angle of 15 degrees, which is towards the lower end of the range available. In both the axial and radial directions, the rate of increase of stiffness is greatest at low preloads. Since high preloads increase bearing friction and limit maximum speed and life, the optimum preload is very application dependent, being that which is judged to give the best trade off between stiffness and the other operating parameters of importance.

The generalised stiffness matrix for this bearing mounting arrangement is complex and if it is to describe a wide range of loading conditions should contain all 5 direct and 20 cross-coupling terms. However, in the case of a precision bearing subject to axial preload, light or no radial

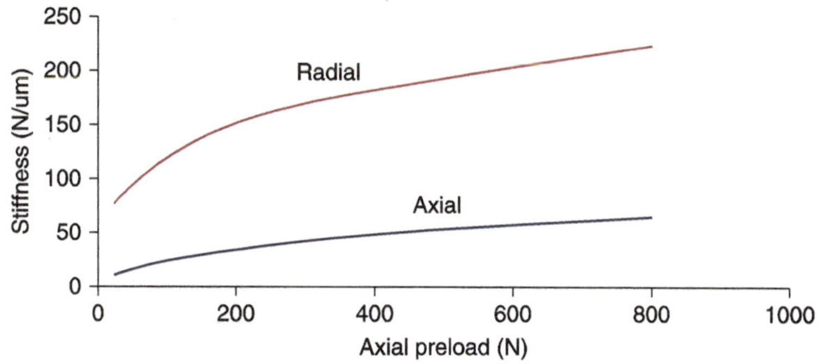

Figure 2.15 **Dependence of bearing stiffness on preload for a 7006 angular contact ball bearing**

loads and for which the alignment of both inner and outer rings is constrained by housing or shaft abutments to be zero, some of the cross-coupling terms may be neglected. The dominant terms are shown in Eq. 2.50, where it should be noted that $\Delta_\theta = \Delta_\phi = 0$ and hence the stiffness terms in columns 4 and 5 have no effect on the system stiffness. However, these terms may be significant if shaft tilt or bending results from, for example, the application of an overhung radial load:

$$\begin{bmatrix} F_x \\ F_y \\ F_z \\ F_\theta \\ F_\phi \end{bmatrix} = \begin{bmatrix} K_{xx} & 0 & K_{xz} & K_{x\theta} & 0 \\ 0 & K_{yy} & K_{yz} & 0 & K_{y\phi} \\ K_{zx} & K_{zy} & K_{zz} & K_{z\theta} & K_{z\phi} \\ K_{\theta x} & 0 & K_{\theta y} & K_{\theta\theta} & 0 \\ 0 & K_{\phi y} & K_{\phi y} & 0 & K_{\phi\phi} \end{bmatrix} \begin{bmatrix} \Delta_x \\ \Delta_y \\ \Delta_z \\ \Delta_\theta \\ \Delta_\phi \end{bmatrix} \qquad [2.50]$$

2.7.2 Spring preloaded bearing with uncontrolled ring alignment

For the case of a spring preloaded bearing, where the alignment of just one ring is controlled by an abutment, the other ring is free to tilt under the action of ball loads. This is the case of a bearing required to take up axial thermal growth. A spring preload is applied to one ring, normally the outer which is free to slide, whilst the other ring, normally the inner is clamped firmly in position so that the preload can be transmitted to the bearing providing axial location of the shaft. Depending on how the spring load is applied, the sliding ring may be subject to some degree of moment restraint. The static stiffness of two different methods of spring loading are considered in Figure 2.16 and compared with that of the bearing having both rings clamped (Section 2.7.1). In Figure 2.16(b), a short spring such as a wavy washer is placed directly in contact with the outer ring and since the spring stiffness is several orders of magnitude less than the stiffness of the bearing, it accommodates any tilt deflection of the outer ring without any significant effect on the preload distribution around the ring. In Figure 2.16(c), the spring preload is applied via a long sleeve that is a close fit in the housing bore; sleeve alignment is therefore controlled and any tilt of the outer ring due to applied loads results in an uneven preload distribution around the ring that has the effect of generating a restraining moment. Allowing a ball bearing ring to tilt under the action of applied loads has a large effect on radial stiffness. In Figure 2.17, the radial stiffness of the two situations just described are compared with the radial stiffness of the bearing with both rings clamped (Section 2.7.1).

For the moment restraint situation, the moment is assumed to be the product of preload and the bearing's outer ring radius. The difference in static stiffness from one situation to another is large – the bearing subject to moment restraint has a stiffness of 5 to 10% of that of the bearing with both rings located, whilst the stiffness of the bearing with one ring completely free to tilt is even lower at only 1 to 2% of the located ring bearing stiffness. In addition to this, the spring preloaded bearing makes no contribution to the axial stiffness of the system of which it is part, as it acts in series with the preloading spring of negligible stiffness.

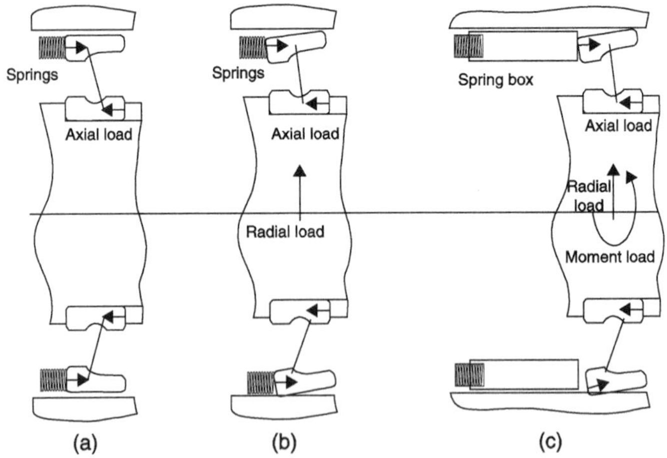

Figure 2.16 Comparison of spring and springbox preloaded bearing arrangements

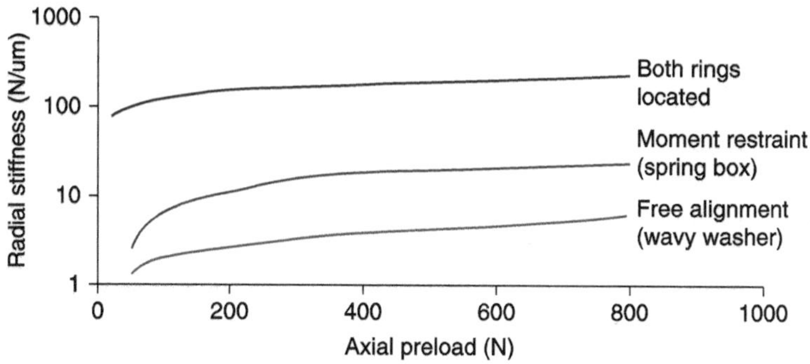

Figure 2.17 The effect of mounting on the radial stiffness of a preloaded 7006 15-degree angular contact ball bearing

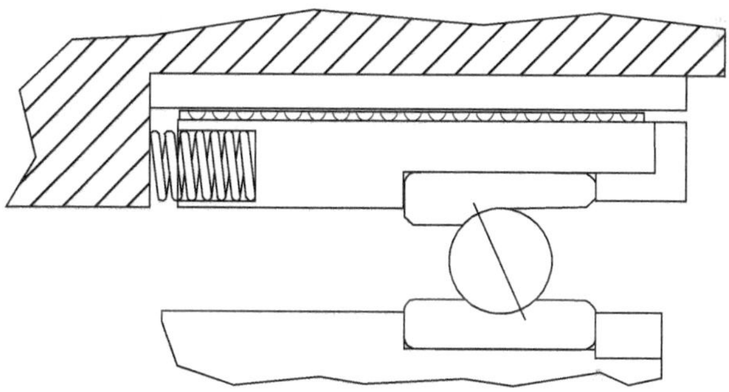

Figure 2.18 Spring preloaded angular contact bearing mounted in a linear bearing

Figure 2.17 shows the particular case of a 7006 15-degree contact angle bearing. The loss in stiffness in spring preloading a bearing is dependent on bearing geometry and may be greater for bearings with higher contact angles and lower for deep groove ball bearings or angular contact ball bearings with lower contact angles. Figure 2.17 also represents a worst case scenario, as it completely neglects any restraint on ring tilt that the housing may produce. In practice, and in particular for precision bearing arrangements, the outer ring–housing clearance is made a minimum consistent with the requirement for sliding and in general friction at the ring–housing interface provides a positive but indeterminate level of restraint to ring tilt. However, if friction is significant, then for a spring preloaded bearing the accuracy to which preload can be controlled is compromised. One solution to this is to mount the spring preloaded bearing in a sleeve supported by a linear bearing (Figure 2.18). The linear bearing uses a large number of small diameter balls in radial preload to maintain a high stiffness for the overall arrangement, yet axial preload may be accurately maintained at an optimum value. This arrangement is frequently used on high speed machining spindles, where the optimum preload is relatively light and critical to performance.

The stiffness matrix for the spring preloaded bearing with a single ring located takes a similar form to Eq. 2.50, but the values of the radial stiffness terms are lower and the axial stiffness of the bearing, K_{zz}, does not contribute to the axial stiffness of the system, as it acts in series with the preload spring of negligible stiffness.

2.7.3 Face step preloaded bearings

The third type of bearing arrangement, having yet another static stiffness characteristic, is the face step preloaded arrangement such as any of those shown in Figures 2.4(a to d). Face step preload applies to sets of two or more angular contact bearings, where inner and outer ring end faces are matched in such a way that clamping outer rings together and inner rings together generates a preload. Bearings are available in sets and with different preloads – light, medium and high.

They are also available 'flush faced', so that the preload can be set by matching the lengths of spacers placed between outer rings and inner rings (Figure 2.19). Face step preload controls the axial position of inner rings with respect to outer rings. In a symmetrical pair of bearings, such as the back–back arrangement shown in Figures 2.4(b) and 2.19, an applied radial load has no effect on the axial position of the rings but instead influences the preload. This contrasts with the spring preloaded bearing described in Section 2.7.1, for which an applied radial load induces an axial displacement between inner and outer rings. The main effect of face step preload is to increase the components of angular and radial stiffness on bearings that experience radial and moment loading. To illustrate the improvement, Figure 2.20 compares the radial load–deflection characteristics of the back–back arrangement shown in Figures 2.4(b) and 2.19, with that of a spring preloaded bearing. The preload for the back–back arrangement is set by a relative axial displacement of the inner rings of 8.6 μm generating 50 N preload. This is a light preload for a 7006 bearing, but the improvement in static stiffness becomes substantial for radial loads exceeding 100 N.

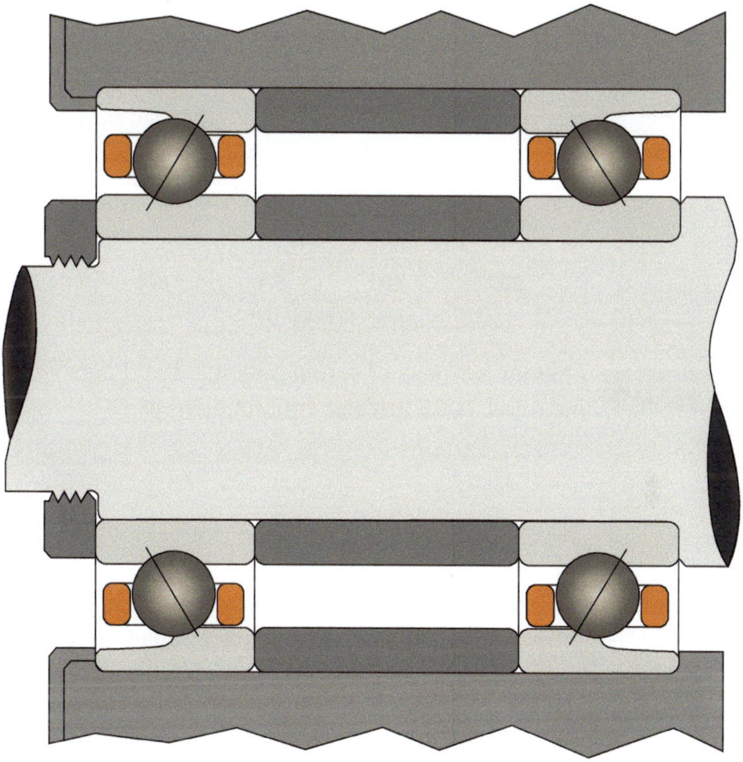

Figure 2.19 Back-back angular contact bearing preloaded with spacers

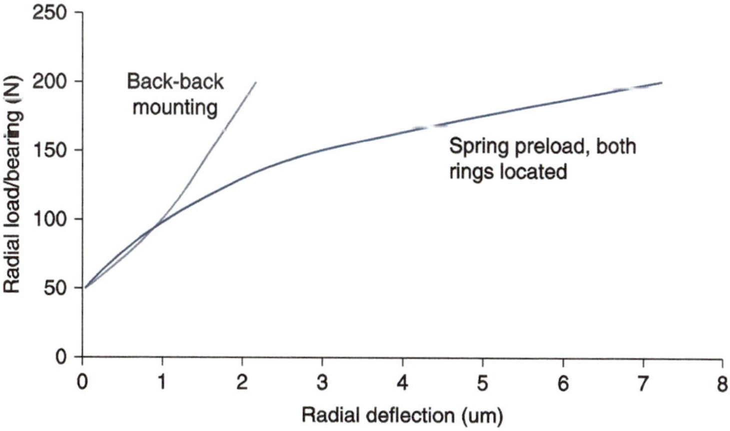

Figure 2.20 Radial load-deflection relationships for 2007 spring preloaded and back-back mounted bearings subject to 50 N preload

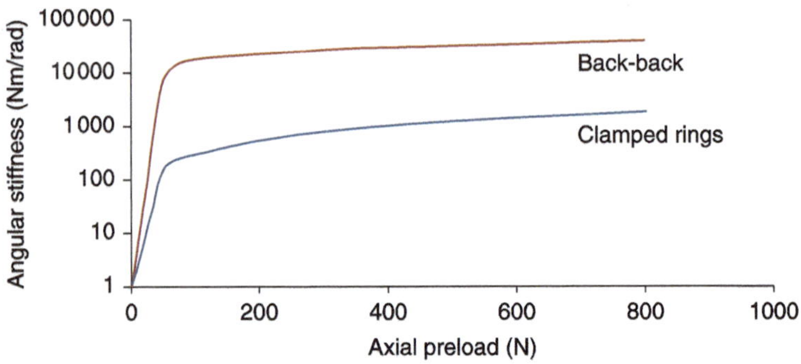

Figure 2.21 Angular stiffness of spring preloaded and back–back mounted 7006 angular contact bearing

The angular stiffness of one bearing mounted in a back–back arrangement is compared to that of the spring preloaded bearing in Figure 2.21 for a range of axial preloads.

For this example, the back–back bearings are located without any axial spacing and the preload is set by a face step. A range of face steps are considered, but for the purpose of comparing angular stiffness with that of a spring preloaded bearing, the face step is converted to an equivalent preload force. The improvement in angular stiffness of the back–back bearing over a spring preloaded bearing is more than an order of magnitude and is due mostly to the tilt of the rings inducing a large increase in the axial preload set by the face step. In contrast, for the spring preloaded bearing, tilt or moment loading produces instead, significant axial and radial displacements of the rings.

For the general case of a face step preloaded bearing subject to combined axial, radial and moment loading, the dominant terms in the stiffness matrix for a pair of bearings mounted back–back is given by:

$$\begin{bmatrix} F_x \\ F_y \\ F_z \\ F_\theta \\ F_\phi \end{bmatrix} = \begin{bmatrix} K_{xx} & 0 & K_{xz} & 0 & K_{x\phi} \\ 0 & K_{yy} & K_{yz} & K_{y\theta} & 0 \\ K_{zx} & K_{zy} & K_{zz} & K_{z\theta} & K_{z\phi} \\ 0 & K_{\theta y} & K_{\theta z} & K_{\theta\theta} & 0 \\ K_{\phi x} & 0 & K_{\phi z} & 0 & K_{\phi\phi} \end{bmatrix} \begin{bmatrix} \Delta_x \\ \Delta_y \\ \Delta_z \\ \Delta_\theta \\ \Delta_\phi \end{bmatrix} \qquad [2.51]$$

However, under no applied load, symmetry means that all cross-coupling terms are negligible so that Eq. 2.51 reduces to:

$$\begin{bmatrix} F_x \\ F_y \\ F_z \\ F_\theta \\ F_\phi \end{bmatrix} = \begin{bmatrix} K_{xx} & 0 & 0 & 0 & 0 \\ 0 & K_{yy} & 0 & 0 & 0 \\ 0 & 0 & K_{zz} & 0 & 0 \\ 0 & 0 & 0 & K_{\theta\theta} & 0 \\ 0 & 0 & 0 & 0 & K_{\phi\phi} \end{bmatrix} \begin{bmatrix} \Delta_x \\ \Delta_y \\ \Delta_z \\ \Delta_\theta \\ \Delta_\phi \end{bmatrix} \quad [2.52]$$

2.7.4 Effect of bearing speed

Centrifugal force acting on the balls in a bearing running at high speed can have a marked effect on static stiffness. Figure 2.22 shows the relationship between speed and stiffness for a 7006 15-degree angular contact bearing subject to a spring preload of 200 N, mounted with both inner and outer rings clamped. The four curves are for axial and radial components of stiffness for standard and hybrid bearing versions. The only difference between the hybrid bearing and the standard bearing is that the former has ceramic (silicon nitride) balls whereas the latter has steel balls. As bearing speed increases, centrifugal force acting on the balls sets up a difference in contact angle between ball–outer and ball–inner raceways, with the contact angle reduced at the outer and increased at the inner raceway (Figure 2.23). Under these conditions, ball equilibrium is determined by both the applied loads and the level of

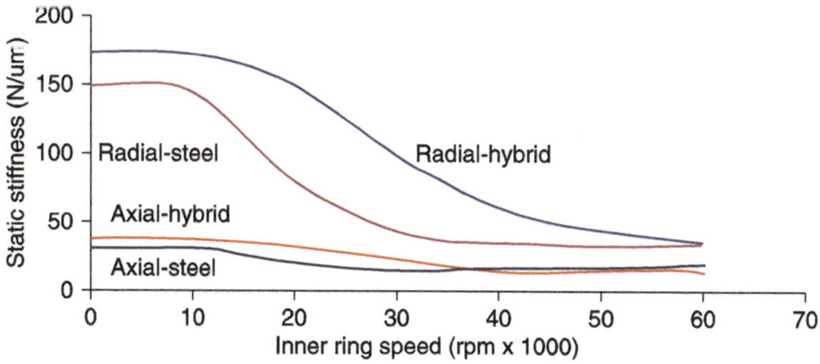

Figure 2.22 Effect of bearing speed on stiffness for a 7006 15-degree angular contact bearing subject to 200 N axial preload

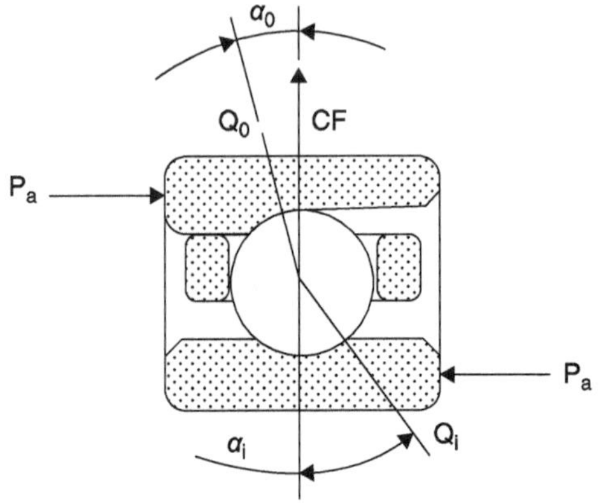

Figure 2.23 **Force equilibrium of ball in a high-speed bearing**

centrifugal force generated by the orbital speed of the ball. Furthermore, when a contact angle difference between inner and outer contacts exists, a small change in axial preload not only produces a change to the elastic deformation at the contacts but also a significant direct change to the contact angles themselves. The net effect is to reduce bearing stiffness so in general the higher the speed of the bearing the lower its stiffness.

For the example of Figure 2.22, ball centrifugal force reduces static stiffness by more than a factor of 3 over its full speed range. The effect is preload dependent, the higher the preload, the higher the speed before loss of stiffness occurs. The hybrid bearing with silicon nitride balls having a density 40% of steel enables considerably higher speeds to be obtained before centrifugal effects become significant. Silicon nitride also has a higher Young's modulus than steel, so a hybrid bearing is inherently more rigid than an equivalent all-steel bearing.

2.7.5 Summary of ball bearing static stiffness characteristics

The static stiffness of a ball bearing varies with bearing size, contact angle and series. But even for a given bearing design, stiffness may vary by more than a factor of 100, depending on how it is mounted and the conditions of load and speed it is operated under. In general, a bearing has five components of stiffness in the axial, radial – x, radial – y, angular – xx and angular – yy directions. Stiffness components may be determined theoretically from a load–deflection analysis taking into account ball–raceway contact mechanics, the bearing's internal geometry, mounting arrangement and operating conditions. Stiffness determined in this way is an upper bound estimate ignoring ring bending and any flexibility at ring interfaces. Nonetheless, since stiffness is so critically dependent on operating and mounting conditions such an analysis enables the most appropriate bearing design and arrangement to be chosen and the preload to be optimised for any given application.

2.8 Damping in ball bearings

Damping levels associated with ball bearings can vary widely, even under apparently identical situations so for the purpose of determining the dynamic response of a rotating system, values based on practical experience or even guesswork are often used. The damping characteristics of ball bearings have been subject to scientific investigation, mostly in relation to machine tool spindles, and the effect of bearing parameters on damping levels observed or measured. Numerical models for predicting damping levels have also been developed. All of this work has helped to improve the understanding of ball bearing damping mechanisms and therefore improving estimates of damping levels in given bearing arrangements.

In this section the dominant sources of damping normally associated with ball bearings are described and some guidance to estimating approximate damping values for different operating conditions given.

2.8.1 Sources of damping

Ring–housing and ring–shaft interfaces, their mounting tolerances and fits are an integral part of system design when utilising ball bearings. These factors may also influence overall bearing damping levels and should be considered in any practical machine design. The major sources of damping normally associated with ball bearings are shown in Figure 2.24. Squeeze films at the ball–raceway contacts are an inherent feature of ball bearings, but squeeze films may also occur at the ring–housing or ring–shaft interfaces, particularly when a small clearance exists as for the outer ring of a bearing that is free to slide to take up axial thermal expansion. Micro-slip may also occur at the ring–housing or ring–shaft interfaces, and again this is particularly significant where rings are free to slide. It is interesting to note that a small radial deflection of a

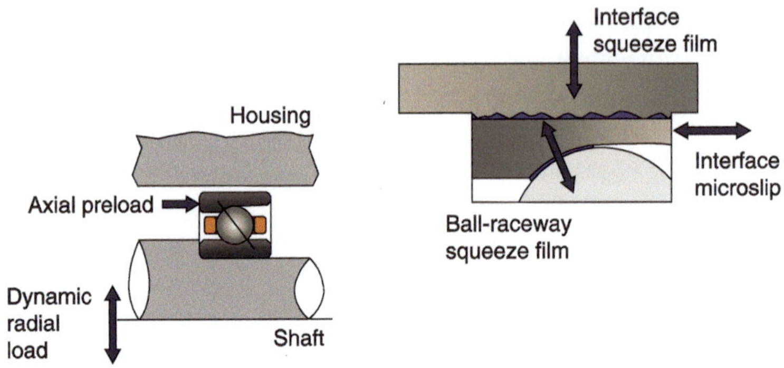

Figure 2.24 **Sources of damping associated with precision ball bearings**

spring preloaded bearing induces the preloaded ring to tilt. For the case of a 7006 15-degree angular contact bearing, a 1 μm radial displacement of the inner ring may create a 4 μm movement between the outer ring and housing.

Squeeze film damping at the rolling element–raceway contacts has perhaps received the most attention from research workers. One investigation, performed by [10], using two discs in rolling contact is worthy of note because of its elegant simplicity, isolating a single rolling contact and providing clear evidence of the importance of the effect of lubrication on contact damping. For complete bearings, [11] published the results of an experimental investigation into damping levels in tapered roller bearings as early as 1976, whilst in 1983 [12] published experimental measurements of damping exhibited by a pair of precision angular contact ball bearings. An approximate analysis of the squeeze films occurring at the ball-raceway contacts was also performed [13], concluding that they are significant sources of damping. A numerical simulation of the dynamic behaviour of ball bearings and ball bearing-shaft systems was undertaken by [14] and again damping was found to be significant. Perhaps the most comprehensive analysis of squeeze film damping at ball–raceway contacts was performed by [15–18], where they formulated both stiffness and damping coefficients for ball bearings subject to mixed elasto-hydrodynamic lubrication. In this series of papers, they developed the analysis, derived an empirical formula for damping and used the results in a non-linear dynamic simulation of a structural system to determine its overall dynamic response.

Unlike ball-raceway contacts, conditions at the inner ring-shaft and in particular the outer ring–housing interfaces are not so clearly defined.

Surface geometries and roughness are not as well controlled as the bearing's rolling surfaces, the fits may vary from a few microns interference to a few microns clearance and rings may be clamped or left free to slide, depending on the bearing mounting arrangement. Furthermore, depending on the direction of applied dynamic loads, the interfaces may be subject to squeeze and/or sliding motions. It is not unreasonable to expect damping at these interfaces to be somewhat inconsistent, varying not only from one bearing arrangement to another, but even on the same arrangement after dismantling and reassembly.

Research into damping at interfaces is either related to structural vibration or high speed machinery, where squeeze film dampers are purposefully designed. Perhaps the most relevant research in precision bearings is by [19], where they examined the effect of surface roughness and lubrication on the damping capacity of a series of flat circular discs clamped together and subjected to a dynamic squeeze film action. When the interfaces were dry, they found negligible levels of damping occurred. However, when lubricant was introduced into the interfaces, damping levels significantly increased but the maximum damping levels were found to occur when the lubricant film thickness was low and the dynamic load transmitted partly by metal–metal contact of the rough surfaces and partly by the lubricant film. They concluded that under these conditions useful levels of energy dissipation could be achieved. The conditions of lubricated rough surfaces and squeeze film loading occur at bearing interfaces and although the damping levels the authors found do not apply directly to ball bearing–housing or –shaft interfaces, they do point towards these interfaces being a useful source of damping in rotating machinery.

2.8.2 Squeeze film damping at ball–raceway contacts

Squeeze film damping at ball–raceway contacts is particularly significant as it must occur in all ball bearings and it enables a lower bound estimate of bearing damping levels to be made and used to determine a worst case dynamic response of a complete system. This source of damping is likely to be the dominant source for ball bearings that have rings clamped or interference fitted. For squeeze film damping at ball–raceway contacts [18], is particularly useful as it provides an empirical formula quantifying the ball–raceway damping coefficients in terms of bearing and lubricant parameters. However, it should be borne in mind that empirical predictions are only accurate for the range of conditions used in deriving the formula.

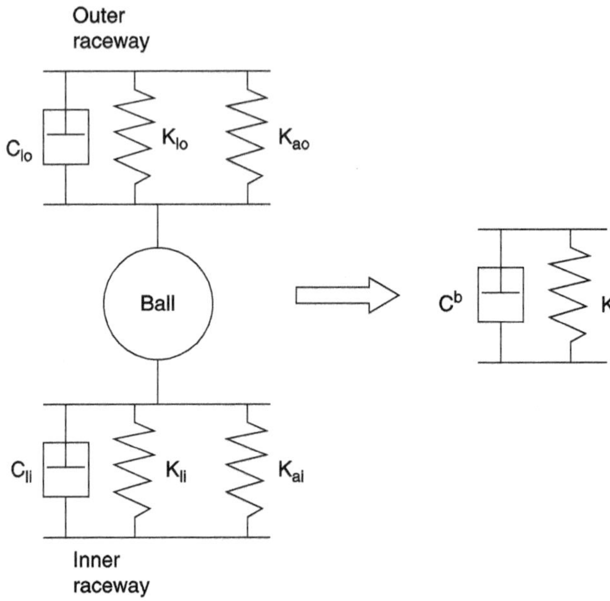

Figure 2.25 Model of ball–race contacts [18]

The damping was considered to be produced entirely within the lubricant film in and around the Hertzian zone at each of the ball–raceway contacts and modelled the film as a linear spring and viscous damper (Figure 2.25) [17]. In this model the normal load acting on the contact is transmitted through the central Hertzian zone, which therefore determines the film stiffness whilst the damping is generated by squeezing of the lubricant surrounding the Hertzian zone. The model takes into account roughness of the ball and raceway surfaces, so within this zone exists a thin elasto-hydrodynamic lubricant film of known average thickness 'h' together with ball and raceway surface asperities. Thus load transmitted across the contact is shared between the lubricant film and the asperities. Lubricant in the space surrounding the Hertzian zone provides a hydrodynamic squeeze film action in the presence of a dynamic load. The dynamic model of the ball and inner and outer contacts is shown in Figure 2.25. Stiffnesses K_{li} and K_{lo} are the stiffnesses of the lubricant film at inner and outer contacts and they act in parallel with stiffnesses K_{ai} and K_{ao} representing asperity contact. Damping acts in parallel with the film stiffness and is represented by the coefficients C_{li} and C_{lo}. The net stiffness and damping of the lubricant films associated with each ball is represented by K^b and C^b respectively where:

$$[K^b + j\omega C^b]^{-1} = [K_{li} + K_{ai} + j\omega C_{li}]^{-1} + [K_{lo} + K_{ao} + j\omega C_{lo}]^{-1} \quad [2.53]$$

where $j = -1^{0.5}$ and ω is the frequency of an oscillating load transmitted by the ball. In a complete bearing model the net film stiffness given above acts in series with the stiffness associated with elastic deformation of the ball–raceway contacts.

The empirical formulae for film stiffness and damping at a single contact were derived from a detailed numerical analysis by curve fitting calculated results. With this approach the results are valid for the range of conditions used for curve fitting. Extrapolating outside of this range inevitably leads to inaccurate predictions. Here, the empirical formulae are quoted as:

$$\underline{K}_l = a_1 G^{a2} \underline{U}^{a3} \underline{W}^{a4} [a_5 - (k\exp(k))^{a6}] K'^{a7} \Lambda^{a8} \exp(-\Lambda^{a9}) \gamma'^{a10} \quad [2.54]$$

$$\underline{C}_l = c_1 G^{c2} \underline{U}^{c3} \underline{W}^{c4} [c_5 - k^{c6} \exp(k^{c7})] K'^{c8} \Lambda^{c9} \exp(-\Lambda^{c10}) \gamma'^{c11} / \underline{r} \quad [2.55]$$

where the coefficients are: $a_1 = -0.4053$, $a_2 = -0.2521$, $a_3 = -0.6995$, $a_4 = 1.1678$, $a_5 = -0.5891$, $a_6 = -0.3102$, $a_7 = -0.0358$, $a_8 = -0.3368$, $a_9 = -0.928$, $a_{10} = -0.0399$, $c_1 = -0.73$, $c_2 = -0.7511$, $c_3 = -0.5871$, $c_4 = 1.1963$, $c_5 = -0.5188$, $c_6 = 0.1275$, $c_7 = -0.4548$, $c_8 = 0.0013$, $c_9 = -0.3521$, $c_{10} = -0.8472$, and $c_{11} = -0.0366$.

The equations are in a non-dimensional form and can be applied to any ball bearing type and size. The symbols are defined to be $\underline{K}_l = K_l/(E'R_x)$, $\underline{C}_l = C_l U/(E'R_x r)$, $G = \alpha E'$ and $U = (U_a + U_b)/2$, the contact's average entrainment speed and $\underline{U} = \eta_o U/(E'R_x)$, W = total contact load and $\underline{W} = W/(E'R_x^2)$, k = a/b, the contact's ellipticity ratio, K′ is the asperity contact stiffness, $\Lambda = h_c/R_x$ is the hydrodynamic roughness parameter, and γ' is a surface pattern parameter and r is ball radius.

E′ is the combined Young's modulus of the ball and raceway materials, R_x is the equivalent contact radius in the direction of rolling, η_o is absolute lubricant viscosity at the operating conditions, α is the lubricant's pressure viscosity coefficient, $\underline{r} = r/R_x$ where r is ball radius and h_c is the central lubricant film thickness of the contact.

An empirical formula for asperity contact stiffness was also derived:

$$\underline{K}_a = b_1 G^{b2} \underline{U}^{b3} \underline{W}^{b4} [b_5 - k^{b6} \exp(k^{b7})] K'^{b8} \Lambda^{b9} \exp(-\Lambda^{b10}) \gamma'^{b11} \quad [2.56]$$

where $b_1 = -0.1828$, $b_2 = 0.1065$, $b_3 = -0.4931$, $b_4 = 0.7239$, $b_5 = 0.0588$, $b_6 = 0.442$, $b_7 = 0.1081$, $b_8 = 1.0418$, $b_9 = 0.5396$, $b_{10} = 1.6221$ and $b_{11} = 0.0861$.

The formulae are theoretical and difficult to verify experimentally as experimental apparatus's that isolate the damping generated by ball–raceway contacts from other sources of damping are highly sophisticated. Nonetheless, the formulae predict the trends observed in practice and damping levels that are realistic and representative of a ball bearing for which both inner and outer rings are interference fitted to a structure.

Equations 2.54 to 2.56 require a detailed knowledge of the ball–raceway contact geometry, surface roughness and lubrication conditions to determine film stiffness and damping levels. However, precision ball bearings are characterised by steel or ceramic materials, well finished rolling surfaces and ball–raceway contact ellipticity ratios between 8 and 10. Assuming the bearings are also lubricated by a mineral oil, then for the purpose of estimating typical damping levels in bearing applications, representative values may be assigned to the parameters Λ, γ, K' and k. These are $\Lambda = 2.0$, $\gamma = 1$, $K' = 0.001$ and $k = 9$. This enables a simple formula for the damping force at a ball–raceway contact to be derived in terms of commonly known bearing and lubricant parameters and operating conditions:

$$\underline{C} = 0.000329\,\underline{U}^{-0.5871}\,\underline{W}^{1.1963}/\underline{r} \qquad [2.57]$$

Equation 2.57 assumes both balls and raceways are made of steel, for which the combined modulus $E' = 2.29 \times 10^{11}\,N/m^2$. For hybrid ball bearings with silicon nitride balls and steel raceways, $E' = 2.73 \times 10^{11}\,N/m^2$ and Eq. 2.57 is modified to:

$$\underline{C} = 0.000288\,\underline{U}^{-0.5871}\,\underline{W}^{1.1963}/\underline{r} \qquad [2.58]$$

For mineral oil, a pressure–viscosity coefficient of $1.88 \times 10^{-8}\,m^2/N$ is assumed for both Eqs 2.57 and 2.58. Thus, although the hybrid bearing offers many advantages over all steel bearings in precision applications, damping is not one of them. In terms of absolute parameters, Eq. 2.57 may be expressed as:

$$C = 3.952 \times 10^{-11}\,R_x^{-0.805}\,U^{-1.5871}\,\eta_o^{-0.5871}\,W^{1.1963}/r \qquad [2.59]$$

and for hybrid bearings, Eq. 2.58 becomes:

$$C = 3.108 \times 10^{-11}\,R_x^{-0.805}\,U^{-1.5871}\,\eta_o^{-0.5871}\,W^{1.1963}/r \qquad [2.60]$$

where C is N/(m/s), R_x is m, U is m/s, W is N, η_o is Ns/m^2 and r is m. It can be seen that damping increases with load (W) but decreases with bearing size (R_x and r), speed (U) and viscosity (η_o).

2.8.3 Ball bearing damping coefficients

Equations 2.59 and 2.60 apply to a single ball–raceway contact. The damping coefficient exhibited by a complete bearing is determined by calculating the damping coefficients associated with each ball using Eq. 2.53 and then summing the coefficients for all balls in the appropriate directions:

$$C_x = \Sigma_1^N C_{bj} \cos\alpha_j \cos\theta_j$$
$$C_y = \Sigma_1^N C_{bj} \cos\alpha_j \sin\theta_j$$
$$C_z = \Sigma_1^N C_{bj} \sin\alpha_j \qquad [2.61]$$
$$C_{xx} = 0.25 D_p^2 \Sigma_1^N C_{bj} \cos\alpha_j \sin\theta_j$$
$$C_{yy} = 0.25 D_p^2 \Sigma_1^N C_{bj} \cos\alpha_j \cos\theta_j$$

where C_{bj} is the damping associated with ball number j and is given by:

$$C_{bj} = C_o C_i / (C_o + C_i)$$

and C_o and C_i are the damping coefficients for outer and inner contacts respectively. D_p is the bearing pitch circle diameter, α is contact angle and θ is the ball's angular position. Equations 2.59 to 2.61 are conveniently integrated into the bearing load–deflection model described in Section 2.8, to give the damping coefficients of bearings under any loading conditions.

2.8.4 Example of ball bearing damping characteristics

The axial damping coefficient of the 7006 15-degree angular contact bearing described in Section 2.9.1 is shown for different preloads and speeds in Figure 2.26. The bearing is lubricated with a 20 cSt mineral oil

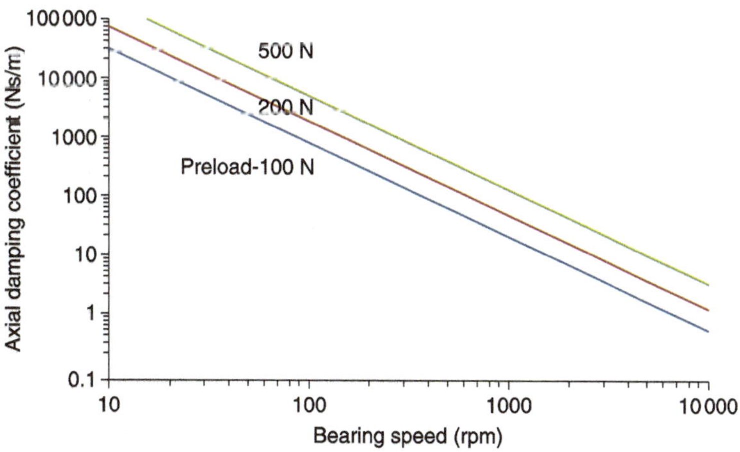

Figure 2.26 Axial damping coefficient for preloaded 7006 15-degree angular contact bearing

and viscosity is assumed to remain constant regardless of bearing speed and temperature. The results are plotted for rotational speeds between 10 and 10000 rpm, and the damping coefficient at speeds between 100 and 3000 rpm are representative of practical bearing–rotor systems. Below 100 rpm the coefficients are unrealistically high, whereas above 3000 rpm they are unrealistically low, illustrating the limited range for which the empirical formula was derived.

2.8.5 Damping at bearing interfaces

Damping at ball bearing interfaces is still quantified empirically. One approach to doing this is briefly described in [20]. Here the authors assumed interface damping was predominantly hysteretic, being due to micro-slip induced by dynamic loads applied to the bearing. Levels of energy dissipation due to hysteresis were determined by measuring the static load–deflection relationship of bearings *in situ* in machine applications. One example of this is shown in Figure 2.27. In this case a spring preloaded 7210 15-degree angular contact bearing is subject to a radial load.

Hysteresis is experimentally determined by measuring deflections for the load increasing and decreasing. The stiffness of the bearing and interface is assumed to take the form:

$$K_s = K(1 + i\,\mathrm{Tan}\delta) \qquad [2.62]$$

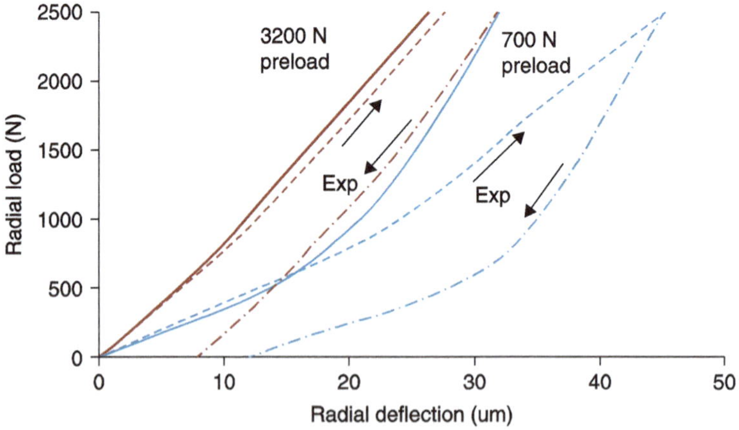

Figure 2.27 Experimental and theoretical radial load deflection curves for preloaded 7210 angular contact bearing

and for small deflections, the load F is related to deflection x by:

$$F = K_s.x$$

where i is the square root of −1 and Tanδ is a loss tangent. The energy dissipated by hysteresis in a loading and unloading cycle is:

$$W = \int_1 F.dx - \int_2 Fdx = 0.5K(x_m^2 - x_h^2)\text{Tan}\delta \qquad [2.63]$$

where x_m is the maximum deflection measured and x_h is a residual deflection left when the bearing is completely unloaded. In Figure 2.27, the experimental measurement for a preload of 700 N shows a complete hysteresis loop for which W is the area contained within the dashed curves. Measuring W and using Eq. 2.63 gives Tanδ = 0.24. Measurements performed over a range of applications, but only in a radial direction, found Tanδ to vary from 0.05 to as high as 0.4, 0.05 to 0.1 being typical of bearings for which the rings were either an interference fit or firmly clamped, whilst for spring preloaded bearings where one ring is free to slide or to tilt under applied radial loads $0.2 < \text{Tan}\delta < 0.4$.

Hysteretic damping is dependent on amplitude of motion but independent of frequency. The damping coefficient is therefore inversely proportional to frequency and given by:

$$C_s = \frac{K}{\omega}.\text{Tan}\delta$$

2.8.6 *Total bearing damping*

Measurements of interface damping obtained by applying a static load as described in Section 2.10.5 do not include the effect of squeeze films at the ball–raceway contacts that are generated when the bearing is rotating. Interface damping acts in series with damping at the rolling contacts and in principle the two should be added using a reciprocal law. This results in a damping coefficient, the complexity of which is not justified by the accuracy to which it can predict damping levels in a given application. Small variations in machine build quality and hence ring fits can significantly affect interface damping levels, whilst the effect of temperature on lubricant viscosity, for example, can have a significant influence on squeeze film damping.

In practice, for ball bearings, approximate damping coefficients are used to estimate the dynamic response of rotating machinery. Either viscous damping coefficients as derived in Section 2.10.2 or hysteretic damping coefficients may be used. Viscous coefficients for a complete

Table 2.7 **Approximate damping loss tangents for ball bearings**

Direction	Loss tangent and mounting type		
	spring preloaded	*clamped rings*	*face step preloaded*
Radial X	0.2–0.4	0.05–0.1	0.05–0.1
Radial Y	0.2–0.4	0.05–0.1	0.05–0.1
Axial Z	N/A	0.05	0.05–0.1
Moment XX	N/A	N/A	0.05–0.1
Moment YY	N/A	N/A	0.05–0.1

bearing may be determined from a load–deflection model modified to include damping effects. For hysteretic damping, the loss tangents in Table 2.7 are suggested.

2.9 Motion error

There is a considerable overlap between the fields of bearing motion error and bearing noise and vibration in as much as many of the factors that need to be controlled to achieve reductions in one also benefit the other. Within the rolling bearing industry, electric motors used in domestic, office and military applications provided the initial driver for reducing noise and vibration. Research work on this subject started around 1950, with Perret [21] and Meldau [22] investigating the variation in elastic deflection of rotating bearings under load. Early progress on the understanding of the effect of bearing surface quality on bearing noise and vibration was published by Tallian and Gustafsson [23] and Sheftel [24] in the mid-1960s and, based on this work, a method was developed to measure bearing surface quality suitable for mass production [25]. The early work developed an empirical understanding between bearing surface quality and vibration. During the 1970s and 1980s, research work aimed at producing a more definitive understanding with the development of analytical models relating surface topography to vibration amplitude and frequency, initially for two discs in rolling contact [26], then for point defects in rolling bearings [27,28] and eventually for thrust loaded ball bearings [29,30]. More recently, researchers have developed bearing models based on dynamic simulation [31–33] that cannot only predict the time signature of ball bearings but can also be used to determine the complete dynamic performance of bearing–machine systems.

During the 1980s and 1990s, the computer disc drive industry emerged as a major application for ball bearings and motion error was recognised as a critical bearing performance parameter directly influencing disc capacity. Unlike the electric motor application, where bearings may operate under a diverse range of conditions, this application was focused on low cost, miniature bearings operating under specific conditions of light axial load and medium speed at near ambient temperatures. Early research work, performed mainly in Japan, developed an understanding of the factors that determine the radial motion error of disc drive ball bearings [34–38] and later focused specifically on reducing the 'Non-Repeatable Run Out' (NRRO) [39–43]. Because in this application bearing speeds are moderate, the NRRO was found to be largely influenced by ball size variation.

In terms of peak–peak motion error amplitudes, ball bearings can achieve a creditable performance. Amplitudes as low as 48 nm have been reported in scientific papers [41], for ball bearings used in computer hard disc drives. This is comparable to the motion error of some types of fluid film, but the disadvantage of ball bearings is that the motion error is predominantly asynchronous whereas for fluid film bearings it is mostly synchronous. The main reason is that for ball bearings, motion error frequencies relate to the orbital and spinning speeds of the balls and these can never be harmonic of shaft speed in a practical bearing design. Ball bearing motion error is influenced by a large number of parameters, some a function of the bearing design and manufacturing processes, others being dependent on application conditions. However, there are relatively few basic mechanisms by which motion error can be generated and by understanding these, the influence of different parameters can be more clearly defined and in many cases, even quantified.

Thus, in this section, theoretical models that explain the basic mechanisms and hence provide a framework for interpreting the effect of bearing and mounting parameters on motion error are first presented. Subsequently, the models are used to examine the influence of important bearing and application variables.

2.9.1 Dynamic model for estimating ball bearing motion error

In an ideal case of a geometrically perfect ball bearing, rotating at constant speed, properly lubricated and subject only to a concentric axial load, there is nothing to generate relative movement between inner and outer rings and hence the bearing should, in theory, exhibit zero motion error. It follows that motion error must arise from imperfections within the bearing or non-asymmetric loading, either of which create fluctuations in ball load that act as a source of excitation. The imperfections may be, for example, out of round raceway surfaces created by the bearing's manufacturing process or induced by application conditions, such as a lobed shaft or housing seating. Non-axisymmetric loading is also a feature of most real ball bearing applications where, more often than not, it is necessary to support a radial or moment load in addition to an axial preload. However, as in high precision bearing applications, axial preload is dominant, a simple analytical model relating motion error to the size and type of imperfections on an axially loaded ball bearing's rolling surfaces is given in Appendix 2A. Here just the essential features of the model are described in order to provide an understanding of the mechanism by which surface imperfections generate vibration and in particular the relationship between amplitude and wavelength of the feature and amplitude and frequency of the vibration it produces.

For a ball bearing rotating at constant speed and subject to axial load, the model (Appendix 2A) derives a series of transfer functions relating the amplitude and wavelength of imperfections on the ball or raceway surfaces to the magnitude and frequency of force or moment generated between inner and outer rings. For simplicity it considers a bearing for which the inner and outer rings remain fixed in position and simple harmonic waveforms are superimposed on ball and raceway surfaces (Figure 2.28).

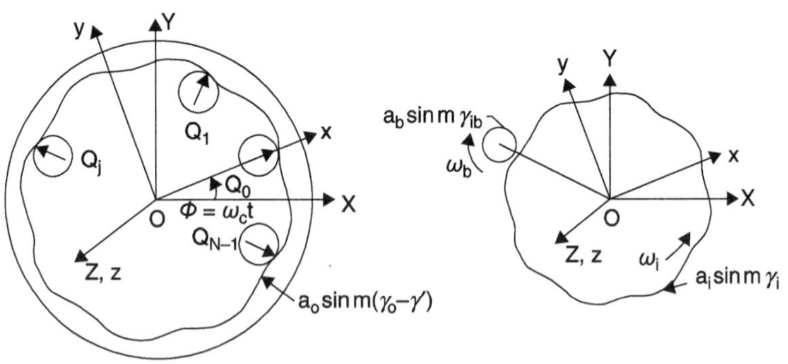

Figure 2.28 Ball bearing with imperfections on ball, inner and outer ring surfaces

As inner and outer rings are fixed in position then ball loads must fluctuate as the imperfections pass through the ball–raceway contacts. The model derives expressions for the fluctuation in each ball load as a function of its angular position, so that the net dynamic force acting between inner and outer rings is found simply by summing all ball loads. Motion error, for the case of a bearing mounted with its rings free to move, is derived from the dynamic force assuming force magnitude and motion amplitude are related by bearing stiffness (Figure 2.29).

Here the bearing is represented by a spring of stiffness K in series with a source of excitation. When inner and outer rings are fixed in position the excitation is force, $F_c \sin\omega t$, and when they are free to move the excitation is displacement, $A_f \sin\omega t$. The two are assumed to be related by:

$$F_c = K\,A_f$$

Transfer functions, expressing dynamic force or moment in terms of imperfection amplitude and wavelength for imperfections on each of the ball, inner and outer raceways are derived. In total, there are 15 transfer functions; however, only those giving axial and radial force are summarised here. These are for the case of a ball bearing with a stationary outer ring and a rotating inner ring, subject to axial load only.

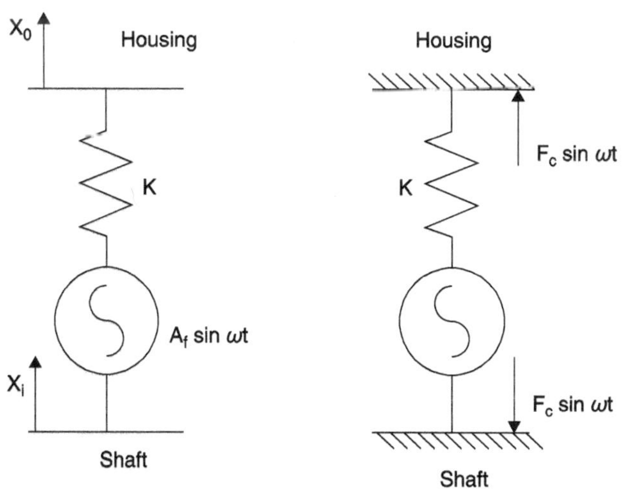

Figure 2.29 **Relationship between dynamic force and displacement**

2.9.2 Transfer functions for outer raceway waviness

The outer raceway imperfection is assumed to be simple harmonic and is referred to as waviness. It is described by:

$$a = a_o \sin m(\gamma_o - \gamma_o') \qquad m = 1, 2, 3 \ldots$$

where a_o is the amplitude of the waviness, γ_o' is a phase angle defining the position of the waviness with respect to a fixed reference frame and m is an integer giving the number of waves/circumference. The axial component of dynamic force is given by:

$$F_a/a_o = N K_a \sin m(\omega_c t + \gamma' + \pi(N-1)/N) + (m = N, 2N, 3N) \ldots$$
$$\tfrac{1}{2} N \sum_{i=1}^{\infty} K_{ai} \{\sin[(m+i)(\omega_c t + \pi(N-1)/N) + m\gamma'] + \quad (m \pm i = N, 2N, 3N) \ldots$$
$$\sin[(m-i)(\omega_c t + \pi(N-1)N) + m\gamma']\} \; (i = 1, 2, 3) \qquad [2.64]$$

where N is the number of balls in the bearing, i is an integer, K_a is the axial component of ball–raceway stiffness taking into account both inner and outer contacts and K_{ai} is a time dependent component of stiffness, as defined in Appendix 2A. The two terms contained in Eq. 2.64 give the component of dynamic force arising directly from the waviness and the component of force due to the bearing varying in stiffness as it rotates. The second term is due to non-linear load–deflection characteristics and is only significant when the bearing load situation or the amplitude of the imperfection produces large fluctuations in ball load.

Forcing frequencies are always harmonic of the ball pass rate $N\omega_c$ and are produced by particular waviness wavelengths. The dominant wavelengths have an integer multiple of N waves per circumference. For this situation, all balls roll over high and low points of the waviness in phase and produce the maximum fluctuation in force. However, other wavelengths can produce vibration at the same frequencies due to the time dependent component of bearing stiffness, K_{ai}. For bearings subject only to axial load and normal levels of waviness, i.e. amplitudes less than 1 μm, K_{ai} is negligible. However, on bearings where, for example, ring distortion may create waviness amplitudes of up to 10 μm in amplitude or significant radial load or misalignment is present, then K_{ai} is significant and should be taken into account. In this situation, waviness wavelengths of m, m + 1, m − 1, m + 2, m − 2 . . ., where m ± i is an integer multiple of N, the number of balls can produce a dynamic force. In this case the magnitude of force at any one frequency is the vector sum of that due to the individual wavelengths. It should be noted that the contact compliance parameter K_{ai} may generally be expected to reduce in magnitude as i increases and in practice may be negligible for values of i greater than N.

Similar expressions may be derived for radial force and moments. However, they contain additional terms that arise due to orbital rotation of the balls with respect to the direction of the measurement. For radial vibration:

$$\begin{aligned} F_{rx}/a_o = \tfrac{1}{2}\, Nk_r\, &\{\sin[(m+1)(\omega_c t+(N-1)\pi/N)+m\gamma']+ \\ &\sin[(m-1)(\omega_c t+(N-1)\pi/N)+m\gamma']\}+ \\ &\tfrac{1}{4}\, N\sum_{i=1} k_{ri}\{\sin[(m+1+i)(\omega_c t+(N-1)\pi/N)+m\gamma']+ \\ &\sin[(m-1+i)(\omega_c t+(N-1)\pi/N)+m\gamma']+ \\ &\sin[(m+1-i)(\omega_c t+(N-1)\pi/N)+m\gamma']+ \\ &\sin[(m-1-i)(\omega_c t+(N-1)\pi/N)+m\gamma']\} \end{aligned} \qquad [2.65]$$

where $m \pm 1 \pm i = pN$ and $p = 1, 2, 3 \ldots$

2.9.3 *Transfer functions for inner raceway waviness*

The mechanism by which waviness on the inner raceway produces dynamic force is the same as that for the outer raceway. However, inner ring rotation must be taken into account and the main effect of this is to change the basic frequency of the force from ball pass rate to ball pass rate relative to the inner ring, i.e. from $N\omega_c$ to $N(\omega_I - \omega_c)$.

For convenience, let $\omega_{ci} = \omega_I - \omega_c$ then substitution of this into Eq. 2.63 and simplifying as in the previous section gives the transfer functions for inner ring waviness:

$$\begin{aligned} F_a/a_i = NK_a &\sin m(\omega_{ci}t+\pi(N-1)/N)+ \\ &\tfrac{1}{2}N\sum_{i=1}^{\infty} K_{ai}\{\sin[(m\omega_{ci}+i\omega_c)t+(m+i)\pi(N-1)/N)]+ \\ &\sin[(m\omega_{ci}-i\omega_c)t+(m-i)\pi(N-1)/N)]\} \qquad m = N, 2N, 3N\ldots \\ F_r/a_i = \tfrac{1}{2}\, NK_r &\{\sin[(m\omega_{ci}+\omega_c)t+(m+1)\pi(N-)/N)]+ \\ &\sin[(m\omega_{ci}-\omega_c)t+(m-1)\pi(N-1)/N)]\}+ \end{aligned}$$

$$\tfrac{1}{4}\, N \sum_{i=1}^{\infty} K_{ri} \{ \sin[(m\omega_{ci} + (1+i)\omega_c)t + (m+1+i)\pi(N-1)/N)] +$$
$$\sin[(m\omega_{ci} + (1-i)\omega_c)t + (m+1-i)\pi(N-1)/N)] +$$
$$\sin[(m\omega_{ci} + (I-1)\omega_c)t + (m-1+i)\pi(N-1)/N)]] + \qquad [2.66]$$
$$\sin[(m\omega_{ci} - (1+i)\omega_c)t + (m-1-i)\pi(N-1)/N)]\}$$
$$M \pm 1 = N,\ 2N,\ 3N\ldots$$

2.9.4 Transfer functions for ball imperfections

Ball motion is complex as it can spin and roll; however, for the purpose of understanding the mechanism of dynamic force generation, the ball is assumed to rotate about one axis so that force is produced by waviness present in a diametrical plane passing through the ball centre, perpendicular to the spin axis. Ball waviness is therefore described by:

$$a = a_b \cos(m\phi) \qquad m = 1, 2, 3\ldots$$

The corresponding variation in the bearing's radial internal clearance experienced by the ball is then:

$$dP_D = 2a_b \cos(2m\omega_b t) \qquad m = 1, 2, 3\ldots$$

where it is assumed that only even numbers of waves/circumference generate a local internal clearance variation. This enables axial force to be expressed as:

$$F_a/a_b = 2K_a \cos(2m\omega_b t) + \sum_{i=1}^{\infty} K_{ai}[\cos(i\omega_c + 2m\omega_b)t + \cos(i\omega_c - 2m\omega_b)t]$$

Similarly, for radial force:

$$F_{rx}/a_b = K_r[\sin(2m\omega_b + \omega_c)t + \sin(2m\omega_b - \omega_c)t] +$$
$$\sum_{i=1}^{\infty} K_{ri}[\sin((i+1)\omega_c + 2m\omega_b)t + \sin((i-1)\omega_c - 2m\omega_b)t + \qquad [2.67]$$
$$\sin((i-1)\omega_c + 2m\omega_b)t + \sin((i+1)\omega_c - 2m\omega b)t]$$

The above equations relate dynamic force to the waviness on just one ball. Assuming all balls in a bearing are manufactured in one batch and that their surface geometries are therefore similar, then the total vibration force for a given waviness wavelength is approximately the root mean squared (RMS) sum of that due to all N balls.

2.9.5 Ball size variation

Ball size variation may be considered as a special case of ball waviness for which the number of waves/circumference, $m=0$. The equations of the preceding section can then be simplified to

$$F_a/a_b = 2K_a + 2\sum_{i=1}^{\infty} K_{ai} \cos i\omega_c t$$

$$F_{rx}/a_b = 2K_r \sin\omega_c t + \sum_{i=1}^{\infty} K_{ri}[\sin(i+1)\omega_c t + \sin(i-1)\omega_c t] \qquad [2.68]$$

where now a_b is the departure of the ball radius from the standard ball radius. Again the above equations relate dynamic force to ball size error for a single ball. Note that the frequencies produced are only harmonic of cage speed, whereas ball waviness produces frequencies harmonic of ball rotational speed modulated by cage speed. The above equations suffice to determine the relative importance of ball size variation on bearing motion error; however, the analysis may in principle be extended to consider the general case of all balls being of different diameter.

2.9.6 Experimental validation of the dynamic model

Experimental measurements of the motion error of a ball bearing fitted to a mechanical system were taken and used to validate the dynamic model described above. A full description of the experimental procedure is given in [30] and only a summary is given here.

The experimental set-up is shown in Figure 2.30. The inner ring of the test bearing is supported by a mandrel fitted to a quiet running hydrostatic spindle, while the outer ring of the test bearing is mounted in a rigid, steel housing. Thrust load is applied to the housing by a pneumatic cylinder via three flexible rubber pads. The spindle and loading mechanism are both attached to a heavy base plate mounted on vibration isolators. Spindle drive is by means of a soft, endless rubber belt from an electric motor mounted on a separate vibration isolated base. Spindle speed is constant at 1800 rpm.

With this set-up, direct measurement of bearing vibration force was not possible because of flexibility within the loading mechanism and spindle. As a result, validation consisted of comparing the experimentally measured amplitude response of the bearing housing with that predicted

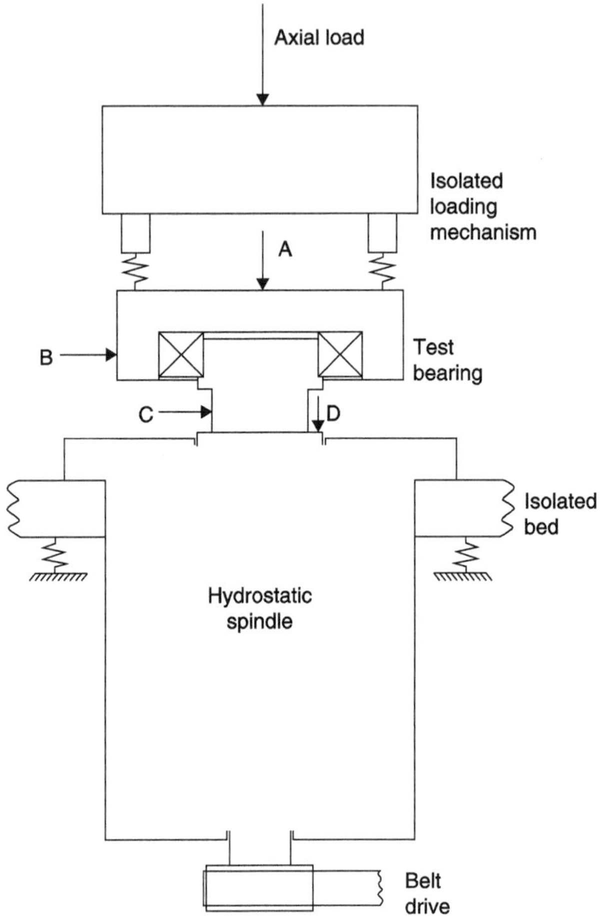

Figure 2.30 Arrangement for measuring bearing vibration

by theoretical force levels and measurements of the system's structural response characteristics. Narrow band spectral analysis was used to measure axial and radial components of motion error and these are compared with the theoretical spectral lines predicted by the model described above.

Bearing motion error, in terms of acceleration, was measured on the housing in an axial direction concentric with the spindle axis and radially in the plane containing the effective centre of the bearing. An essential requirement of the arrangement was for housing motion produced by the test bearing to be clearly measurable over and above any motion produced by extraneous sources. Spindle generated motion was therefore an important parameter that was kept to a minimum by using a precision boring spindle fitted with hydrostatic bearings.

In the frequency range of interest, 20 to 1000 Hz, the response characteristics of the experimental arrangement contained a number of resonances. Thus in order to compare motion errors measured on the bearing housing with those deduced from measurements of surface waviness, it was necessary to measure the response characteristics of the experimental arrangement. Receptance of the spindle and the bearing housing in axial and radial directions was therefore measured and used to factor the theoretically predicted motion error spectrum.

Waviness of the inner and outer raceway and ball surfaces of the test bearing were measured using a roundness tester coupled to a spectrum analyser to generate a narrow band spectrum of RMS waviness amplitude versus wavelength. Figure 2.31 is an example of the outer raceway amplitude spectrum.

The relevant wavelengths, i.e. those predicted by the model described above, were identified and used to calculate a theoretical motion error spectra. These were compared with the spectra measured directly on the bearing housing in axial and radial directions. The results are shown in Figure 2.32.

Good agreement between experimentally measured and theoretically predicted vibration spectra is achieved over the range of wavelengths and frequencies considered. The experimental spectra contain all of the predicted frequencies and levels generally compare to within ±6 dB. Additional frequencies present in the experimental spectra are either harmonic of spindle running speed or correspond to odd harmonics of ball rotational speed. The former was attributed to motion error of the

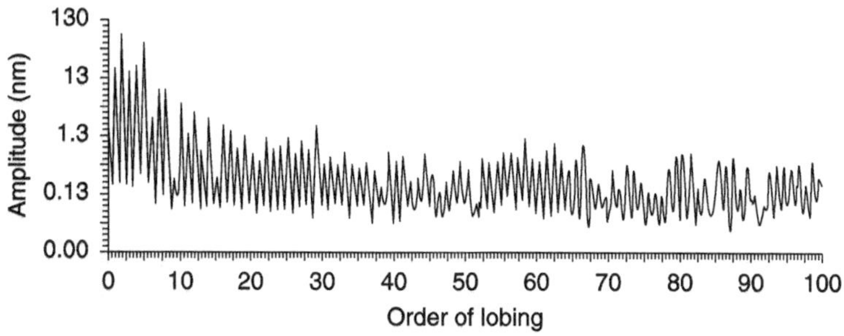

Figure 2.31 Spectral analysis of outer raceway surface

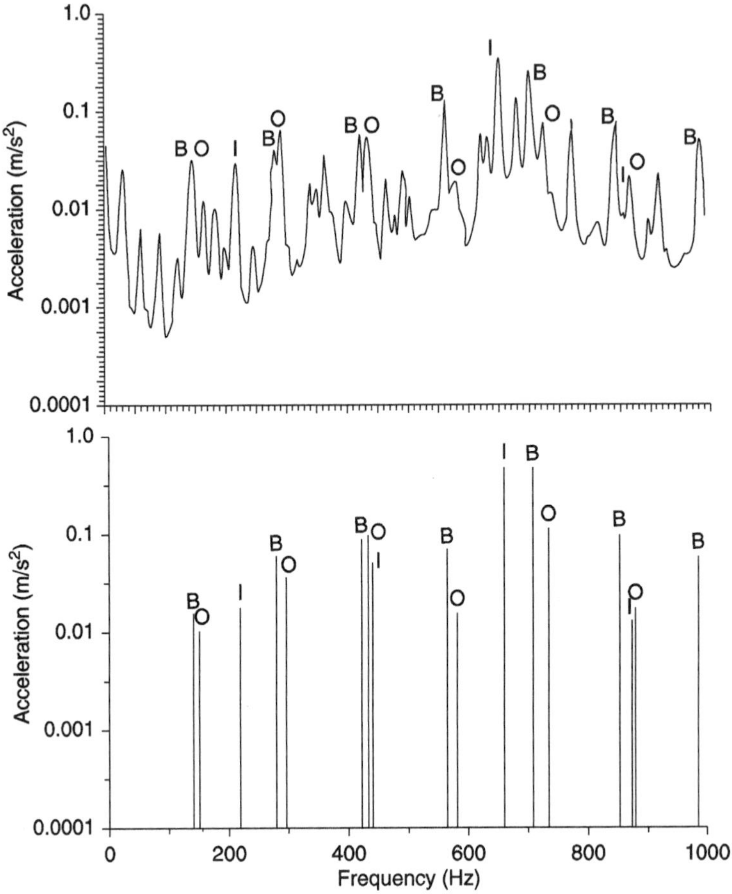

Figure 2.32 **Comparison of measured and theoretical axial acceleration spectra**

spindle, whereas the absence of the latter from the theoretical spectra is a limitation of the model.

2.9.7 Review of sources of ball bearing motion error

The model described in the previous section can be used to relate many types of bearing surface imperfection to amplitudes and frequencies of motion error. In this section, a review of the main sources of motion error is presented for the case of axially preloaded ball bearings. Whilst the model can be used to assess a wide range of conditions, it does not cover all sources of motion error. Cage related issues fall completely outside the scope of the model and impulsive vibration due to short wavelength surface features such as roughness or damage are not fully explained. The review starts with cage related issues, as in the vast majority of applications these are symptomatic of non-optimum bearing set-up. This is followed by sources of motion error related to bearing surface quality and finally the influence of application conditions.

Cage related motion error

The function of the cage, or separator as it is sometimes referred to, is to keep balls equally spaced and to prevent ball–ball contact, a condition that generates high sliding speeds and for which satisfactory lubrication cannot be achieved. Normally the cage rotates about a fixed centre at a constant speed, but under some operating conditions it can interact with the balls or the ring riding surface, if it is ring located, and create high dynamic forces. Cage related issues fall into two main categories – cage instability and cage binding.

Cage instability

Usually this is associated with an audible noise – cage rattle, hammer or squeal, depending on the nature of the instability. Although an audible noise is emitted, cage instability produces high impulsive forces between itself and the balls, resulting in high amplitude motion errors of the bearing as a whole.

Whilst different cage designs may exhibit different stability characteristics, for preloaded, precision ball bearings, cage instability is very seldom a bearing design or manufacturing quality issue. In practice, it is experienced more often in very lightly loaded bearings and is associated with insufficient traction at the ball–raceway contacts to overcome frictional drag within the bearing. Loss of traction results in a 'stick–slip' motion of the balls that is highly erratic, causing them to intermittently make and break contact with the cage. Lightly loaded grease lubricated bearings frequently exhibit cage instability, particularly in the first few hours of running when the grease is not optimally distributed.

The effect of cage instability can be clearly seen by measuring motion error or vibration level as a function of bearing preload using the test described in Section 1.3.6. Figure 2.33 shows vibration level in terms of

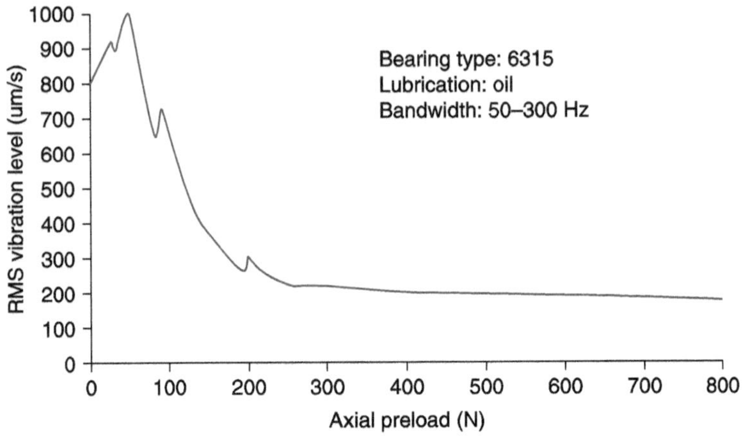

Figure 2.33 Examples of the effect of axial preload on the low frequency vibration of a single-row radial ball bearing

RMS velocity in the 50 to 300 Hz frequency band for a single-row radial ball bearing.

The relationship shown is typical. Vibration level varies very little over a wide range of preloads, but as preload is reduced a threshold value is reached at which there is a sudden and large increase in vibration accompanied by audible noise. It follows that cage instability can be avoided by subjecting the bearing to a minimum axial preload [44]. A guide to the value of this preload for single-row radial ball bearings is given in Figure 2.34. The figures are conservative, taking into account different bearing sections over a broad range of operating conditions of lubrication, speed and temperature. It is suggested they should only be used as a starting point to finding the optimum preload in any given application.

Cage binding

The phenomenon of cage binding should not occur in precision bearing applications, but is included here for the sake of completeness. It most commonly occurs on preloaded ball bearings subject to excessive misalignment of either or both inner and outer rings. Under this condition, the ball–raceway contact angle varies with position around the bearing and as the orbital speed of the balls is dependent on contact angle, so they accelerate and decelerate as they orbit, varying the ball spacing. When the variation in spacing exceeds the clearance between the ball and the cage pocket, high ball–cage forces are generated and the balls are forced to skid.

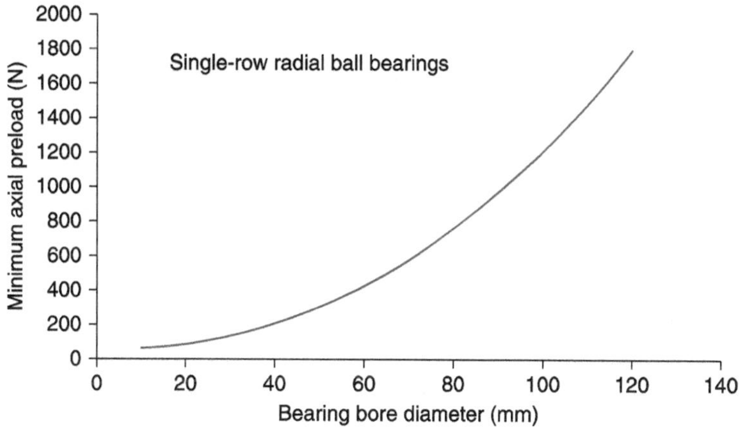

Figure 2.34 **Suggested minimum axial preload to prevent cage instability in grease lubricated single-row radial ball bearings**

In addition to generating high motion errors, the bearing generates high friction and fluctuations in torque. The condition sometimes emits an audible noise referred to as 'squeal'.

Levels of misalignment to cause cage binding are high, being greater than the limits recommended by manufacturers, so in high precision machinery it is essentially a fault condition. Misalignment is usually the cause of cage binding, but excessive ring distortion can also generate the condition.

Variable compliance

Variable compliance is the term used to describe a variation in a component of a bearing's flexibility (inverse of stiffness) that occurs when it is rotating. The mechanism was first identified as early as 1959 [34]. If a bearing is subject to static loads and its flexibility is changing with time, it will produce motion error even if the bearing is geometrically perfect. The most obvious example is a ball bearing subject to radial load only (Figure 2.35). This example shows a 6206 deep groove ball bearing subject to a 200 N radial load for the case in which one ball is in line with the applied load vector and the case for which two balls are equi-spaced about it. The maximum ball load of 112.8 N and minimum approach of the inner and outer rings occurs for the former case whereas for the latter, the ball loads reduce to 100.2 N but the ring approach is a maximum.

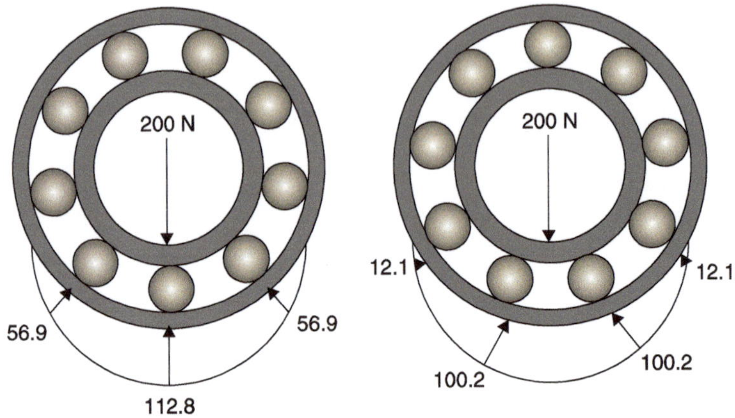

Figure 2.35 **Ball load distribution in a radially loaded ball bearing**

Thus as the balls rotate, a cyclic motion is set up with a basic frequency equal to the ball pass rate. In the example of Figure 2.35, the difference in ring approach in line with the load is 0.1 μm. However, in a radial direction normal to the applied load vector, the ring approach varies by ±0.9 μm, giving a peak–peak amplitude of 1.8 μm.

Fundamental to the mechanism by which variable compliance generates motion error is a variation in ball load with angular position around the bearing. The variation is obviously large in a ball bearing subject to radial load only, but on a geometrically perfect bearing can be reduced to zero by subjecting it to axial preload only. In high precision applications, ball bearings will be subject to axial preload but in general they may also experience radial and/or moment loading that will cause a degree of ball load variation and hence some variable compliance motion error.

The analysis described earlier in this section can be used to investigate how levels of motion error due to variable compliance vary with load. Furthermore, the motion error can be expressed in terms of harmonics of the basic ball pass frequency. For the case of the 6206 single-row radial ball bearing of Figure 2.35 subject to radial load only, the dynamic component of variable compliance is shown in Figure 2.36 in terms of force for applied loads of 500, 1000 and 1500 N. Dynamic force levels increase with bearing load and although they are dominated by the fundamental ball pass frequency, significant harmonics occur at up to five times the fundamental frequency.

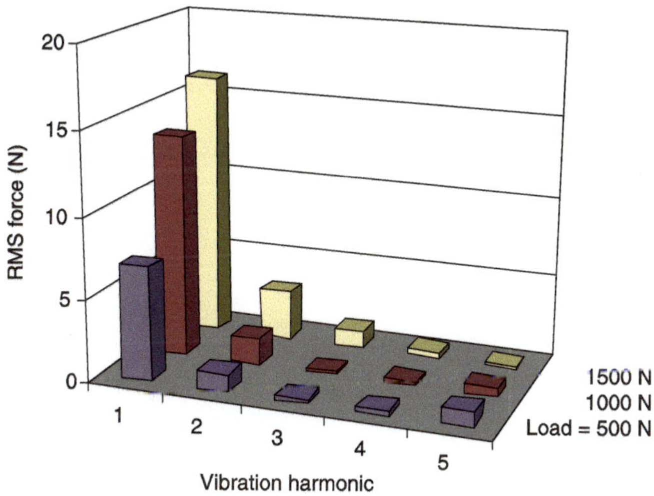

Figure 2.36 Variable compliance vibration produced by radially loaded bearing

For the 6206 bearing considered, a representative radial stiffness in the load range 500 to 1500 N is 116 N/μm. Forcing levels of 5 to 15 N rms thus represent motion error amplitudes of 0.43 to 1.29 μm rms, which are unacceptably high for ultra precision machine applications.

The effect of axial preload on variable compliance error motion for the 6206 single-row radial bearing considered above is shown in Figure 2.37. In this case an axial preload of 500 N is superimposed on the radial loads of 500, 1000 and 1500 N considered earlier. The dynamic components of force are greatly attenuated and for radial loads of 500 and 1000 N are below 0.1 N rms, which corresponds to an error motion amplitude of 8.6 nm rms. For the radial load of 1500 N, the dynamic force of 2 N is still significant and is due to the fact that 500 N preload is insufficient for this level of radial load. At the lower radial loads of 500 and 1000 N, the preload of 500 N enables all balls to contact both inner and outer raceways, whereas for 1500 N radial load, 4 of the 9 balls in the bearing are unloaded. It should be noted that the more uniform the ball load distribution, the lower the variable compliance motion error.

Misalignment between inner and outer rings may also produce variable compliance motion error, although for levels of misalignment within manufacturer's recommendations it is generally much lower in level than that produced by radially loaded bearings and usually negligible. Figure 2.38 is an example of the radial component of dynamic variable compliance force

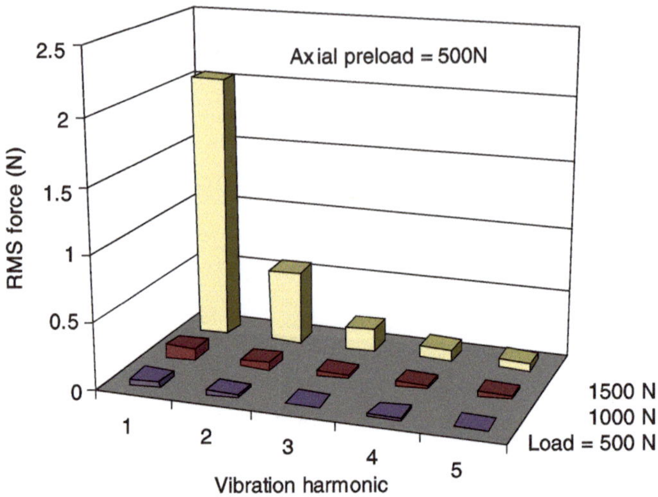

Figure 2.37 Variable compliance produced by axially preloaded ball bearing

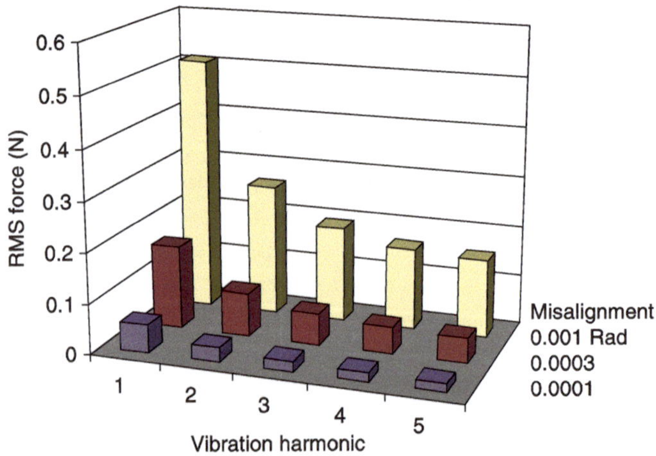

Figure 2.38 Dynamic component of variable compliance force due to misalignment

produced by a 7006 angular contact ball bearing subject to 200 N axial preload and misalignments of 0.0001, 0.0003 and 0.001 radians. The acceptable limit of misalignment for this type of bearing recommended by most bearing manufacturers is 0.0003 radians. The largest component of dynamic force occurs at the fundamental ball pass frequency but the force components corresponding to the higher-order harmonics decay slowly in magnitude. However, at the maximum recommended level of misalignment, the components of dynamic force are only 0.15 N or less. A representative radial stiffness for this bearing is 140 N/μm, so in terms of motion amplitude, levels are around 1 nm rms.

Even geometrically perfect ball bearings will produce variable compliance motion error. This can be unacceptably high in magnitude on ball bearings that are subject to radial load only. However, axially preloading ball bearings can reduce variable compliance error motions to a negligible level, even in the presence of radial load and normal levels of misalignment. For a geometrically perfect ball bearing subject only to axial preload, variable compliance error motion is theoretically zero. Apart from radial load and misalignment, long wavelength geometry errors of the inner and outer raceways such as, for example, 2- or 3-point lobing, can also produce variable compliance error motion. This is described in the following section.

Effect of bearing surface geometry on motion error

Modern ball bearings from reputable manufacturers are manufactured to very high standards of finish but no surface is perfectly round or perfectly smooth and as the levels of motion error required by some ultra precision machine builders are now well below 50 nm peak-peak, even the smallest departures from perfection can be significant.

Rolling surface characterisation

From the point of view of relating surface topography to bearing motion error, it is convenient to consider the rolling surfaces to contain three types of feature: lobing, waviness and roughness. Each type of feature generates motion error by a slightly different mechanism.

Lobing refers to surface features, in the direction of rolling, that have a longer wavelength than the ball spacing and is normally present on inner and outer raceways. Across the raceway, lobing is usually of constant height, this being a function of the raceway machining processes. In the example of Figure 2.39 the outer raceway is shown to have 3-point lobing.

Waviness is the name given to surface features, in the direction of rolling, with wavelengths less than the ball spacing but greater than the width of the ball–raceway contact. As with lobing, waviness is usually of constant height across raceways. Waviness may also be present on ball surfaces in which case height may vary both in and normal to the direction

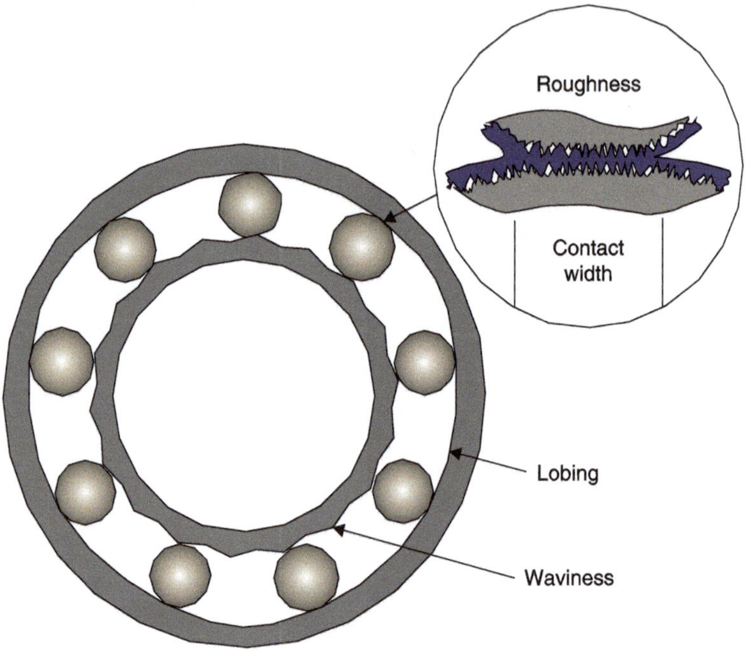

Figure 2.39 Lobing, waviness and roughness of rolling surfaces

of rolling. Figure 2.39 shows waviness to be present on the inner raceway and one of the balls.

Roughness refers to surface features that have a wavelength comparable or less than the width, in the direction of rolling, of the elastically deformed ball–raceway contact. Contact widths vary with bearing, size, geometry and load but in most cases may be expected to be in the range 0.1 to 1 mm. Roughness varies in amplitude both in and normal to the direction of rolling. Raceway machining processes tend to produce surfaces that may be described as having a series of fine furrows around the ring, whereas ball surfaces may be described as having a random distribution of fine scratches or pits. Figure 2.39 illustrates roughness within a lubricated ball–raceway contact where, in this case, the film thickness is sufficient to prevent metal–metal contact.

Motion error produced by lobing

The longer wavelength raceway surface features, referred to as lobing affect the distribution of ball load within axially preloaded bearings and may produce motion error in a manner similar to the variable compliance

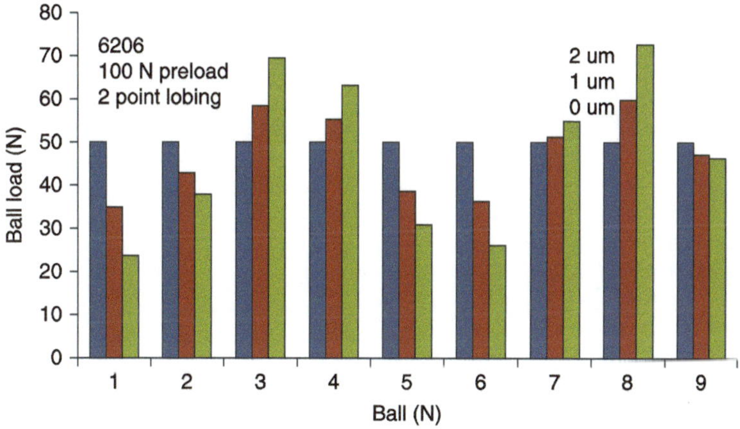

Figure 2.40 **Example of ball load variation due to 2-point lobing**

mechanism described in Sub-section 2.9.7. Lobing errors are easily transferred to raceways from out of round shafts and housings or induced by uneven clamping forces, and for these reasons can have a significant effect on motion error amplitudes.

It is not unreasonable to expect fitted rings to have 2- or 3-point lobing of 1 μm or more in amplitude transferred through to the raceway. Figure 2.40 gives an example of the effect of this upon ball load distribution in a single-row radial ball bearing with 100 N axial preload. Blue represents the ball loads in a geometrically perfect bearing, red is for 2-point lobing of 1 μm in amplitude and green is for 2-point lobing of 2 μm in amplitude present on just one of the raceways. The variation is large and unsurprisingly creates a significant net dynamic force (Figure 2.41).

Depending on whether the lobing is on the outer or inner raceway, it will produce motion error at frequencies harmonic of ball–outer or ball–inner pass rate. The motion is predominantly in an axial direction (Figure 2.41). This bearing has an axial stiffness of approximately 10 N/μm, so the motion errors associated with Figure 2.41 may be as large as 1 μm in amplitude.

The effect of lobing is even more significant when it is present on both inner and outer raceways at the same time, where now the motion error will be modulated by the speed of the rotating ring.

Motion error produced by ball size variation

The ring lapping process used to finish machine balls provides excellent control over ball diameter and variation in diameter. Furthermore,

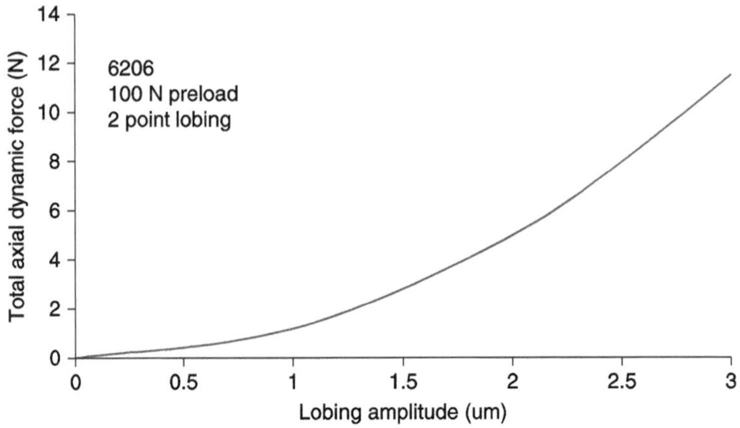

Figure 2.41 **Dynamic force produced by 2-point lobing in a preloaded ball bearing**

subsequent size grading processes enable balls from a given batch to be subdivided into even finer diameter tolerance bands, with the net result that ball diameter variation within a single bearing can be controlled to better than 0.1 μm in precision applications. Whilst ball grade standards exist, specifications for ball size variation in a particular bearing are proprietary to manufacturers.

The effect of ball size variation on bearing motion error has been studied in relation to hard disc drive bearings by [43]. The investigation considered bearings with different numbers of balls and different levels of ball diameter variation. Bearings in which just one ball differed in size and bearings with all balls varying in size were experimentally measured for motion error.

Figure 2.42 shows the averaging effect of the number of balls in a bearing upon the radial motion error generated by one over-sized ball. Five is a practical minimum number of balls and the bearing motion error at 0.045 μm is just less than the ball oversize error of 0.05 μm. Bearing motion error decreases exponentially with increasing ball complement to about 30% of the ball size error at 15 balls. Although not investigated, higher numbers of balls would appear to offer little extra benefit.

When there is a random size distribution of ball diameters, the bearing motion error is, on average, substantially less than the maximum ball diameter difference (Figure 2.43). The results are influenced by the order in which balls of different diameters are placed in the bearing but [43] concluded that it was reasonable to expect a bearing motion error of approximately 20% of the maximum ball diameter difference for a production lot of balls.

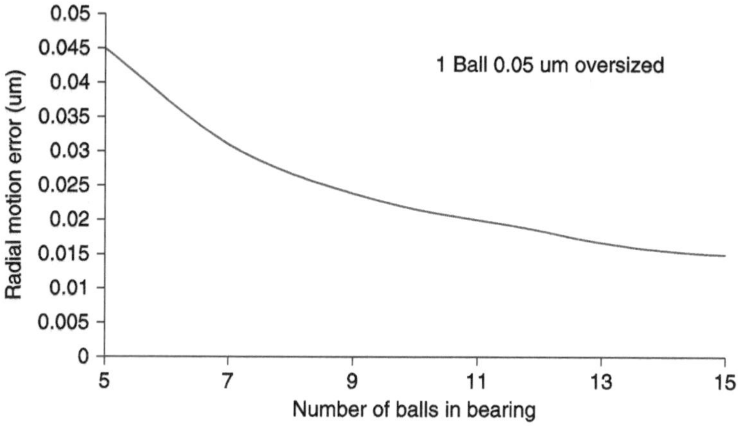

Figure 2.42 Effect of one large ball on bearing motion error

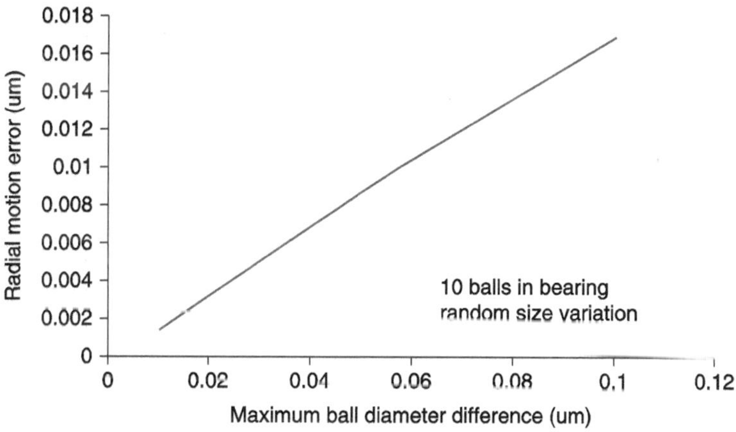

Figure 2.43 Effect of random ball size variation on motion error

Motion error produced by waviness

As for lobing, waviness of the raceway and ball surfaces are a function of the bearing's manufacturing process. In the case of the raceways, grinding machine vibration is a common cause of waviness. In fact, the quality and condition of the bearings (usually ball bearings) in the grinding spindle are critically important. Ideally, any waviness generated in raceway grinding is removed by a subsequent honing process.

The sensitivity of motion error or dynamic force to waviness amplitude can be determined using the model presented in Section 2.9.1. Figures 2.44 and 2.45 show the axial and radial components of dynamic force generated by outer raceway waviness on a 7006 angular contact ball bearing subject to 200 N axial and 500 N radial load. The amplitude of the waviness is 0.1 μm and waviness wavelengths between 4 and 35 waves/circumference are considered. Both figures show that particular waviness wavelengths have a critical effect on dynamic force. For axial force, waviness orders that are an integer multiple of the number of balls in the bearing produce maximum forcing levels, whereas for radial force, waviness orders that are an integer multiple of the number of balls plus or minus one produce the highest levels of force. In this case, the number of balls in the bearing is 14. Maximum

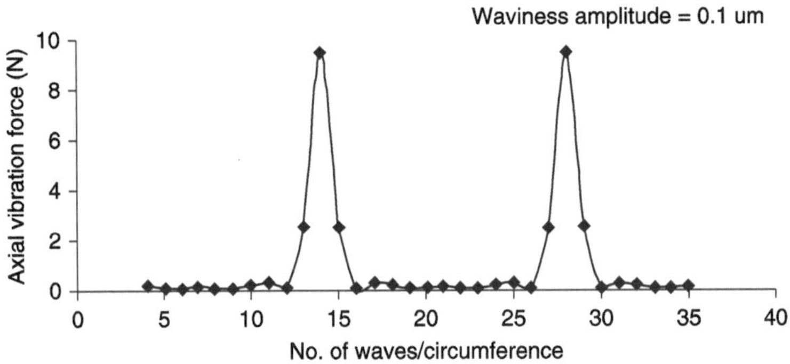

Figure 2.44 Example of axial dynamic force due to waviness

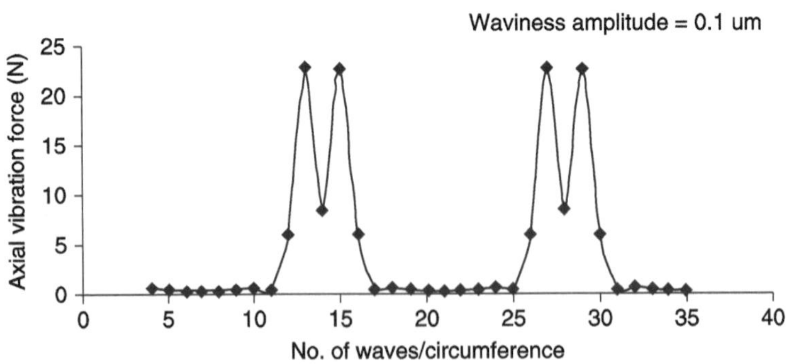

Figure 2.45 Example of radial dynamic force due to waviness

force levels of 9.5N axial and 22.9N radial are produced by waviness of just 0.1 μm in amplitude and illustrate just how critical the intermediate wavelength rolling surface features are. In the previous section, lobing of 3 μm in amplitude was shown to produce similar dynamic force levels to the 0.1 μm waviness considered in Figure 2.44.

Figures 2.44 and 2.45 are for a bearing subject to combined axial and radial loading, where there is a variable compliance effect superimposed upon the direct effect of waviness. Under these conditions, all waviness wavelengths create some degree of motion error. In contrast, a bearing subject to axial preload only does not exhibit variable compliance and only waviness orders that are an exact integer multiple of the number of balls produce axial motion error, whilst waviness orders that are an exact integer multiple of the number of balls plus or minus one produce radial motion error.

The effect of surface roughness on motion error

Surface roughness describes short wavelength surface features, comparable to the width of the ball raceway contact in the direction of rolling or less. Its effect on motion error cannot be considered without taking into account lubrication of the ball–raceway contacts. Even though the stresses in the central region of such contacts are extremely high, usually between 1 and 3 GPa in real bearing applications, remarkably, a lubricant film is formed with the potential to prevent metal–metal contact. This is explained by a relatively new mechanism of lubrication, Elastohydrodynamic Lubrication (EHL) that emerged between the 1950s and 1980s. An explanation of the lubricant films formed in the elliptical contacts characteristic of ball bearings and a means of calculating their thickness is provided by [45]. They also provide a comprehensive set of references describing the early research work on the subject.

The realisation that a lubricant film is formed in such highly stressed contacts has had a profound effect on contact design, particularly in the case of ball bearings. For practical operating conditions, the films are thin, typically in the range 0.1 to 0.5 μm, and at the time EHL theory was being developed, bearing surface roughness was of a comparable or even higher level so that some fraction of the load applied to a ball–raceway contact was supported by metal–metal contact at a number of highly localised asperities. Furthermore, asperity contact also served to further raise the maximum stress within the contact field. Developments within the rolling bearing industry have since been aimed at reducing the incidence and severity of asperity contact with the aim of improving bearing fatigue life in highly loaded bearings [35], or reducing noise and vibration in precision and quiet running bearing applications. For the latter, the goal has been to achieve complete separation of the ball–raceway surfaces by reducing their roughness. This has required the development of new or improved surface finishing processes, but the result is that modern bearings have a much higher standard of finish than their equivalents of 30 or 40 years ago.

Achieving full separation of ball–raceway contacts under all operating conditions is a demanding objective and in practice a low incidence of asperity contact may occur.

Thus the mechanism by which roughness generates motion is predominately impulsive, as illustrated in Figure 2.46, where an asperity breaks through the lubricant film and contacts the opposing surface. As the width of the contact is typically 0.1 to 1 mm and rolling speeds are usually more than 1 m/s, the asperity contact time is short, being less than 1 ms. The impulse created excites high frequency resonances within the

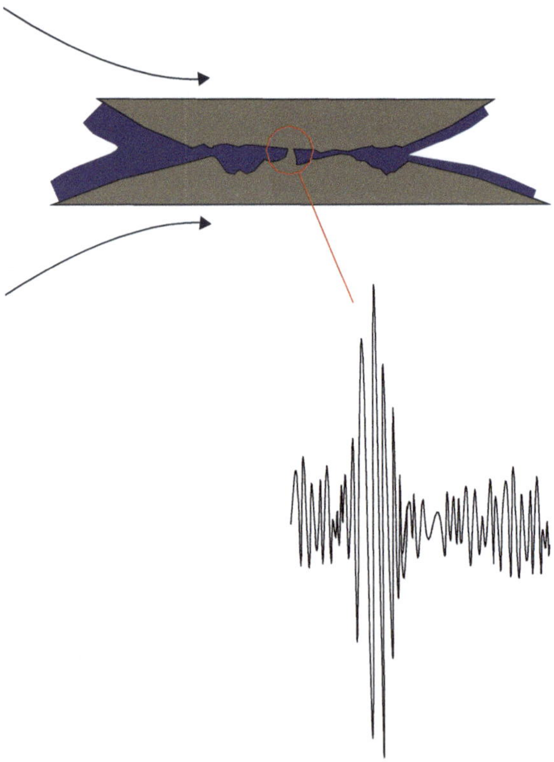

Figure 2.46 Surface roughness in relation to a ball–raceway contact

bearing and machine structure, so the bearing's amplitude–time signature is characterised by a random sequence of decaying transients. Because the motion occurs predominantly at high frequencies, amplitudes are low, being typically in the range 1 to 10 nm for normal quality bearings. However, the influence on bearing noise is far more significant.

The effect of contamination on bearing motion error

The subject of bearing contamination has been widely researched, mostly in relation to its effect on bearing life. A review of the work that has had a major influence on the way bearing fatigue life is estimated is provide by [46]. As lubricant films at ball–raceway contacts are typically less than 0.5 μm in thickness, it is clear that even sub-micron sized particles may bridge the film with risk of producing permanent damage to the rolling surfaces. One of the objectives of the past research has been to quantify the relationship between particle size and hardness on the level of damage inflicted to rolling surfaces and this confirms that hard particles (similar to bearing hardness levels) as small as one micron in diameter may in fact result in permanent indentation. However, the rolling contacts can survive particles of 30 μm diameter or more, provided they are soft enough to deform as they pass through the contact.

Thus there are two mechanisms by which contamination can influence motion error. Very small or very soft particles that pass through the contact without damaging the bearing surfaces produce a random sequence of impulses. However, hard particles will create permanent surface damage that will result in a regular sequence of impulses. If this occurs, bearing motion errors may be expected to be high compared to that of an undamaged bearing. The importance of bearing cleanliness in precision bearing applications cannot be over-stated.

2.10 Power consumption and thermal distortion

The purpose of this section is to provide a basic awareness of the sources of friction within a ball bearing and an approximate means of estimating its overall power consumption. The detailed dynamic models described in [1–4,6] determine the overall torque and power loss within a ball bearing, by considering the shear stresses and levels of slip at and around all of the interacting internal surfaces. The models generally provide a good understanding of the relative importance of the individual sources of friction and how they relate to bearing design and operating conditions. But when used to predict the torque and power loss in a given application, are still limited in accuracy by not being able to satisfactorily quantify the relationship between applied lubrication conditions and the state of lubricant within the bearing cavity. A much simpler empirical approach to estimating bearing friction is more appropriate to this book and in the case of most precision bearing applications will provide a reasonable estimate of overall torque and power loss.

2.10.1 Overview

Figure 2.47 illustrates the main sources of friction in a ball bearing subject to load and rotating at constant speed.

Within the Hertzian zone at the ball–raceway contacts, pure rolling can only occur at one or two points at most; at all other positions within the contact, rolling motion is accompanied by small amounts of sliding (micro-slip) due in part to the conformal geometry of the contact and in part to spinning motion of the ball. The lubricant film that separates ball and raceway surfaces is thin, less than 1 μm, and subject to a high hydrostatic stress, usually more than 1 GPa in the centre of the contact. This high stress

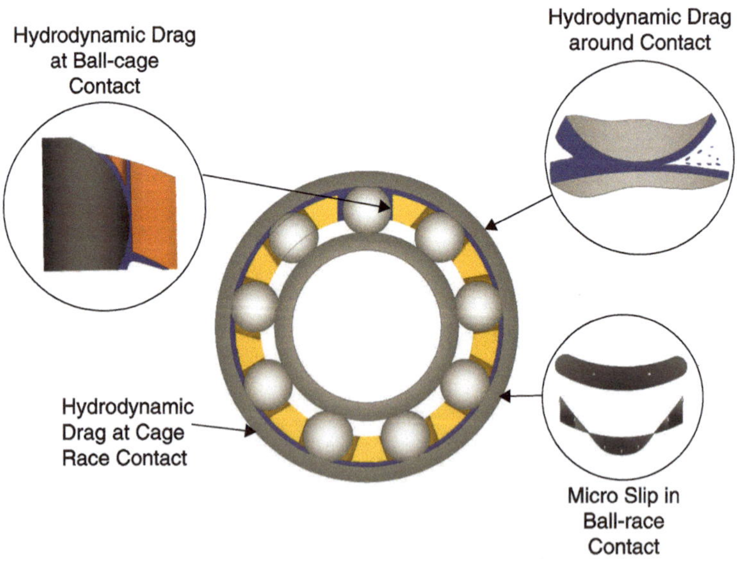

Figure 2.47 Major sources of power loss in ball bearing

dramatically increases lubricant viscosity [47], so that within the contact the lubricant behaves like a low shear strength solid such as, for example, silver. The small amounts of sliding create significant traction forces within the lubricant film and significant energy dissipation. This source of friction is highly dependent on bearing load. Increasing load not only increases the size of the Hertzian contacts and therefore the level of micro–slip, but also the level of stress and the magnitude of the traction forces.

Outside of the Hertzian contacts, a film of lubricant adheres to ball and raceway surfaces. Under normal lubrication conditions, the thickness of this film is usually much greater than the film thickness within the contact zones. Some of the lubricant is dragged into inlet zones, ahead of the Hertzian contacts, by the motion of the balls and raceways. But only a small fraction of this can pass through, leaving the remainder to be displaced by ball motion, creating viscous shearing and churning losses in the regions surrounding the contacts. These losses are essentially hydrodynamic and increase with bearing speed and the quantity of lubricant present within the bearing cavity. The latter is dependent on the quantity and method by which the lubricant is supplied to the bearing.

The cage is another source of friction. The clearances between the cage pocket and ball and the cage riding surface and ring are low. Furthermore, sliding speeds at these interfaces are high so that any lubricant in these gaps creates significant drag forces even though the normal forces between the cage and the ring or ball surfaces are low. Drag forces increase with bearing speed and the quantity of lubricant present in the gaps.

Analytical models for predicting bearing torque and power loss exist, but are complex and fall outside the scope of this text. For example, to determine friction at cage interfaces, it is first necessary to determine cage loads and this requires a full dynamic analysis of the bearing [1]. Even with a rigorous analysis it is difficult to account for the quantity of lubricant within the bearing cavity. Practical experience shows bearing friction to generally increase with the amount of lubricant present and although some account of this can be made by defining the percentage of the bearing cavity filled with lubricant, the relationship of this parameter to the quantity and mode of lubricant supply still needs to be determined empirically.

2.10.2 Bearing torque

Detailed analytical models are of value in understanding and optimising bearing design. For the purpose of estimating overall friction torque and power loss to assess power requirements and thermal distortion in precision machine design, an empirical approach is favoured. Such an approach was published by Palmgren [48] in 1959 and is well-established within the rolling bearing industry. He developed equations describing friction torque and determined the empirical factors for all types of rolling bearing. Here we are concerned with ball bearings and only the relevant data is included. He also considered that ball bearing torque was the sum of two components, one load dependent, the other speed dependant:

$$M = M_l + M_v \qquad [2.69]$$

In the case of roller bearings a third component representing roller end-flange contact was included but is omitted here. The load dependant torque in a ball bearing is associated with micro-slip within the ball–raceway contacts, whilst the speed dependant torque is due to viscous shearing and churning of the lubricant everywhere else within the bearing cavity.

Micro-slip within the Hertzian contacts subjects the lubricant film to shear stress. The response of lubricant to shear stress in a rolling contact is generally referred to as its traction characteristics, with the ratio of shear stress/normal stress defined to be the traction coefficient. A full explanation of lubricant traction behaviour can be found in [49], and typical traction coefficients as a function of the slide/roll ratio in Figure 2.48. Under the higher pressures reached in a ball–raceway

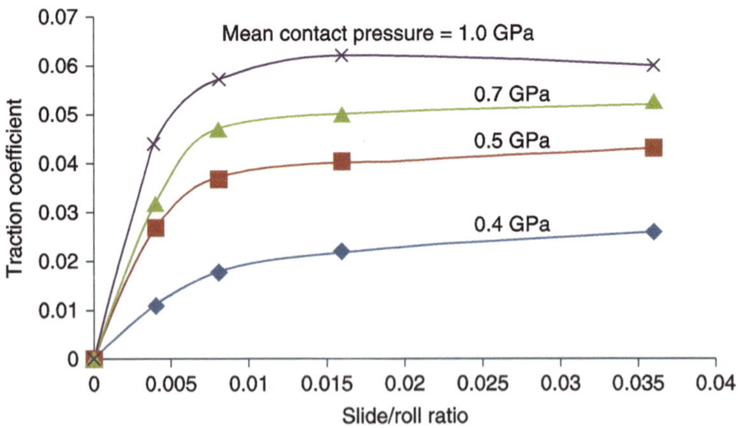

Figure 2.48 Typical lubricant traction characteristics

contact, traction forces reach a limiting value and the form of the traction–slip curve appears somewhat similar to a Coulomb friction law. This complex reality was simply represented by [48] as:

$$M_l = f_1 \; F_b \; D_p \qquad [2.70]$$

where f_1 is a factor dependant on bearing design and relative load. For ball bearings:

$$f_l = z(F_s/C_s)^y \qquad [2.71]$$

C_s is the basic static load rating of the bearing, as given in the manufacturer's catalogue, and F_s is the static equivalent applied load:

$$F_s = X_s F_r + Y_s F_a \qquad [2.72]$$

F_r and F_a are radial and axial applied loads respectively and for single-row ball bearings (as in Table 2.8).

The factors z and y have been determined experimentally and are given in Table 2.9.

Table 2.8 Equivalent load factors for ball bearings

Bearing Type	X_s	Y_s
Single row radial	0.6	0.5
Angular contact, α=15°	0.5	0.47
α=20°	0.5	0.42
α=25°	0.5	0.38
α=30°	0.5	0.33
α=35°	0.5	0.29
α=40°	0.5	0.26

Table 2.9 Ball bearing load torque factors

Bearing type	z	y
Single row radial	0.0007	0.55
Angular contact α=15 to 40°	0.001	0.33

The factor F_b in Eq. 2.70 depends on the magnitude and direction of the applied load and is given by:

$$F_b = 0.9\ F_a \text{ ctn } \alpha - 0.1\ F_r \qquad [2.73]$$

or:

$$F_b = F_r$$

whichever gives the larger result.

Viscous shearing of the oil within a ball bearing occurs at various positions within the bearing cavity (Figure 2.47). Palmgren [48] found that the combined effect of these could be adequately described by:

$$M_v = 1.42 \times 10^{-5}\ f_0\ (\nu_0\ n)^{2/3}\ D_p^{\ 3} \qquad \text{for } \nu_0 n > 2000 \qquad [2.74]$$

or:

$$M_v = 2.3 \times 10^{-3}\ f_0\ D_p^{\ 3} \qquad \text{for } \nu_0 n \leq 2000 \qquad [2.75]$$

where ν_0 is the kinematic oil viscosity in centistokes, n is bearing speed in rpm, D_p is pitch circle diameter in inches and f_0 is a factor depending on bearing type and mode of lubrication.

Table 2.10 highlights the importance of the effect of mode of lubrication on friction torque. For a single bearing, friction levels may vary by 6 to 1, depending upon whether the bearing is supplied with minimal or copious amounts of oil. For grease lubrication, the factor f_0 applies to the first few hours of running and provides an upper bound estimate of friction torque. The viscosity to be used for grease is that of its base oil.

The above approach to estimating friction torque was developed prior to 1960 and applies to open bearings operating under low and moderate speeds. It does not account for ball spinning friction in high speed bearings nor the effect of seals in closed bearings. It provides an estimate of friction

Table 2.10 **Bearing type and lubrication factor, f_0**

Bearing type	Oil mist or oil injection	Oil bath or grease	Oil bath (vertical shaft) or oil jet
Single row radial	0.7–1.0	1.5–2.0	3.0–4.0
Angular contact single row	1.0	3.0	6.0
Angular contact double row	2.0	6.0	9.0

torque sufficiently accurate for determining drive power and heat generation in bearing arrangements but is not refined enough to distinguish between different bearing designs. For example, the hybrid ball bearing developed during the 1980s offers significantly lower friction levels than its standard equivalent, yet the above formulae suggests it to be the same.

2.10.3 Application to precision bearings

In the vast majority of precision bearing applications, the dominant bearing load is axial preload and the lubrication is minimal, either air–oil or grease. This means the preceding expressions for friction torque can be simplified. For load dependent torque:

$$M_l = 0.9.z[Y_s.F_s/C_s]^y ctn(\alpha)d_m \qquad [2.76]$$

where z = 0.0007 for single-row radial or 0.001 for angular contact ball bearings and y = 0.55 for single-row radial or 0.33 for angular contact ball bearings. For viscous or speed dependent friction torque, the lubrication factor is 0.7 to 1 for air–oil lubrication or 1 to 3 for grease lubrication, depending on bearing type.

2.10.4 Lubrication

Table 2.10 highlights the effect of lubrication on bearing power consumption. Not only the viscosity of the lubricant but the quantity and mode of supply affects both the power consumption and running temperature. For oil lubrication, torque and hence power loss generally continues to increase with the rate at which lubricant is supplied to the bearing (Figure 2.49). However, the dependence of bearing running temperature on lubricant flow rate exhibits a characteristic curve, whereby it initially increases with flow rate to a maximum and then decreases again as flow rate continues to rise. The decrease is due to an oil cooling effect which, at very high flow rates, more than offsets the large power losses produced. Figure 2.49 covers several different modes of lubrication, whereby air–oil systems deliver minute quantities of lubricant and cover the lower end of the curve. Oil bath and low pressure oil feed systems flood the bearing with lubricant and operate in the intermediate range of flow rates, whilst high pressure oil feed systems operate at the high flow rate end of the curve.

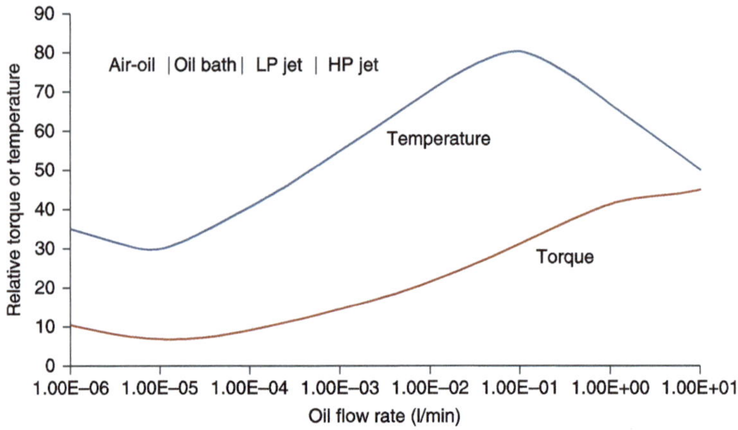

Figure 2.49 Characteristic effect of oil flow rate on torque and temperature

Minimising bearing running temperature, and hence the thermal distortion in adjacent machine parts, is an important aspect of precision bearing selection and application. High pressure oil feed systems, although offering good temperature control, are seldom used in high precision machines, as the overall cost of the system and the bearing power losses are high, and the oil supply is also relatively sophisticated having a high pressure pump, means of regulation, filter and oil cooler. In addition, the bearing chambers require elaborate sealing arrangements to prevent excessive leakage. For these reasons a minimal lubrication approach is preferred in a majority of precision bearing applications. This means either grease lubrication or the use of air–oil systems. The latter provides a limited amount of cooling, enabling significantly lower temperatures than grease lubrication at very high speeds.

Air–oil lubrication systems inject small, precise quantities of oil into an air stream. The oil moves along the walls of the feed tube and at the end of the tube is blown into the bearing by the air stream. The optimum amount of oil to inject is found by monitoring bearing temperature for a range of oil flow rates. Figure 2.50 shows the characteristic bearing temperature–oil flow rate curve for air–oil lubrication systems. The curve exhibits a minimum temperature at a given oil flow rate.

At lower flow rates, the bearing operates starved of lubrication and there is a risk of wear occurring at the cage interfaces. At higher flow rates, the bearing operates with enough lubricant to generate churning losses. For reliable bearing operation, the optimum lubricant flow rate is

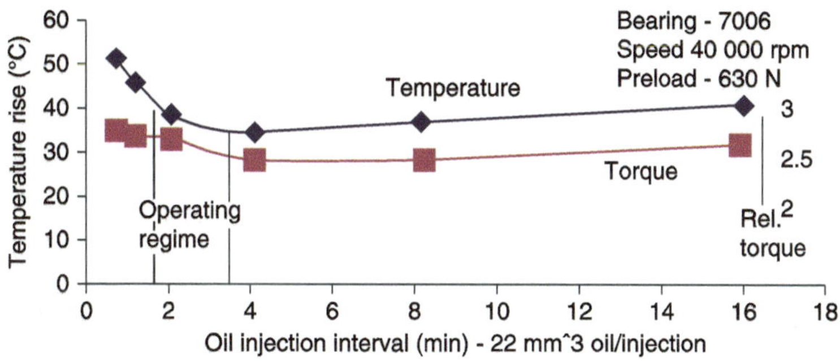

Figure 2.50 **Bearing temperature and torque versus oil flow rate for air–oil lubrication**

chosen to be just greater than that to give minimum temperature. Oil flow rates are typically 1 to 2 cc/hr and as the flow rates are so low, the oil is not usually recovered. In addition to minimising bearing friction losses, air–oil systems provide a small but useful amount of air cooling.

Grease is a popular alternative to oil for lubricating precision bearings. A single charge of grease can often last the required design life of the bearings, so there is no need for any form of delivery system. Furthermore, it places no demands on sealing, which is generally designed to keep contamination out of the bearing rather than grease in. Thus, unlike air–oil lubrication, there is no contamination of the environment.

Nowadays both angular contact and single-row radial ball bearings manufactured in precision grades are available pre-greased and sealed or shielded. This is a relatively recent development in the ball bearing industry. Fitting seals or shields to precision bearings may distort the outer ring, making it difficult to achieve the class 2 specification ring tolerances and this has hindered the introduction of such products. The advantages of greased and sealed bearings are that there is no risk of contaminating the bearing during assembly and they are supplied with the correct type, amount and distribution of grease. They should be the preferred choice for most precision machine builders.

Where it is necessary to use open bearings and provide grease lubrication separately, care is needed in designing the bearing cavity and charging the bearing with the correct amount of grease. Guidance is provided in bearing manufacturer's catalogues on grease type and fill to be used for precision bearings. The issue is to provide sufficient grease to properly lubricate the bearing but not so much that it is continually churned. Churning of the grease soap structure can lead to excessive heat generation and even

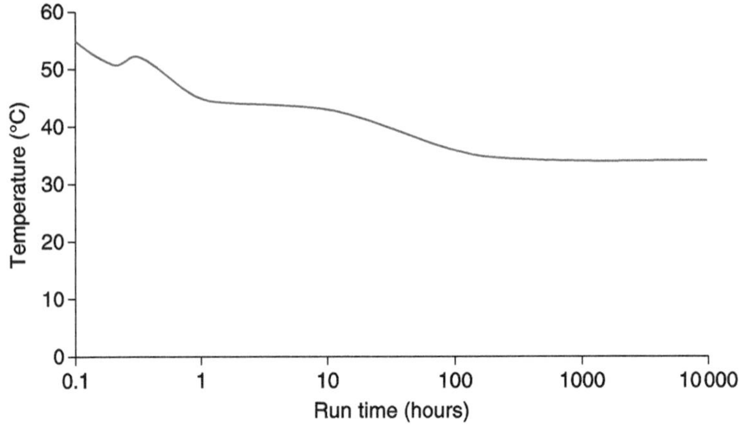

Figure 2.51 Characteristic temperature–time relationship for grease lubricated ball bearing

thermal instability in some precision bearing arrangements. The charge of grease is based on the internal volume of the bearing, being typically between 20 and 40%. Sufficient space, either side of the bearing, must be available for surplus grease to be expelled during the early stages of bearing running.

Whether open or sealed bearings are used, the grease must be run in and a running in procedure should be adopted. Figure 2.51 illustrates how bearing running temperature reduces with running time for a newly greased bearing. Depending on operating conditions, as much as 200 hours can be required for the temperature to stabilise.

2.10.5 Thermal distortion and instability

For grease and air–oil lubrication, the frictional power generated by a ball bearing is dissipated mostly by conduction through adjacent machine components and convection from their external surfaces. The associated temperature rise creates thermal growth of the machine and adversely affects the overall dimensional precision of the machine. But in addition to this, conduction rates through the housing, surrounding the outer ring, are usually far higher than those through the shaft upon which the inner ring is mounted. The consequence is that not only is there an increase in bearing temperature but a temperature gradient is developed across the bearing where in general the inner ring runs hotter than the outer ring. This reduces the internal clearance of the bearing, affecting ball load distribution. In face step preloaded bearing arrangements, for example, the back–back arrangement shown in Figure 2.4(b), the reduction in clearance increases ball loads and induces a thermal preload that further increases the friction power generated by the bearing. At a sufficiently high speed a runaway situation occurs, where preload and temperature continue to increase until overheating or seizure irretrievably damage the bearings.

This effect was clearly observed in [50]. A pair of 7013, angular contact bearings were mounted in a back–back arrangement with a 12.5 μm face step preload, corresponding to 125 N. Both conventional all-steel bearings and hybrid bearings were investigated.

Thermal preload tests consisted of running the bearings at a set speed until outer ring temperatures stabilized. At this point preload and outer ring temperature measurements were recorded. The procedure was repeated at speed increments until a limiting speed was reached at which stable running conditions could not be obtained. Thermal preload was determined mainly for grease lubrication conditions. Figure 2.52 shows the results. For the hybrid bearings, induced preload increased gradually with speed to 930 N at 17 000 rpm, whilst for the conventional bearings, preload increased at a higher rate to 2150 N at 15 000 rpm.

The bearing outer ring temperatures corresponding to Figure 2.52 are shown in Figure 2.53. The increase in preload from 125 to 2150 N in the conventional bearings corresponds to a 35 °C increase in outer ring temperature. For the same preload, the hybrid ball bearing has a smaller ball–raceway contact area and produces less friction than a conventional bearing. Also with the relatively light ceramic balls, centrifugal loads generated at speed are lower. This results in lower running temperatures,

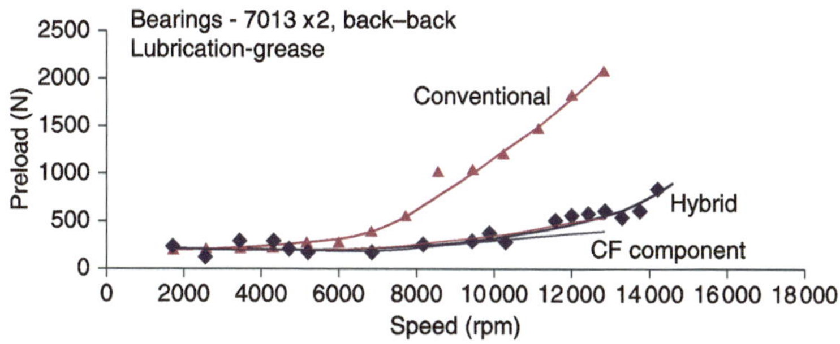

Figure 2.52 Thermally induced preload in a back–back bearing arrangement

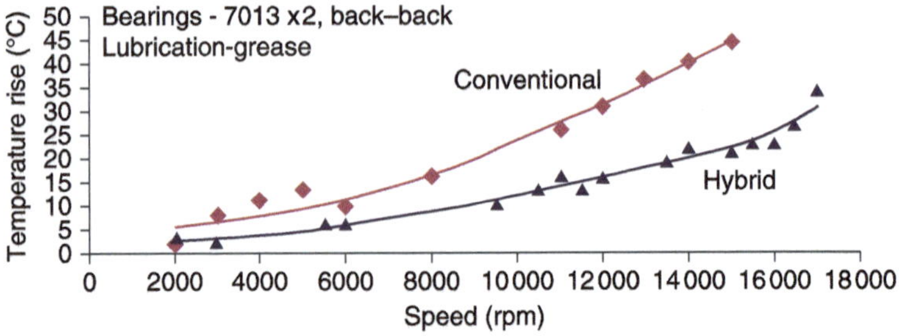

Figure 2.53 Running temperature versus speed for the back–back arrangement of Figure 2.52

lower thermally induced preload and a higher running speed before thermal instability sets in.

2.10.6 Bearing thermal model

The preceding section provides an approximate empirical calculation of overall bearing power loss. The loss is due to several sources of friction, occurring at different positions within the bearing. For the purpose of estimating temperature distribution at the bearing and in the adjacent machine structure, this can be split into two equal sources of magnitude P/2 acting at the bearing's inside and outside diameters (Figure 2.54).

The thermal resistance (inverse of thermal conductivity) of ball bearings has not been extensively researched compared to other aspects of performance, but has received some attention mostly in relation to operating in a vacuum such as in outer space. Stevens and Todd [51] performed experimental measurements of thermal conductivity of low speed ball bearings in 1977, whilst more recently Takeuchi et al. [52] performed further experiments on P9 (class 2) precision bearings of 20 mm

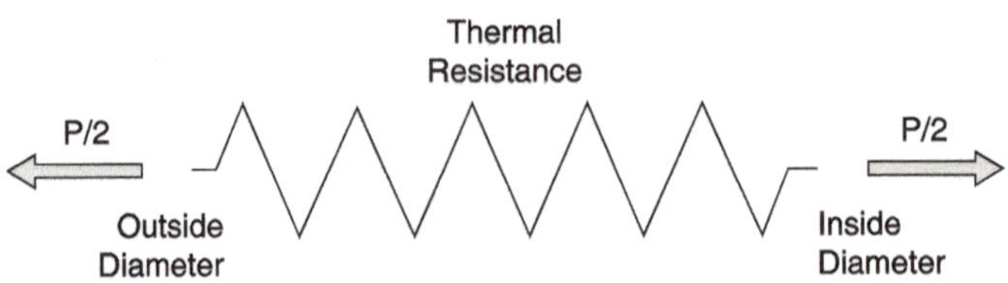

Figure 2.54 Thermal model of ball bearing

bore, extending maximum speed to 10 000 rpm. They found that thermal conductance depended on bearing speed and a running in process. At static and very low speeds, the thermal conductance was primarily through the ball–raceway contacts, whereas at higher speeds mass transport of the lubricant increased the bearing's conductivity. However, measured values at between 0.092 and 0.18 W/°C were very low. Thus for the purpose of estimating temperature distributions and thermal distortion in structures of which ball bearings are part, the bearing's thermal conductivity may be neglected.

2.11 Life and reliability

Bearing life is normally accepted to be the duration that it can safely support load under rotating conditions, and bearing manufacturer's life calculations centre upon the mechanism of rolling contact fatigue. Here the high cyclic stresses experienced by the rolling surfaces may lead to the development of localised cracks at or just under the surface. These grow and lead to a flake of material breaking away from the surface leaving a pit. The first indication of a fatigue pit is usually accepted to be the end of the bearing life. In many applications the development of a fatigue pit is clearly evident from noise or vibration or some other departure from normal operation, yet the bearing can continue to run for a significant time before it fails catastrophically, i.e. seizes. Thus the failure mode is safe. To the end user the bearing appears to operate perfectly well for a long period and then suddenly gives warning that it is beginning to deteriorate and the warning period is sufficiently long for remedial action to be taken.

In a precision engineering context, bearing life can take on a slightly different perspective. Here the bearing may be expected to meet performance criteria related to, for example, running accuracy, noise, temperature or rigidity and deterioration in any of these parameters may cause the machine function to fall outside of specification. At this point the useful life of the bearing has expired and bearing renewal is required to remedy matters. The mechanisms that produce this type of deterioration are not often related to fatigue but more to external influences or to non-optimal bearing set-up. As such it is difficult to define a theoretical bearing life but an awareness of common mechanisms of deterioration can help minimise risk of failure.

Nonetheless, in designing bearing arrangements for precision engineering applications, a fatigue life calculation should be part of the overall design procedure to ensure that the selection of size and type is appropriate and the bearing can meet reliability requirements.

2.11.1 Fatigue life

Within the bearing industry there is continual improvement in materials and manufacturing processes, with the result that bearing fatigue life is also continually improving. Alongside these developments, the understanding of rolling contact fatigue has undergone radical rethinking. Up until the 1980s, fatigue was accepted as an inherent mode of rolling element bearing failure that would occur even on bearings operating under ideal conditions. For ball bearings, fatigue life was accepted to be inversely proportional to applied load cubed. Lightly loaded bearings would have long fatigue lives, but would always be expected to fail. In 1985, Harris [53] proposed the 'New Life Theory' based upon the notion that in common with structural materials, rolling bearing materials exhibit an endurance limit and that if used under moderate loads, kept correctly lubricated and free from all types of contamination, rolling bearings would last indefinitely. Controversial at the time, the theory is now widely accepted and has since reshaped bearing application engineering and bearing development.

Fatigue life theory is based on the hypothesis that the bearing material contains microscopic inclusions and metallurgical dislocations that act as local weaknesses. Those defects just beneath the rolling surfaces are subject to high cyclic stress fields and can act as sites for micro-crack initiation, the time to initiation being dependent on the size and type of defect, its position and the magnitude of the stress field it experiences. As the material is subject to a distribution of defect types and sizes, the fatigue lives of bearings in a batch run under identical conditions, are not the same but also exhibit a distribution. For this reason, fatigue life is defined in statistical terms as a probability of survival. The parameter L_{10} is used to define the basic rating life of a population of bearings, L_{10} life being the life in revolutions that 90% of the population will reach or exceed before developing a fatigue pit or spall.

Fatigue Life Theory, first developed by Lundberg and Palmgren [54], considered the probability of an elemental volume of material dV surviving N stress cycles but failing in the next dN cycles to be dependent on N and the location and size of the volume element. Integrating over the total stressed volume gives the probability of survival of the complete component. They simplified this integral to:

$$\mathrm{Ln}\ \{(1/S(N)\} = \tau_o^{\,c}\ N^e\ V/z_o^{\,h} \qquad [2.77]$$

where τ_o is the maximum orthogonal shear stress developed within the volume at risk calculated from Hertzian contact analysis, z_o is the depth below the surface at which this stress occurs and V is a measure of the effective volume for which the shear stress has a significant value. The exponents c, e and h are constants determined empirically from test data. Ioannides and Harris extended this model in two respects. First, they introduced a fatigue limit, σ_u, patterned on structural fatigue such that any stress less than or equal to σ_u makes no contribution to the risk of fatigue initiation. Second, τ_o was replaced by a more general fatigue criterion, σ, and interpreted locally in order to allow for effects such as surface traction to be accounted for. Lundberg and Palmgren's equation was first rewritten as:

$$\text{Ln}\,\{(1/S(N)\} = A\,(\sigma - \sigma_u)^c\,N^e\,H(\tau - \tau_u)\,dV/z_o^{\,h} \qquad [2.78]$$

where H(x) is the Heaviside step function and then approximated to:

$$\text{Ln}\,\{(1/S(N)\} = \frac{AN}{V_R}\int (\sigma - \sigma_u)^c / z'^h\,dV \qquad [2.79]$$

where A is the volume averaged value of A and the integration is performed only over the region where the stress criterion exceeds the threshold value. The fatigue limit may also assume local values. The factor z_o in the Lundberg–Palmgren equation that accounted for the interval between crack initiation and its appearance on the surface is changed to z′ and regarded as a local stress weighted average depth. The exponents are the same as those used by Lundberg–Palmgren.

Interpreting Eq. 2.79 locally has the advantage of being able to quantify the effects of a number of variables on fatigue life hither to unaccounted for. Surface roughness, residual stress, surface traction and damage due to over rolling of contamination were some of the important factors investigated. Being able to separate out the effects of parameters such as these from material effects and observing their effect on fatigue life strengthened the support for the hypothesis of an endurance limit and this is now generally accepted as being applicable to rolling contacts.

Bearing life calculation using the above equation requires considerable computing resources and test data. Thus for the purpose of simplifying life calculation for practical bearing applications, manufacturers have condensed the results of the above into a simple equation and tabulated adjustment factors and product endurance limits. For SKF, the life equation takes the form:

$$L_{naa} = a_1\,a_{skf}\,(C/P)^p \qquad [2.80]$$

where now the factor a_1 accounts for reliability and a_{skf} for fatigue limit and operating conditions. C is basic dynamic load rating, P is equivalent dynamic load and the exponent p = 3 for ball bearings.

2.12 Closing remarks

Ball bearings can meet the requirements of a majority of precision bearing applications, providing excellent static stiffness, adequate damping and low friction at a relatively low cost. They are also available in a wide range of types and sizes and can be mounted in different arrangements to meet a variety of requirements. The one limitation is perhaps the non-synchronous motion error inherent in a rolling bearing geometry. Whilst peak–peak motion errors as low as 50 nm have been achieved in computer disc drive bearings, the motion error is predominantly asynchronous and is still considerably higher in level than can be achieved with most fluid film bearing types.

2.13 Appendix 2A, Vibration forces generated by waviness on the rolling surfaces of thrust bearing ball bearings

The analysis presented here is a development of [30] considering the vibration of thrust loaded ball bearings. The purpose is to define transfer functions relating the amplitude and wavelength of waviness on the rolling surfaces to the magnitude and frequency of vibration forces generated.

2A.1 Dynamic model of ball bearing

Figure 2A.1 shows a model of the bearing and loading situation to be analysed. The bearing is subject to a centrally applied axial load, P_a, that produces an axial approach between inner and outer rings of Δ_a. Both the applied load and the associated ring approach are assumed to remain constant.

Only inner ring rotation is considered and is assumed to rotate at constant speed, ω_i. Cage speed is also assumed to be constant at ω_c and can be determined from the kinematic relationships given in Section 2.6.3:

$\omega_c = \frac{1}{2}\, \omega_i(1 - D\cos\alpha/D_p)$

The bearing has N equi-spaced balls that act as flexible elements interspaced between rigid inner and outer rings. Ball flexibility is, in general, a function of inner and outer raceway contact compliance, contact load and bearing geometry. Ball mass and centrifugal force is neglected so the contact angles and loads at inner and outer contacts are equal and each ball can be represented simply as a non-linear spring without any inertia.

Waviness is assumed to be present on ball, inner and outer raceway surfaces. It is assumed to be simple harmonic in the direction of rolling and independent of position across the surface, normal to the direction of rolling. For inner and outer raceways, these assumptions are reasonable for the longer wavelength features greater than about three times the ball–raceway contact width. For balls, surfaces are isotropic in form, so the above assumptions are an approximation. As rings are fixed in position waviness on the rolling surfaces creates a fluctuation in elastic approach as it passes through the ball–raceway contacts. Lubricant film thickness at the contacts is assumed to remain constant, so the variation in elastic approach is equal to the waviness amplitude.

2A.2 Ball load – deflection characteristics

In the bearing model, each ball is represented by a linear spring of stiffness K. The spring rate is determined from a consideration of the ball–inner

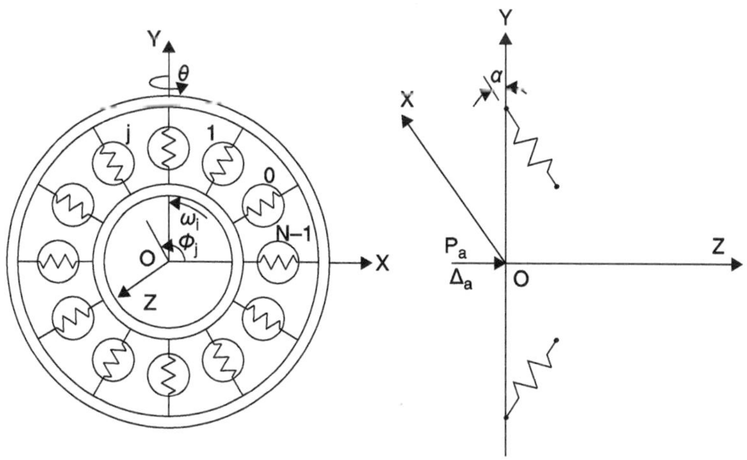

Figure 2A.1 Dynamic model of ball bearing

and ball–outer contact characteristics. For a single contact, load and deflection obey the Hertzian equation (Section 2.7.2):

$$Q = \kappa\delta^{3/2}$$

For inner and outer contacts acting in series:

$$\delta = \delta_o + \delta_i \qquad \text{and} \qquad 1/\kappa^{2/3} = 1/k_o^{\,2/3} + 1/\kappa_i^{\,2/3}$$

The spring rate representing each ball is then:

$$K = dQ/d\delta = 1.5\kappa\delta^{1/2} \qquad [2A.1]$$

With waviness superimposed upon the rolling surfaces and as the bearing rotates, each ball will experience a time dependent component of deflection in addition to the steady component produced by the applied bearing load. For ball number j:

$$\delta_j = \delta + \delta_j(t) \qquad (2A.2)$$

Assuming $\delta_j(t)$ is small compared to δ, then Eq. 2A.2 can be substituted into Eq. 2A.1 and simplified using a binomial expansion to give:

$$K_j = 1.5\kappa\delta^{1/2}[1 + \delta_j(t)/2\delta + \delta_j^{\,2}(t)/4\delta^2] \qquad [2A.3]$$

where first- and second-order terms in the expansion are preserved.

Equation 2A.3 states that waviness superimposed on rolling surfaces not only affects the ball load but also ball stiffness, the variation in stiffness being dependent on the amplitude of the waviness compared to the steady state deflection produced by bearing load, $\delta_j(t)/\delta$. For good quality precision bearings, $\delta_j(t)$ may typically be in the range $0.1 < \delta_j(t) < 1\,\mu m$ and δ to be $1 < \delta < 10\,\mu m$, depending on applied load and ball size. Thus, in general, ball stiffness variation is significant. In this model ball stiffness is considered to consist of steady and time dependent components:

$$K_j = K_{oj} + \Delta K_j(t)$$

where $K_{oj} = 1.5\,\kappa\delta^{1/2}$ and $K_{oj} = 1.5\,\kappa\delta^{1/2}$ and $\Delta K_j \simeq K_{oj}\left(\frac{\delta_j}{2\delta}\right)$. Furthermore, ball stiffness is considered to have axial and radial components given by:

$$K_{aj} = K_{oj}\sin\alpha_j + \Delta K_j \sin\alpha_j \quad \text{and} \quad K_{ij} = K_{oj}\cos\alpha_j + \Delta K_j \cos\alpha_j \qquad [2A.4]$$

2A.3 Contact deflections

The steady component of spring deflection is dependent upon the axial approach of inner and outer rings and the internal geometry of the bearing, while the time dependent component is a function of waviness amplitude and bearing geometry. Waviness is described by defining the bearing's radial internal clearance to have a constant component equal to the designed clearance and a time varying component representing waviness amplitude:

$$P_D = P_D + P_D\{\phi(t)\}$$

Ball–race contact angle and hence the line of action through centres of inner and outer contacts is given by (Figure 2A.2):

$$\text{Tan } \alpha = \frac{\delta_A + (r_o + r_i - D) \sin \alpha'}{r_i + r_o - D - P_D/2}$$

where α' is the no load contact angle:

$$\cos \alpha' = 1 - P_D/[2(r_o + r_i - D)]$$

and the steady component of spring deflection is given by:

$$\delta \cos \alpha = (r_o + r_i - D)(1 - \cos\alpha) - P_D/2$$

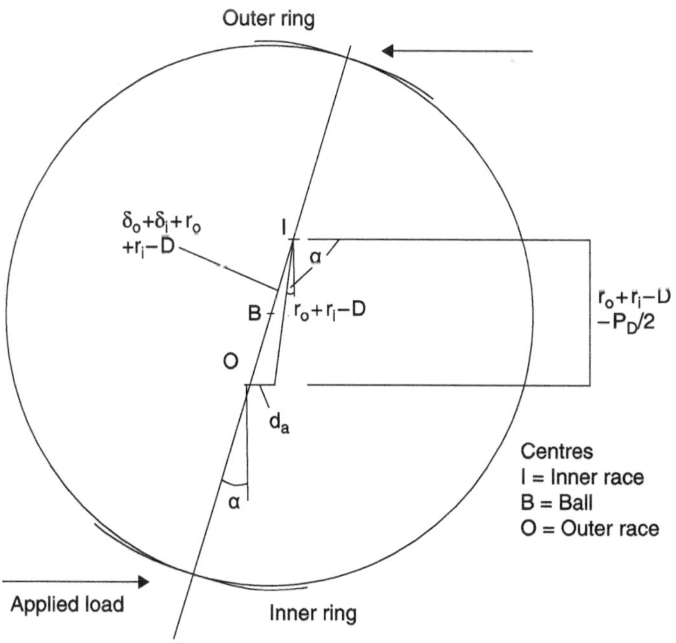

Figure 2A.2 Relationship between contact deformations and ball-raceway geometry

The time-dependent component of spring deflection and radial internal clearance are related by:

$$\delta_j(t)\cos\alpha = -dP_D\{\phi_j(t)\}/2 \qquad [2A.5]$$

where $\delta_j(t)$ is the deflection of spring number j and $\phi_j(t)$ is its angular position. The negative sign indicates that a reduction in internal clearance produces an increase in spring compression and hence spring force. Substitution of Eq. 2A.5 into Eq. 2A.3 enables the spring stiffness to be determined.

2A.4 Force analysis

The total vibration force generated by the bearing is the vector sum of that generated at each ball. In general, five components of vibration may be of interest: axial, radial in X and Y directions, and moment about X- and Y-axes. Each ball will generate axial and radial components of vibration given by:

$$F_{aj} = K_{aj}\frac{dP_D(\phi_j)}{2} \quad \text{and} \quad F_{rj} = K_{rj}\frac{dP_D(\phi_j)}{2}$$

where $dP_D(\phi_j)$ is a time varying component of radial internal clearance due to waviness of the bearing's rolling surfaces. The waviness may be present on the outer raceway, inner raceway or ball surfaces and there may also be ball size variation. Each error yields a different expression for ϕ, so it is convenient to consider them separately.

The waviness model is shown in Figure 2A.3. OXYZ is a fixed frame of reference with OZ coincident with the bearing axis, Oxyz is a rotating

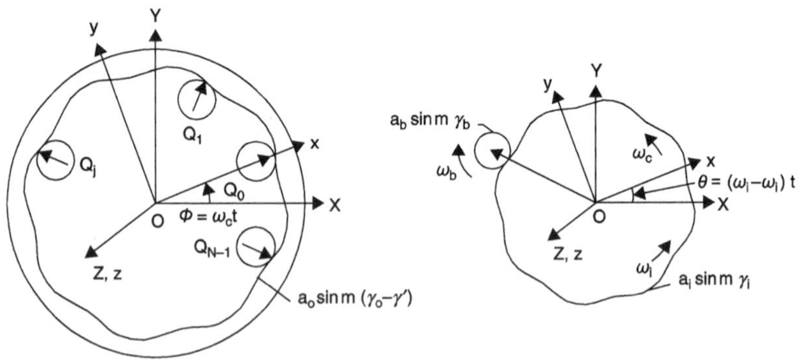

Figure 2A.3 Waviness model

frame with Oz also coincident with the bearing axis and Ox and Oy rotating with the ball complement at an angular speed ω_c. Waviness or lobing errors are present on ball, inner and outer raceway surfaces and are described by the simple harmonic functions:

Inner	$a = a_i \sin m\gamma_I$	$m = 1, 2, 3, \infty$
Outer	$a = a_o \sin m(\gamma_o - \gamma_o')$	
Ball	$a = a_b \sin m\gamma_b$	

Relative to the rotating frame Oxyz net bearing forces and moments are given by:

$$F_a' = \sum_{j=0}^{N-1} K_{aj} \frac{dP_D}{2}(\phi_j)$$

$$F_{rx}' = \sum_{j=0}^{N-1} K_{rj} \frac{dP_D}{2}(\phi_j)\cos\phi_j$$

$$F_{ry}' = \sum_{j=0}^{N-1} K_{rj} \frac{dP_D}{2}(\phi_j)\sin\phi_j$$

$$M_{xx}' = D_p/2 \sum_{j=0}^{N-1} K_{aj} \frac{dP_D}{2}(\phi_j)\sin\phi_j$$

$$M_{yy}' = D_p/2 \sum_{j=0}^{N-1} K_{aj} \frac{dP_D}{2}(\phi_j)\cos\phi_j$$

As frame Oxyz is rotating, then relative to the fixed frame OXYZ:

$$F_a = \sum_{j=0}^{N-1} K_{aj} \frac{dP_D}{2}(\phi_j) \qquad [2A.6]$$

$$F_{rx} = F_{rx}'\cos\omega_c t - F_{ry}'\sin\omega_c t = \sum_{j=0}^{N-1} K_{rj} \frac{dP_D}{2}(\phi_j)\cos(\omega_c t + \phi_j) \qquad [2A.7]$$

$$F_{ry} = F_{rx}'\sin\omega_c t - F_{ry}'\cos\omega_c t = \sum_{j=0}^{N-1} K_{rj} \frac{dP_D}{2}(\phi_j)\sin(\omega_c t + \phi_j) \qquad [2A.8]$$

$$M_{xx} = M_{xx}'\cos\omega_c t - M_{yy}'\sin\omega_c t = D_p/4 \sum_{j=0}^{N-1} K_{rj} \frac{dP_D}{2}(\phi_j)\sin(\phi_j - \omega_c t) \qquad [2A.9]$$

$$M_{yy} = M_{xx}'\sin\omega_c t - M_{yy}'\cos\omega_c t = D_p/4 \sum_{j=0}^{N-1} K_{rj} \frac{dP_D}{2}(\phi_j)\cos(\phi_j - \omega_c t) \qquad [2A.10]$$

It should be noted that the influence coefficients K_{aj} and K_{rj} are time dependent and therefore the forces F_{rx} and F_{ry} are not necessarily the same in magnitude. This is also the case for M_{xx} and M_{yy}.

2A.5 Transfer functions for outer raceway waviness

Outer raceway waviness (Figure 2A.3) is assumed to be described by:

$$a = a_o \sin m(\gamma_o - \gamma_o') \qquad m = 1, 2, 3...$$

where a_o is the amplitude of the waviness, γ_o' is a phase angle defining the position of the waviness with respect to the fixed frame OXYZ and m is an integer giving the number of waves/circumference. As each ball orbits around the bearing, it experiences a time dependent elastic deflection given by $\gamma_o = \omega_c t + \phi_j$. Since $\phi_j = 2\pi j/N$, then the change in radial internal clearance experienced by ball number j may be written as:

$$\frac{dP_D}{2}(\phi_j) = a_o \sin m(\omega_c t + \gamma' + 2\pi j/N) \qquad j = 1, 2, 3N - 1 \qquad [2A.11]$$

For a bearing subject to both axial and radial loads the axial component of ball stiffness, K_{aj}, is a function of angular position and time. It is convenient to express K_{aj} in the form:

$$K_{aj} = K_a + \sum_{i=1}^{\infty} K_{ai} \cos i\left(\omega_c t + \frac{2\pi j}{N}\right)$$

Substituting for dP_D and K_{aj} in Eq.2A.6 and using Eqs 2A.5 and 2A.11 gives:

$$F_a = a_o \sum_{j=0}^{N-1}\left[K_a + \sum_{i=1}^{\infty} K_{ai} \cos i(\omega_c t + 2\pi j/N)\right] \sin m(\omega_c t + \gamma' + 2\pi j/N)$$

which may be written in the form:

$$F_a/a_o = K_a S_1 + \sum_{i=1}^{\infty} K_{ai} S_2$$

The summation S_1 may be simplified [55] to:

$$S_1 = \sin m(\omega_c t + \gamma' + \pi(N - 1)/N)\sin m\pi/\sin m\pi/N$$

For which the term $\sin m\pi \sin m\pi/N = N$ when m is an integer multiple of N and zero for all other values of m. Thus:

$$S_1 = N \sin m(\omega_c t + \gamma' + \pi(N - 1)/N) \qquad m = N, 2N, 3N...$$

Similarly, S_2 can be simplified to:

$$S_2 = \tfrac{1}{2} \sum_{i=1}^{\infty} K_{ai} \sum_{j=0}^{N-1} \{\sin[(m+i)(\omega_c t + 2\pi j/N) + m\gamma'] +$$

$$\sin[(m-i)(\omega_c t + 2\pi j/N) + m\gamma']\}$$

$$S_2 = \tfrac{1}{2} N \sum_{i=1}^{\infty} K_{ai} \{\sin[(m+i)(\omega_c t + \pi(N-1)/N) + m\gamma'] +$$

$$\sin[(m-i)(\omega_c t + \pi(N-1)N) + m\gamma']\} \qquad m \pm i = N,\ 2N,\ 3N\ldots$$

Thus the transfer function for axial vibration is:

$$F_a/a_o = N\,K_a \sin m(\omega_c t + \gamma' + \pi(N-1)/N) + \qquad m = N,\ 2N,\ 3N\ldots$$

$$\tfrac{1}{2} N \sum_{i=1}^{\infty} K_{ai} \{\sin[(m+i)(\omega_c t + \pi(N-1)/N) + m\gamma'] + \qquad m \pm i = N, 2N,\ 3N\ldots$$

$$\sin[(m-i)(\omega_c t + \pi(N-1)N) + m\gamma']\} \qquad [2A.12]$$

The above equation gives the axial vibration force F_a resulting from waviness of the outer raceway. Vibration frequencies are always harmonic of the ball pass rate, $N\omega_c$, and are produced by particular waviness wavelengths. The dominant wavelengths have an integer multiple of N waves per circumference. For this situation all balls roll over high and low points of the waviness in phase and produce the maximum fluctuation in force. However, other wavelengths can produce vibration at the same frequencies due to the variation in ball contact compliance with orbital position. On bearings only subject to axial load and with normal levels of waviness, i.e. amplitudes less than 1 μm, variation in contact compliance is negligible. However, on radially loaded or misaligned bearings, variation in contact compliance is large and must be taken into account. In this situation waviness wavelengths of m, m + 1, m − 1, m + 2, m − 2 . . ., where $m \pm i$ is an integer multiple of N, the number of balls can produce vibration. In this case the magnitude of vibration force at any one frequency is the vector sum of that due to the individual wavelengths. It should be noted that the contact compliance parameter K_{ai} may generally be expected to reduce in magnitude as i increases and in practice may be negligible for values of i greater than N.

Similar expressions may be derived for radial and moment vibration. They contain additional terms that arise due to orbital rotation of the balls with respect to the direction of measurement. These expressions are summarised as follows.

For radial vibration:

$$\frac{F_{rx}}{a_o} = \sum_{j=0}^{N-1}\left[K_r + \sum_{i=1}^{\infty} K_{ri}\cos i(\omega_c t + 2\pi j/N)\right]\sin m(\omega_c t + \gamma' + 2\pi j/N)$$
$$\cos(\omega_c t + 2\pi j/N)$$

$$\frac{F_{rx}}{a_o} = \tfrac{1}{2}\sum_{j=0}^{N-1}\{\sin[(m+1)(\omega_c t + m\gamma' + 2\pi j/N)] + \sin[(m-1)(\omega_c t + m\gamma' + 2\pi j/N)]a_o$$
$$\left[K_r + \sum_{i=1}^{\infty} K_{ri}\cos i(\omega_c t + 2\pi j/N)\right]$$

$$\begin{aligned}\frac{F_{rx}}{a_o} = &\tfrac{1}{2}NK_r\{\sin[(m+1)(\omega_c t + (N-1)\pi/N) + m\gamma'] +\\ &\sin[(m-1)(\omega_c t + (N-1)\pi/N) + m\gamma']\} +\\ &\tfrac{1}{4}N\sum_{i=1}^{\infty}K_{ri}\{\sin[(m+1+i)(\omega_c t + (N-1)\pi/N) + m\gamma'] +\\ &\sin[(m-1+i)(\omega_c t + (N-1)\pi/N) + m\gamma'] +\\ &\sin[(m+1-i)(\omega_c t + (N-1)\pi/N) + m\gamma'] +\\ &\sin[(m-1-i)(\omega_c t + (N-1)\pi/N) + m\gamma']\}\end{aligned} \qquad [2A.13]$$

where:

$m \pm 1 \pm i = pN$ and $p = 1, 2, 3 \ldots$

F_{ry} only differs from F_{rx} in respect of phase so that:

$$\begin{aligned}\frac{F_{ry}}{a_o} = &\tfrac{1}{2}NK_r\{\cos[(m+1)(\omega_c t + (N-1)\pi/N) + m\gamma'] +\\ &\cos[(m-1)(\omega_c t + (N-1)\pi/N) + m\gamma']\} +\\ &\tfrac{1}{4}N\sum_{i=1}^{\infty}K_{ri}\cos[(m+1+i)(\omega_c t + (N-1)\pi/N) + m\gamma'] +\\ &\cos[(m-1+i)(\omega_c t + (N-1)\pi/N) + m\gamma'] +\\ &\cos[(m+1-i)(\omega_c t + (N-1)\pi/N) + m\gamma'] +\\ &\cos[(m-1-i)(\omega_c t + (N-1)\pi/N) + m\gamma']\}\\ &m \pm 1 \pm i = pN \text{ and } p = 1, 2, 3\ldots\end{aligned} \qquad [2A.14]$$

Similarly for moment vibration:

$$\frac{2M_{xx}}{D_p a_o} = \tfrac{1}{2} NK_r \{\cos[(m+1)(\omega_c t+(N-1)\pi/N)+m\gamma']+$$
$$\cos[(m-1)(\omega_c t+(N-1)\pi/N)+m\gamma']+$$
$$\tfrac{1}{4} N \sum_{i=1}^{\infty} K_{ri} \{\cos[(m+1+i)(\omega_c t+(N-1)\pi/N)+m\gamma']+$$
$$\cos[(m-1+i)(\omega_c t+(N-1)\pi/N)+m\gamma']+$$
$$\cos[(m+1-i)(\omega_c t+(N-1)\pi/N)+m\gamma']+ \quad [2A.15]$$
$$\cos[(m-1-i)(\omega_c t+(N-1)\pi/N)+m\gamma']\}$$
$$m \pm 1 \pm i = pN \quad \text{and} \quad p=1, 2, 3\ldots$$

and:

$$\frac{2M_{yy}}{D_p a_o} = \tfrac{1}{2} NK_r \{\sin[(m+1)(\omega_c t+(N-1)\pi/N)+m\gamma']+$$
$$\sin[(m-1)(\omega_c t+(N-1)\pi/N)+m\gamma']\}+$$
$$\tfrac{1}{4} N \sum_{i=1}^{\infty} K_{ri} \{\sin[(m+1+i)(\omega_c t+(N-1)\pi/N)+m\gamma']+$$
$$\sin[(m-1+i)(\omega_c t+(N-1)\pi/N)+m\gamma']+ \quad [2A.16]$$
$$\sin[(m+1-i)(\omega_c t+(N-1)\pi/N)+m\gamma']+$$
$$\sin[(m-1-i)(\omega_c t+(N-1)\pi/N)+m\gamma']\}$$
$$m \pm 1 \pm i = pN \quad \text{and} \quad p=1, 2, 3\ldots$$

2A.6 Transfer functions for inner raceway waviness

The mechanism by which waviness on the inner raceway produces vibration is the same as that for the outer raceway and the analysis follows the same route. Inner ring rotation must be taken into account and the main effect of this is to change the basic frequency of the vibration from ball pass rate to ball pass rate relative to the inner ring, i.e. from $N\omega_c$ to $N(\omega_i - \omega_c)$.

The time varying component of deflection experienced by ball number j is:

$$\frac{dP_D}{2}(\phi_j) = a_i \sin m((\omega_i - \omega_c)t + 2\pi j/N) \qquad j = 1, 2, 3 \ldots N-1$$

For convenience, let $\omega_{ci} = \omega_i - \omega_c$, then substitution of this into Eq. 2A.12 and simplifying as in the previous section gives the transfer functions for inner ring waviness:

$$
\begin{aligned}
F_a/a_i = NK_a \Big\{ & \sin m\left(\omega_{ci}t+\pi(N-1)/N\right) + \\
& \tfrac{1}{2}N\sum_{i=1}^{\infty}K_{ai}\sin[(m\omega_{ci}+i\omega_c)t+(m+i)\pi(N-1)/N)] + \\
& \sin[(m\omega_{ci}-i\omega_c)t+(m-i)\pi(N-1)/N)]\Big\} \qquad m=N,\ 2N,\ 3N\ldots
\end{aligned}
\qquad [2A.17]
$$

$$
\begin{aligned}
\frac{F_{rx}}{a_i} = \tfrac{1}{2}\ NK_r \Big\{ & \sin\left[\left(m\omega_{ci}+\omega_c\right)t+(m+1)\pi(N-1)/N\right)\Big]+ \\
& \sin\left[\left(m\omega_{ci}-\omega_c\right)t+(m-1)\pi(N-1)/N\right)\Big]\Big\}+ \\
& \tfrac{1}{4}\ N\sum_{i=1}^{\infty}K_{ri}\Big\{\sin\left[\left(m\omega_{ci}+(1+i)\omega_c\right)t+(m+1+i)\pi(N-1)/N\right)\Big]+ \\
& {}^{i=1}\sin\left[\left(m\omega_{ci}+(1-i)\omega_c\right)t+(m+1-i)\pi(N-1)/N\right)\Big]+ \\
& \sin\left[\left(m\omega_{ci}+(1-i)\omega_c\right)t+(m-1+i)\pi(N-1)/N\right)\Big]\Big]+ \\
& \sin\left[\left(m\omega_{ci}-(1+i)\omega_c\right)t+(m-1-i)\pi(N-1)/N\right)\Big]\Big\} \qquad m\pm i\pm 1 = N,\ 2N,\ 3N\ldots
\end{aligned}
\qquad [2A.18]
$$

$$
\begin{aligned}
\frac{F_{ry}}{a_i} = \tfrac{1}{2}\ NK_r \Big\{ & \cos\left[\left(m\omega_{ci}+\omega_c\right)t+(m+1)\pi(N-1)/N\right)\Big] + \\
& \cos\left[\left(m\omega_{ci}-\omega_c\right)t+(m-1)\pi(N-1)/N\right)\Big]\Big\}+ \\
& \tfrac{1}{4}\ N\sum_{i=1}^{\infty}K_{ri}\Big\{\cos\left[\left(m\omega_{ci}+(1+i)\omega_c\right)t+(m+1+i)\pi(N-1)/N\right)\Big]+ \\
& {}^{i=1}\cos\left[\left(m\omega_{ci}+(1-i)\omega_c\right)t+(m+1-i)\pi(N-1)/N\right)\Big]+ \\
& \cos\left[\left(m\omega_{ci}+(1-i)\omega_c\right)t+(m-1+i)\pi(N-1)/N\right)\Big]\Big]+ \\
& \cos\left[\left(m\omega_{ci}-(1+i)\omega_c\right)t+(m-1-i)\pi(N-1)/N\right)\Big]\Big\} \qquad m\pm i\pm 1 = N,\ 2N,\ 3N\ldots
\end{aligned}
\qquad [2A.19]
$$

$$
\begin{aligned}
\frac{2M_{xx}}{a_iD_p} = \tfrac{1}{2}\ NK_a \Big\{ & \sin\left[\left(m\omega_{ci}+\omega_c\right)t+(m+1)\pi(N-1)/N\right)\Big] + \\
& \sin\left[\left(m\omega_{ci}-\omega_c\right)t+(m-1)\pi(N-1)/N\right)\Big]\Big\}+ \\
& \tfrac{1}{4}\ N\sum_{i=1}^{\infty}K_{ai}\Big\{\sin\left[\left(m\omega_{ci}+(1+i)\omega_c\right)t+(m+1+i)\pi(N-1)/N\right)\Big]+ \\
& {}^{i=1}\sin\left[\left(m\omega_{ci}+(1-i)\omega_c\right)t+(m+1-i)\pi(N-1)/N\right)\Big]+ \\
& \sin\left[\left(m\omega_{ci}+(1-i)\omega_c\right)t+(m-1+i)\pi(N-1)/N\right)\Big]\Big]+ \\
& \sin\left[\left(m\omega_{ci}-(1+i)\omega_c\right)t+(m-1-i)\pi(N-1)/N\right)\Big]\Big\} \qquad m\pm i\pm 1 = N,\ 2N,\ 3N\ldots
\end{aligned}
\qquad [2A.20]
$$

$$\frac{2M_{yy}}{a_i D_p} = \tfrac{1}{2} NK_a \{\cos[(m\omega_{ci} + \omega_c)t + (m+1)\pi(N-1)/N)] +$$
$$\cos[(m\omega_{ci} - \omega_c)t + (m-1)\pi(N-1)/N)]\} +$$
$$\tfrac{1}{4} N\sum_{i=1}^{\infty} K_{ai}\cos[(m\omega_{ci} + (1+i)\omega_c)t + (m+1+i)\pi(N-1)/N)] + \qquad [2A.21]$$
$$\cos[(m\omega_{ci} + (1-i)\omega_c)t + (m+1-i)\pi(N-1)/N)] +$$
$$\cos[(m\omega_{ci} + (1-i)\omega_c)t + (m-1+i)\pi(N-1)/N)]] +$$
$$\cos[(m\omega_{ci} - (1+i)\omega_c)t + (m-1-i)\pi(N-1)/N)]] \qquad m \pm i \pm 1 = N, 2N, 3N\ldots$$

2A.7 Transfer functions for ball waviness

For the general case of a bearing subject to combined loading rotating at high speed, ball motion is complex and a precise transfer function relating waviness amplitude to vibration force is difficult to define. The main difficulties arise from:

- changes in ball rotational speed in the unloaded zone of the bearing;
- differences between inner and outer contact angles influencing the effective amplitude of ball waviness;
- spinning motion of the ball.

However, these effects are small and in this analysis are ignored. The ball is assumed to rotate about one axis and vibration is produced by waviness present in a diametrical plane passing through the ball centre perpendicular to the spin axis. Ball waviness is therefore described by:

$$a = a_b \cos(m\phi) \qquad m = 1, 2, 3 \ldots$$

The corresponding variation in the bearing's radial internal clearance experienced by the ball is then:

$$dP_D = 2a_b \cos(2m\omega_b t) \qquad m = 1, 2, 3 \ldots$$

where it is assumed that only even numbers of waves/circumference generate a local internal clearance variation. Substitution of the above equation into Eq. 2A.12 and removing the summation corresponding to N balls gives the vibration force generated by one rolling element. For example, consider axial vibration:

$$F_a / a_b = 2K_a \cos(2m\omega_b t) + \sum_{i=1}^{\infty} K_{ai}[\cos(i\omega_c + 2m\omega_b)t + \cos(i\omega_c - 2m\omega_b)t] \qquad [2A.22]$$

Similarly for radial and moment vibration:

$$F_{rx} / a_b = K_r \left[\sin(2m\omega_b + \omega_c)t + \sin(2m\omega_b - \omega_c)t\right] +$$
$$\sum_{i=1}^{\infty} K_{ri}\left[\sin((i+1)\omega_c + 2m\omega_b)t + \sin((i-1)\omega_c - 2m\omega_b)t + \right. \quad [2A.23]$$
$$\left. \sin((i-1)\omega_c + 2m\omega_b)t + \sin((i+1)\omega_c - 2m\omega_b)t\right]$$

$$F_{ry} / a_b = K_r \left[\cos(2m\omega_b + \omega_c)t + \cos(2m\omega_b - \omega_c)t\right] +$$
$$\tfrac{1}{2}\sum_{i=1}^{\infty} K_{ri}\left[\cos((i+1)\omega_c + 2m\omega_b)t + \cos((i-1)\omega_c - 2m\omega_b)t + \right. \quad [2A.24]$$
$$\left. \cos((i-1)\omega_c + 2m\omega_b)t + \cos((i+1)\omega_c - 2m\omega_b)t\right]$$

$$\frac{M_{xx}}{a_b D_p} = -\tfrac{1}{2} K_a \left[\cos(2m\omega_b + \omega_c)t + \sin(2m\omega_b - \omega_c)t\right] +$$
$$\tfrac{1}{2}\sum_{i=1}^{\infty} K_{ri}\left[\sin((i+1)\omega_c + 2m\omega_b)t + \sin((i-1)\omega_c - 2m\omega_b)t + \right. \quad [2A.25]$$
$$\left. \sin((i-1)\omega_c + 2m\omega_b)t + \sin((i+1)\omega_c - 2m\omega_b)t\right]$$

$$\frac{M_{yy}}{a_b D_p} = -\tfrac{1}{2} K_a \left[\cos(2m\omega_b + \omega_c)t + \cos(2m\omega_b - \omega_c)t\right] +$$
$$\tfrac{1}{2}\sum_{i=1}^{\infty} K_{ri}\left[\cos((i+1)\omega_c + 2m\omega_b)t + \cos((i-1)\omega_c - 2m\omega_b)t + \right. \quad [2A.26]$$
$$\cos((i-1)\omega_c + 2m\omega_b)t + \cos((i+1)\omega_c - 2m\omega_b)t$$
$$m = 1, 2, 3, \quad i = 1, 2, 3\ldots$$

Phase terms are of no practical significance for ball vibration and have been dropped from the above equations. The latter relate bearing vibration force to the waviness on just one ball. Assuming all balls in a bearing are manufactured in one batch and that their surface geometries are therefore similar, then the total vibration force for a given waviness wavelength is approximately the RMS sum of that due to all N balls.

2A.8 Ball size variation

Ball size variation may be considered as a special case of waviness for which the number of waves/circumference, $m = 0$. The equations of the preceding section can then be simplified to:

$$F_a / a_b = 2K_a + 2\sum_{i=1}^{\infty} K_{ai} \cos i\omega_c t \quad [2A.27]$$

$$F_{rx}/a_b = 2K_r \sin\omega_c t + \sum_{i=1}^{\infty} K_{ri}\left[\sin(i+1)\omega_c t + \sin(i-1)\omega_c t\right] \qquad [2A.28]$$

$$\frac{M_{xx}}{a_b D_p} = -K_a \sin\omega_c t + \sum_{i=1}^{\infty} K_{ri}\left[\sin(i+1)\omega_c t + \sin(i-1)\omega_c t\right] \qquad [2A.29]$$

$$\frac{M_{xx}}{a_b D_p} = -K_a \sin\omega_c t + \sum_{i=1}^{\infty} K_{ri}\left[\sin(i+1)\omega_c t + \sin(i-1)\omega_c t\right] \qquad [2A.30]$$

$$\frac{M_{yy}}{a_b D_p} = -K_a \sin\omega_c t + \sum_{i=1}^{\infty} K_{ri}\left[\cos(i+1)\omega_c t + \cos(i-1)\omega_c t\right] \qquad [2A.31]$$

$i = 1, 2, 3\ldots$

where now a_b is the departure of the ball radius from the standard ball radius. Again the above equations relate vibration force to ball size error for a single ball. Note that now vibration frequencies produced are only harmonic of cage speed, whereas ball waviness produces vibration frequencies harmonic of ball rotational speed modulated by cage speed. The above equations suffice to determine the relative importance of ball size variation on bearing motion error; however, the analysis may in principle be extended to consider the general case of all balls being of different diameter.

2.14 References

1. Gupta, P.K. (1979), 'Dynamics of rolling element bearings, Part III: Ball bearing analysis', *Transactions of the ASME*, **101**: 312–18.
2. Harris, T.A. (1984), *Rolling Bearing Analysis*, 2nd edition, New York: John Wiley & Sons.
3. Walters, C.T. (1971), 'The dynamics of ball bearings', *Transactions of the ASME, Journal of Lubrication Technology*, **13(4)**: 229.
4. Gentle, C.R. and Boness, R.J. (1998), 'Prediction of ball motion in high-speed thrust-loaded ball bearings', *Transactions of the ASME, Journal of Lubrication Technology, 75-Lub-1*, 463–9.
5. Schulze, D.R. and Dill, J.F. (1976), 'Measurement and analysis of three-dimensional cage motion & bearing performance of a 100 mm bore bearing', *Lubrication Engineering*, **38(7)**: 439–48.
6. Kannel, J.W. and Bupara, S.S. (1978), 'A simplified model of cage motion in angular contact bearings operating in the EHD lubrication regime', *Transactions of the ASME, Journal of Lubrication Technology*, **100**: 395–403.
7. Adam, G.J. and Jones, J.R. (1976), 'The effect of retainer geometry on the stability of ball bearings', *Transactions of the ASLE*, **19(2)**: 95–107.
8. Hertz, H. (1882), 'Ber die Beruhrung fester elastischer Korper', *Journal reine und angewandte Mathematik*, **92**: 156–71.

9. Johnson, K.L. (1985), *Contact Mechanics*, Cambridge: Cambridge University Press.
10. Dareing, D.W. and Johnson, K.L. (1975), 'Fluid film damping of rolling contact vibrations', *Proceedings of the Institute of Mechanical Engineers, Part C: Journal of Mechanical Engineering Science*, **17(4)**: 214–18.
11. Elsermans, M., Hongerloot, M. and Snoeye, R. (1976), 'Damping in taper roller bearings', *Proceedings of the 16th MTDR Conference*, London, 223–9.
12. Walford, T.L.H. and Stone B.J. (1983), 'The source of damping in rolling element bearings under oscillating conditions', *Proceedings of the Institute of Mechanical Engineers, Part C: Journal of Mechanical Engineering Science*, **197**: 225–32.
13. Walford, T.L.H. (1981), 'Some stiffness and damping characteristics of ball bearings', PhD Thesis, University of Bristol, UK.
14. Wensing, J.A. (1998), 'On the dynamics of ball bearings', PhD Thesis, University of Twente, The Netherlands.
15. Sarangi, M., Majumdar, B.C. and Sekhar, A.S. (2004), 'Stiffness and damping characteristics of lubricated ball bearings, Part 1: Theoretical formulation', *Proceedings of the Institute of Mechanical Engineers, Part J: Journal of Engineering Tribology*, **218**: 529–38.
16. Sarangi, M., Majumdar, B.C. and Sekhar, A.S. (2004), 'Stiffness and damping characteristics of lubricated ball bearings, Part 2: numerical results and application', *Proceedings of the Institute of Mechanical Engineers, Part J: Journal of Engineering Tribology*, **218**: 539–47.
17. Sarangi, M., Majumdar, B.C. and Sekhar, A.S. (2004), 'On the dynamics of elasto-hydrodynamic mixed lubricated ball bearings. Part I: Formulation of stiffness and damping coefficients', *Proceedings of the Institute of Mechanical Engineers, Part J: Journal of Engineering Tribology*, **219**: 411–21.
18. Sarangi, M., Majumdar, B.C. and Sekhar, A.S. (2004), 'On the dynamics of elasto-hydrodynamic mixed lubricated ball bearings. Part II: Non-linear structural vibration', *Proceedings of the Institute of Mechanical Engineers, Part J: Journal of Engineering Tribology*, **219**: 423–33.
19. Andrew, C. and Coburn, J.A. (1967), 'Metal surfaces in contact under normal forces: Some dynamic stiffness and damping characteristics', *Proceedings of the Institute of Mechanical Engineers*, **182(11)**: pp 92–100.
20. Wardle, F.P., Lacey, S.J. and Poon, S.Y. (1983), 'Dynamic and static characteristics of a wide speed range machine tool spindle', *Precision Engineering*, **5(4)**: 175–83.
21. Perret, H. (1950), 'Elastischen Spielschwingunen onstant belaster Walzlager', *Werkstatt und Betrieb*, **83(C5)**: 354–8.
22. Meldau, E. (1951), 'Die Bewechung der Achse vonWalzlagern bei geringen Drehzahlen', *Werkstatt und Betrieb*, **84(C5)**: 308–13.
23. Tallian, T.E. and Gustafsson, O.G. (1965), 'Progress in roller vibration research and control', *Transactions of the ASLE*, **8(3)**: 195–207.
24. Sheftel, B.T. et al. (1965), 'The effect of race groove waviness on bearing vibration', *Russian Engineering Journal*, **7**: 50–2.
25. Yhland, E. (1968), 'Waviness measurement: An instrument for quality control in the rolling bearing industry', *Proceedings of the Institute of Mechanical Engineers*, **182(3K)**: 438.

26. Gray, G.G. and Johnson, K.L. (1972), 'The dynamic response of elastic bodies in rolling contact to random roughnesses of their surfaces', *Journal of Sound Vibrations*, **22(3)**: 323–42.
27. McFadden, P.D. and Smith, J.D. (1984), 'Model for the vibration produced by a single point defect in a rolling bearing', *Journal of Sound Vibrations*, **96**: 69–82.
28. McFadden, P.D. and Smith, J.D. (1985), 'The vibration produced by multiple point defects in a rolling element bearing', *Journal of Sound Vibrations*, **98(2)**: 263–73.
29. Rahnejat, H. and Gohar, R. (1985), 'The vibration of radial ball bearings', *Proceedings of the Institute of Mechanical Engineers, Part C*, **199(C3)**: 181–93.
30. Wardle, F.P. (1988), 'Vibration forces produced by waviness of thrust loaded ball bearings, Parts I and II', *Proceedings of the Institute of Mechanical Engineers*, **202**: C5.
31. Wensing, J.A. (1998), 'On the dynamics of ball bearings', PhD Thesis, University of Twente, The Netherlands.
32. Wensing, J.A. and Nijen, G.C.V. (1996), '2-D computational model for vibration analysis of waviness in rolling bearing applications', *Proceedings of the Institute of Mechanical Engineers 6th International Conference of Vibration in Rotating Machinery*, 371–82.
33. Wensing, J.A. (1998), 'Dynamic behaviour of ball bearings on vibration test spindles', *16th International Modal Analysis Conference*, Santa Barbara, USA.
34. Fujii, Y. (2004), 'Bearing for computer magnetic disc spindle', *NSK Bearing Journal*, **644**: 14–23 (in Japanese).
35. Noguchi, S., Tsukada, T. and Sakamoto, J. (1995), 'Evaluation method to determine radial accuracy of high precision rotating spindle units', *Precision Engineering*, **17(4)**: 266–73.
36. Ono, K. and Takahashi, K. (1996), 'Theoretical analysis of shaft vibration supported by a ball bearing with small sinusoidal waviness', *IEEE. Trans. Magn.*, **32(3)**: 1709–15.
37. Momono, T. and Noda, B. (1996), 'Sound and vibration in rolling bearings', *NSK Technical Journal*, **661**: 13–22 (in Japanese).
38. Kubota, Y. (1992), 'NRRO on miniature ball bearing', *Proceedings of the Japanese Society of Mechanical Engineers*, Tokai Branch, Mie District, **923(2)**: 294–97 (in Japanese).
39. Noguchi, S. and Ono, K. (1998), 'Reduction of NRRO in a ball bearing for HDD spindle motors, 2nd report: Experimental analysis of ball bearing with 12 balls)', *Transactions of the Japanese Society of Mechanical Engineers*, **64(624)**: 3149–54 (in Japanese).
40. Noguchi, S., Tanaka, K. and Ono, K. (1999), 'Theoretical analysis of a ball bearing used in HDD spindle motors for reduction of NRRO', *IEEE Trans. Magn.*, **35(2)**: 845–50.
41. Sakaguchi, T. and Akamatsu, Y. (2000), 'Simulation for ball bearing vibration', *Proceedings of the International Tribology Conference*, Nagasaki.
42. Noguchi, S. and Ono, K. (2004), 'Reduction of NRRO in ball bearings for HDD spindle motors', *Precision Engineering*, **28**, 409–18.

43. Noguchi, S., Hiruma, K., Kawa, H. and Kanada, T. (2005), 'The influence of location of balls and ball diameter difference in rolling bearings on the non-repetitive runout (NRRO) of Retainer Revolution', *Precision Engineering*, **29**: 11–18.
44. Wardle, F.P. and Poon, S.Y. (1993), 'Rolling bearing noise: cause and cure', *Institute of Mechanical Engineers Seminar*, London.
45. Hamrock, B.J. and Dowson, D. (1981), *Ball Bearing Lubrication, The Elasto-hydrodynamics of Elliptical Contacts*, Wiley & Sons.
46. Sayles, R.S., Hamer, J.C. and Ioannides, E. (1990), 'The effects of particulate contamination in rolling bearings: A state of the art review', *Proceedings of the Institute of Mechanical Engineers*, **204**: 29–36.
47. Jacobson, B.O. (1991), *Rheology and Elasto-hydrodynamic Lubrication*, Elsevier.
48. Palmgren, A. (1959), *Ball and Roller Bearing Engineering*, 3rd edition, Philadelphia: Burbank.
49. Johnson, K.L. and Tevaarwerk, J.L. (1977), 'Shear behaviour of elasto-hydrodynamic oil films', *Proceedings of the Royal Society of London*, Series *A*, **356**: 215–36.
50. Wardle, F.P., Siu, S.W. and Synnott, A.T. (1989), 'Ceramic rolling elements for machine tool spindle bearings', *Tribology Open Day, Institute of Mechanical Engineers*, December.
51. Stevens, K.T. and Todd, M.J. (1977), 'Thermal conductance across ball bearings in vacuum', Risley, UK: National Centre of Tribology, Report No. ESA-ESTL-25, 51 p.
52. Takeuchi, Y.R., Dickey, J.T., Demsky, S.M., Lue, K.K., Kirsch., J.J. and Frantz, P.P. (2006), 'A methodology for measuring thermal properties of bearings in motion', Los Angeles: The Aerospace Corporation.
53. Ioannides, E. and Harris, T.A. (1985), 'A new fatigue life model for rolling bearings', *ASME Journal of Tribology*, **107(3)**, 367–78.
54. Lundberg, G. and Palmgren, A. (1947), 'Dynamic capacity of rolling bearings', *Acta Polytechic Mechanical Engineering Series 1*, R.S.A.E.E., No. 3: 7.
55. Dwight, H.B. (1968), *Tables of Integrals and other Mathematical Data*, 4th edition, New York: The MacMillan Company.

2.15 Notation

A_f – bearing displacement due to motion error
a – Hertzian contact, half length
b – Hertzian, contact half width
a_o – dynamic displacement
c – constant
C – Damping coefficient
C_{li}, C_{lo} – Damping coefficients associated with lubricant film at inner and outer contacts
C_x, C_y C_z C_{xx}, C_{yy} – Components of damping coefficient
C_d – dynamic capacity

C_s – static capacity
D – ball diameter
D_i – bore diameter of bearing
D_p – pitch circle diameter of bearing
D_o – outside diameter of bearing
d_i, d_o – inner and outer raceway diameters
E – Young's modulus
E′ – combined Young's modulus for ball and raceway
E(e) – elliptic integral of the first kind
e – argument of elliptic integral
F – force
F_b – bearing load
F_c – centrifugal force
F_x, F_y, F_z, F_{xx}, F_{yy} – component of force
f_1 – load factor
G – non-dimensional material parameter
H – Heaviside step function
H – lubricant film thickness
h_c – central lubricant film thickness
L – life
L_c – contact parameter
L_{10} – 90% survival life
$I_i - r_i - D/2 - \delta_i$, distance between inner raceway and ball centres
$I_o - r_o - D/2 + \delta_o$, distance between outer raceway and ball centres
J – square root of –1
K – a/b
K – static stiffness
K′ – non-dimensional contact stiffness
K'_I, K′ – inner and outer contact compliance factors
K_a – stiffness associated with asperity contact
K_o, K_{oj} – static stiffness of one ball
K_x, K_y, K_z, K_{xx}, K_{yy} – component of static stiffness
K(e) – elliptic integral of the second kind
M – total bearing torque
M_c – contact parameter
M_l – component of torque due to load
M_v – component of torque due to speed
M_{xx}, M_{yy} – moments about *x*- and *y*-axes
N – number of balls
N_c – contact parameter
O – origin of coordinate system
P – power
P_d – radial internal clearance
P_e – end play
p_o – maximum stress in Hertzian contact
Q – ball load
Q_a, Q_r – axial and radial components of ball load
Q_i, Q_o – ball load at inner and outer contacts

R – ball radius
r_i, r_o – inner and outer raceway radius normal to the direction of rolling
R – radius
R_x – raceway radius in rolling direction
R′, R″ – principal radii of curvature
S – D/D_p
S(N) – stressed volume
Tanδ – loss tangent associated with hysteretic damping
U – non-dimensional speed parameter
u_1 – elastic deformation
u_2 – elastic deformation
V_i, V_o – rolling speeds at inner and outer contacts
V_m – $(V_i + V_o)/2$, mean rolling speed
V_r – stressed volume
W – non-dimensional load parameter
X, Y, Z, x, y, z – coordinates
z_o, z′ – depth into Hertzian contact

Symbols

α – contact angle
α_i, α_o – contact angles at inner and outer contacts
α' – no load contact angle
γ – $D\cos\alpha/D_p$
γ' – surface texture parameter
δ – Hertzian contact deflection
δ_i, δ_o – contact deflections at inner and outer contacts
λ – surface roughness parameter
√ – Poisson's ratio
$√_o$ – kinematic viscosity
η_o – dynamic viscosity
σ – normal stress
ϕ – angular coordinate
ϕ_i – $D/2r_i$, osculation at inner contact
ϕ_o – $D/2r_o$, osculation at outer contact
ψ – angular coordinate
θ – angular coordinate
τ – shear stress
ω – angular speed
ω_i, ω_o – angular speed of inner and outer rings

Subscripts

i, o – inner and outer contacts
i, j, k, m, n – integers
x, y, z – coordinates

3

Hydrostatic bearings

DOI: 10.1533/9780857092182.147

Abstract: Hydrostatic bearings operate with relatively thick oil films of large area and therefore have the potential for extremely low motion errors coupled with high damping, high stiffness and high load capacity. They are available from specialist bearing companies, generally in the form of sub-assemblies such as spindles, rotary tables and slides, but as dimensional and geometric tolerances are not unusually precise many machine builders also choose to design and manufacture their own bearing systems. The purpose of this chapter is to provide basic calculations that can be used to evaluate bearing performance, define bearing specifications or to aid bearing design. These cover calculations for the load capacity, static stiffness, damping capacity and power consumption of popular hydrostatic bearing geometries and the mechanisms by which hydrostatic bearings generate motion error and estimation of the levels of bearing surface roundness and flatness necessary to achieve low motion errors.

Key words: externally pressurised bearings, hydrostatic bearings, oil lubricated bearings, bearings, low motion error.

3.1 Introduction

Hydrostatic bearings rely on an external source of high pressure liquid, normally oil, to generate load carrying capacity. The liquid is incompressible and can be supplied at high pressure, hundreds or occasionally, even thousands of Bar to provide the bearing with high load carrying capacity. Oil is also available in a wide range of viscosities, so unlike the air bearing it is not essential to maintain small bearing clearances to limit fluid leakage. In practice, bearing clearances are generally in the range 10 to 100 μm and the manufacturing tolerances required to achieve clearances in the mid- to high end of the range are well within the capability of many manufacturing companies. Thus many machine builders requiring ultra precision motion choose to design and manufacture their own hydrostatic bearings. There are no standards for the pressure at which oil is supplied to a hydrostatic bearing, so this is usually chosen to suit the particular requirements of the application – if load capacity is important then a high supply pressure is an advantage, whereas if bearing speed or temperature is of concern, lower pressures coupled with higher flow rates may be more appropriate. Thus machines utilising hydrostatic bearings generally require dedicated high pressure oil supplies matched to the bearing requirements.

The main advantages in lubricating a bearing with an incompressible liquid supplied at high pressure is high load carrying capacity, high static stiffness, high damping and good averaging of the effects of bearing surface imperfections on the distribution of pressure within the bearing. As a result of the latter, hydrostatic bearings achieve motion errors well below 100 nm with bearing surfaces that are not unusually precise, i.e. having roundness or flatness values of several microns. An externally pressurised fluid film bearing also has an advantage of not having any free play, so the shaft or moving components are accurately centralised, even at zero speed. However, a limitation is high friction at moderate and high sliding speeds. The losses are hydrodynamic in nature, so at low speeds they can be small or even negligible, making hydrostatic bearings ideal for linear slide and rotary table applications on machine tools [1]. Hydrostatic bearings are also used in machining spindles where speeds are moderate or high; however, the need to minimise power losses, not only due to sliding friction but also the high pressure supply pump, has a large influence on the bearing design.

In this chapter, the basic types of hydrostatic bearing in common use are described and their operating principles explained. A method for calculating their load capacity, static stiffness damping force, torque and power loss is given and the mechanisms that generate motion error are also described. Hydrostatic bearings are highly specialised machine components and can be designed in a wide variety of configurations to gain useful performance advantages under specific conditions. The purpose of this chapter is not to cover a comprehensive range of hydrostatic bearing designs but just those geometries in popular use and to focus on aspects of performance relevant to ultra precision machinery.

3.2 Bearing description

A schematic arrangement of a hydrostatic bearing is shown in Figure 3.1. The part of the bearing that supports load and enables relative movement simply comprises one or more relatively deep pockets machined into a housing, each surrounded by a land. Each pocket is fed with high pressure oil through some form of compensation element, such as a capillary and the oil is allowed to leak out of the pocket across a land where the gap between the land and the opposing bearing surface is sufficiently small to create a resistance to oil flow. Typically, pocket depths are in the range 0.5 to 5 mm, whilst bearing gaps are 10 to 100 μm. As pocket depths are large compared to land gaps, the oil pressure in the pocket is constant and pressure gradients are confined to the bearing lands. Since most of the resistance to motion occurs in the narrow gap over the lands, the advantage of this type of construction is that pockets can be increased in area to gain load capacity and stiffness with only a modest increase in friction. Since the load capacity is principally dependent on oil supply pressure and bearing area, any increase in load capacity with speed due to hydrodynamic effects is usually neglected. There is a class of hydrostatic

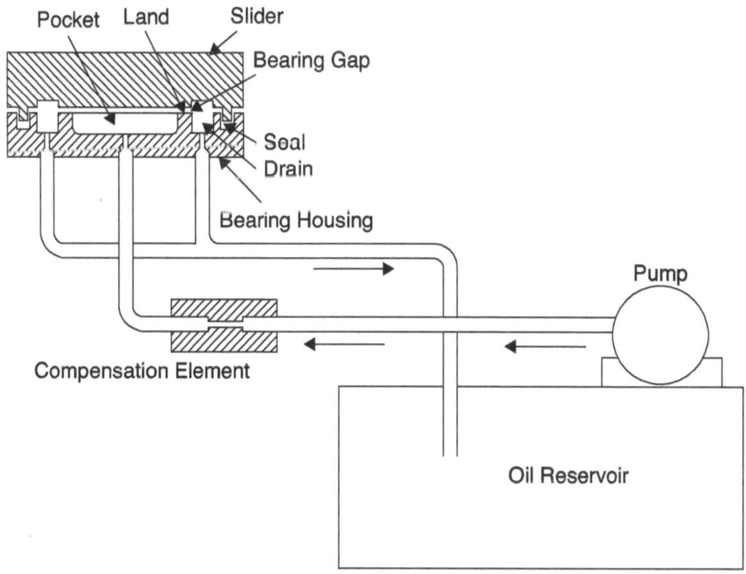

Figure 3.1 Schematic arrangement of a hydrostatic bearing

bearings where plain surfaces are used in place of pockets to generate hydrodynamic lift, sometimes referred to as hybrid bearings [2], but they fall outside the scope of this book.

In practical designs of hydrostatic bearings, the leaking oil must be contained and must be returned to the oil supply to form a closed loop system. The oil may be allowed to fall into a sump but more usually some form of drainage channel and sealing arrangement surrounds the bearing land(s). Oil drained away from the bearing is returned to the oil supply under the action of gravity or if necessary with the aid of an auxiliary pump. With a well designed drainage system, non-contacting labyrinth seals are sufficient to prevent oil from leaking outside of the bearing housing as well as to prevent contamination from entering the system.

Hydrostatic bearing designs are based on the fluid supplied to them being incompressible. Most liquids suitable for use with hydrostatic bearings approximate this requirement provided there is no air entrainment. In the case of hydrostatic bearing systems, this requires some attention to the design of the oil supply, for example, passing the returned oil through baffles and allowing it to dwell in the reservoir for several minutes before it is pumped back to the bearing. As the oil is incompressible, then it is not essential for the compensation elements to be built into the bearing housing. They may be located at any convenient position upstream of the bearing, where they may be easily accessed for adjustment or maintenance. However, if the compensation elements are remote from the bearing, then the piping between the two must be rigid in order to maintain bearing stiffness.

One exception to the bearing arrangement described above is the porous water bearing [3]. It is supplied with high pressure water and one bearing surface is manufactured from porous ceramic to provide compensation. The bearing does not use pockets and has, instead, plain surfaces. This enables cylindrical journal bearing types to generate high hydrodynamic lift that enhances load carrying capacity and stiffness at speed. For reliable bearing operation, the water must be maintained free of chemical or bacterial contamination.

3.3 Basic operating principles

Figures 3.2(a and b) show the parts of a hydrostatic bearing that generates load carrying capacity and static stiffness, namely the pocket and surrounding land. In this example, two identical pockets are machined

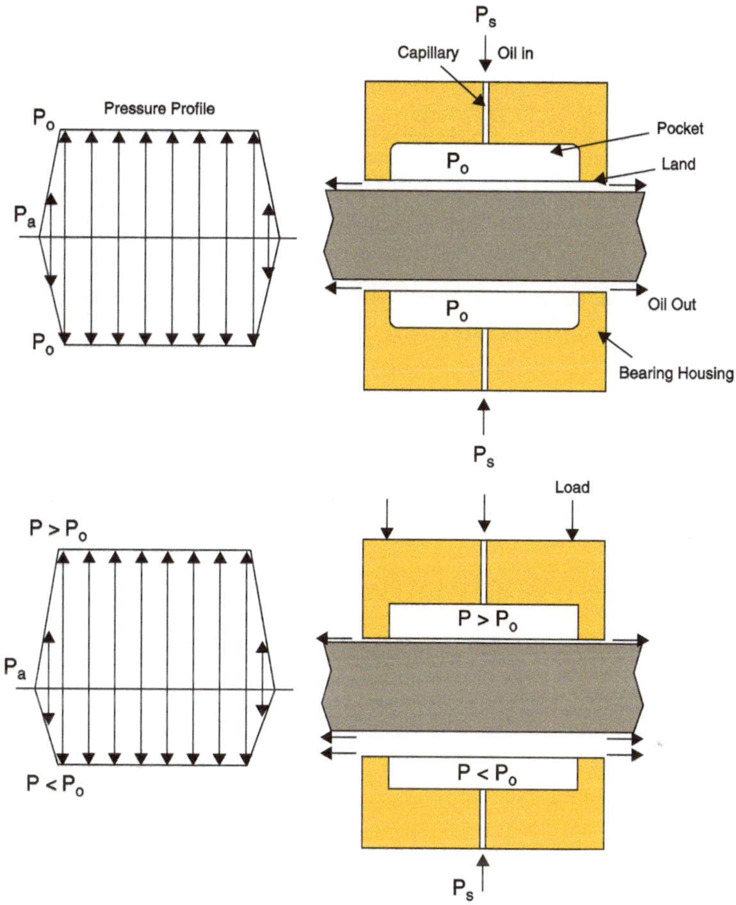

Figure 3.2 **Operating principle of a basic design of hydrostatic bearing**

into a single housing and arranged to act on opposite sides of a flat bar that is fixed in position. Each bearing has an identical compensation element – a capillary, positioned in the housing just upstream of the pocket. The bearings are designed to have a small gap under their lands to allow oil to leak out in a controlled manner, driven by the difference in pressure between oil in the pocket and oil at the external edge of the land, which is at or close to ambient pressure, P_a. Oil is supplied to the capillaries at a high pressure, P_s, and as the capillaries have a finite resistance to oil flow they create a pressure drop so that under no load

(Figure 3.2(a)) oil enters the pockets of both bearings at some pressure, P_o, where $P_o < P_s$. In practice, P_o is about half the supply pressure P_s, which means that the resistance to oil flow of the capillary and the land surrounding the pocket are approximately the same.

The application of a load to the bearing housing displaces it, reducing the gap on one side of the plate and increasing it on the other (Figure 2.1(b)). The reduced gap has a higher resistance to oil flow, so that a greater proportion of the oil supply pressure occurs across the land increasing the oil pressure in the pocket. Conversely, the increased gap on the other side of the bearing has a lower resistance to oil flow and therefore a lower proportion of the oil supply pressure occurs across the land reducing pocket pressure. The difference in pressure between the two pockets, integrated over the bearing area, supports or compensates for the applied load.

3.4 Pocket geometries

3.4.1 Circular and annular thrust bearings

Pocket geometries are simple shapes that are easily manufactured. Figure 3.3 gives some examples of pockets that employ a circular geometry in common usage. The circular pad (Figure 3.3(a)) has a single oil inlet and an annular land around its outside diameter, whereas the annular pad (Figure 3.3(b)) also has a single oil inlet but lands around both its outside and inside diameters. These bearings are used to support centrally applied axial loads only. An offset load or any moment loading may generate a variation in land gap with angular position around the bearing, but as the

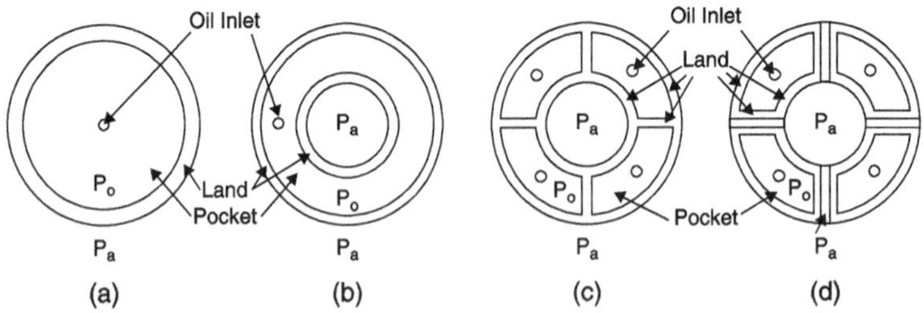

Figure 3.3 Circular pockets: (a) circular pad; (b) annular pad; (c) annular pad with radial lands; and (d) annular pad with radial lands and drainage channel

pocket is relatively deep oil pressure gradients in the pocket are negligible, resulting in pressure remaining constant with respect to angular position. Thus these bearings have zero moment load capacity and zero tilt stiffness.

The bearing designs shown in Figures 3.3(c and d) comprise four pockets separated by radial lands and each pocket has its own oil feed. These geometries enable the bearing to support moment loading as now the radial lands allow differential pressures to be developed between pockets. With a plain radial land (Figure 3.3(c)), the pressure gradient across it is that due to the difference in pressure between adjacent pockets, whereas in Figure 3.3(d) radial lands have a central drainage channel maintained at atmospheric pressure so much larger pressure gradients are developed across them. In principle, this enables the latter to generate higher moment load capacity and static stiffness but at the expense of a slightly more complex bearing geometry. Figures 3.3(c and d) show a bearing design with four pockets but any number of pockets can be used, as increasing the number of pockets gives the potential to generate higher moment load capacity and tilt stiffness but at the expense of complexity and cost.

3.4.2 Rectangular pockets

Rectangular pockets are commonly used in linear slides. In such applications a number of individual pockets, each having their own compensation element and surrounding land, are used (Figure 3.4).

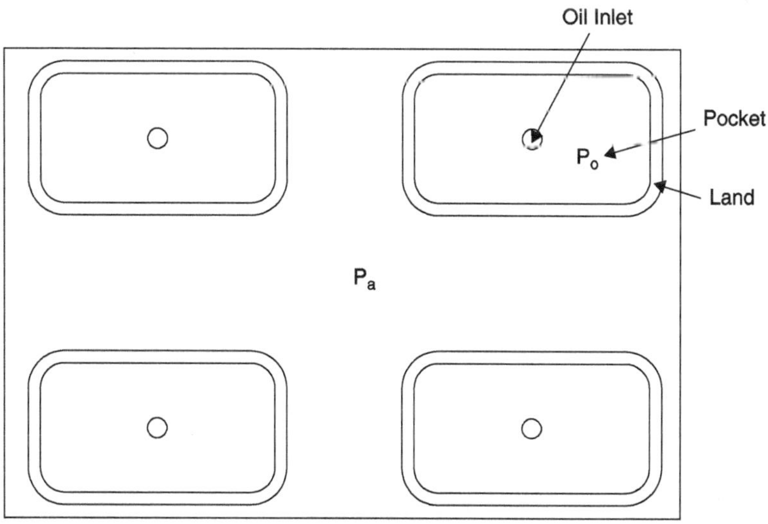

Figure 3.4 Rectangular pockets

Whilst individual pockets cannot support moment loads, the fact that several pockets are used and suitably spaced apart enables the complete bearing arrangement to support both offset loads and moments.

3.4.3 Cylindrical journal pockets

Cylindrical journal bearings must comprise three or more pockets separated by axial lands, in order to support radial load. Figure 3.5 shows a basic journal design with four axial lands and four oil inlets. Again each pocket has its own compensation element and its resistance to oil flow is matched to that of the circular lands at each end of the bearing. Under zero radial load all pocket pressures are equal, and there is no flow across the axial lands so the circular lands form the primary resistance to oil flow. However, radial displacement of the shaft causes the gap under the circular lands to vary with angular position around the bearing, thus creating an imbalance in resistance to oil flow from one pocket to another. Where the gap is low, pocket pressure is increased and visa versa. The difference in pressure between pockets sets up oil flow in a circumferential direction and the axial lands provide a resistance to oil flow that is essential in maintaining the pressure difference, hence enabling the bearing to support load. The axial lands shown in Figure 3.5 are plain surfaces. On some cylindrical journal bearing designs an axial drainage channel is machined into the centre of the land to maintain

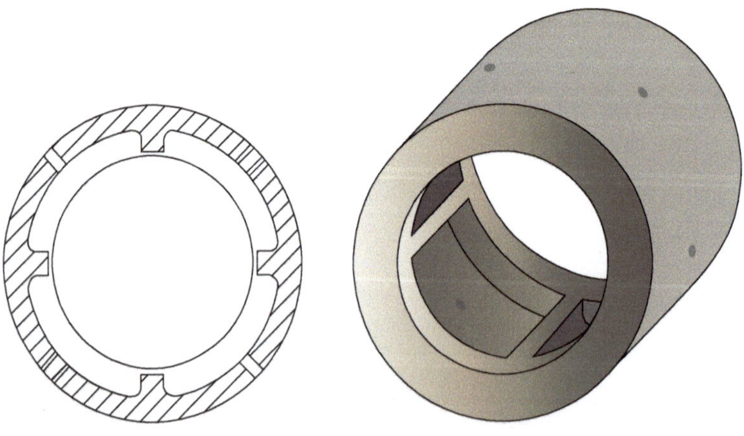

Figure 3.5 Cylindrical journal pockets

ambient pressure at this point. This effectively isolates the pockets and generates a higher bearing stiffness, but at the expense of complexity and a higher oil consumption.

3.4.4 Spherical and conical bearings

Conical and spherical pocket geometries are not covered fully in this book. They offer specific advantages but are more difficult to manufacture. Spherical bearings are used where a pivotal motion is required, such as in scientific instrumentation. Apart from the difficulty in manufacturing them, they do not compete with conical or combined cylindrical journal–annular thrust bearing arrangements in terms of load capacity and stiffness. Conical bearings are more widely used, offering high load capacity and stiffness in a compact arrangement. They are favoured in applications such as rotary tables where they can be set up in a back–back arrangement (Figure 3.6(b)) to achieve a low table height. Here thermal expansion has little effect on performance; however, where it is necessary to space two conical bearings apart, particular care is needed in setting them up as any axial thermal growth of say, a shaft, can affect the bearing clearance and hence performance.

3.5 Hydrostatic bearing arrangements

The pocket geometries described in Section 3.4 can be used to configure a wide range of bearing arrangements used on linear slides, rotary tables and spindles. Examples of common bearing arrangements using flat annular, rectangular, conical and journal pockets described are given in this section.

Examples of popular hydrostatic bearing arrangements used in ultra precision machinery are shown in Figure 3.6, (a) a rotary table comprising two annular thrust and a single journal bearing. The annular thrust bearings each have multiple pockets to enable the bearing arrangement to support moment or offset axial loading. The combination of annular thrust and cylindrical journal bearings is extremely versatile and can be configured in many ways to support different loading situations. For example, Figure 3.6(c) shows a machining spindle and uses a double-sided annular thrust bearing and two cylindrical journal bearings to support offset radial loads and axial loads. The size and the spacing of the journal bearings determine the magnitude of the radial load that can

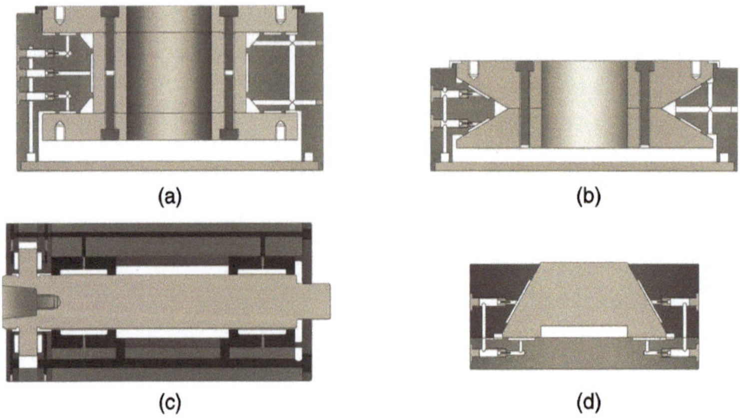

Figure 3.6 Examples of hydrostatic bearing arrangements

be supported and its permissible overhang from the front of the spindle. A double-sided annular thrust bearing provides axial location of the shaft and supports axial load in two directions. Placing the thrust bearings together means that any thermal expansions occurring in the system have only a small effect on bearing clearance and by placing them at the front of the spindle, axial thermal expansion of the shaft between the bearing and the working position is minimised.

Conical bearings provide an alternative to the annular thrust–cylindrical journal combination. Figure 3.6(b) shows another rotary table based on a back–back conical bearing arrangement. This table has a similar load capacity and static stiffness as the rotary table of Figure 3.6(a), but has the advantage of a lower height. Where they can be used, conical bearings do generally save space but they are not as versatile as the annular thrust–cylindrical journal bearing combination. In particular, they are not well suited to applications where high offset radial loads occur or indeed where high speeds are required. The axial spacing between a pair of conical bearings is limited by differential axial thermal expansion, which has the effect of changing the bearing clearance and hence bearing performance. For this reason, conical bearings cannot be spaced apart to support overhung radial loads in the same way as cylindrical journal bearings can. High speeds generally create high running temperatures, and axial differential expansions between the bearing cones and the housing can be significant even if the bearings are abutted together.

Figure 3.6(d) shows a linear bearing arrangement in which a trapezoidal section beam is used. Four pockets are positioned at each end of the carriage as shown to enable it to support vertical and horizontal loads, as well as moment loads in yaw, pitch and roll directions. The inclined pockets on the left- and right-hand sides of the carriage are paired to take horizontal loads, whilst inclined and horizontal pockets are paired to take vertical loads. This arrangement minimises the number of pockets required to support general loading situations. There are many other slide bearing arrangements, with pockets arranged in 'T' or 'H' or in a 'Box' configuration being commonplace and each having specific advantages in different applications.

3.6 Types of compensation

The way in which high pressure oil supplied to a hydrostatic bearing is regulated can have a large effect on bearing performance. As a result, a wide variety of compensation devices of varying complexity have been developed and reported in the scientific press. The trade-off between bearing performance and overall system complexity is an important consideration in selecting the type of compensation element to use. Even with the simplest element designs there is a need to match restriction levels to those of the associated bearing pockets, so that a uniform or constant bearing gap is achieved and this may require time-consuming adjustment on assembly or increased precision of manufacture.

Compensation elements are categorised in terms of:

- fixed restrictors;
- constant flow devices;
- pressure dependent restrictors.

In this section, examples of each type are described and their relative advantages given.

3.6.1 Fixed restrictors

Popular fixed restriction elements are shown in Figure 3.7. They all have the advantage of being compact and therefore easily integrated into the bearing housing, making for a simple overall bearing system design. The capillary, annular and radial slot restrictors have the additional advantage of generating a bearing load capacity and static stiffness that is independent of oil temperature. In both cases, the mass flow rate of oil

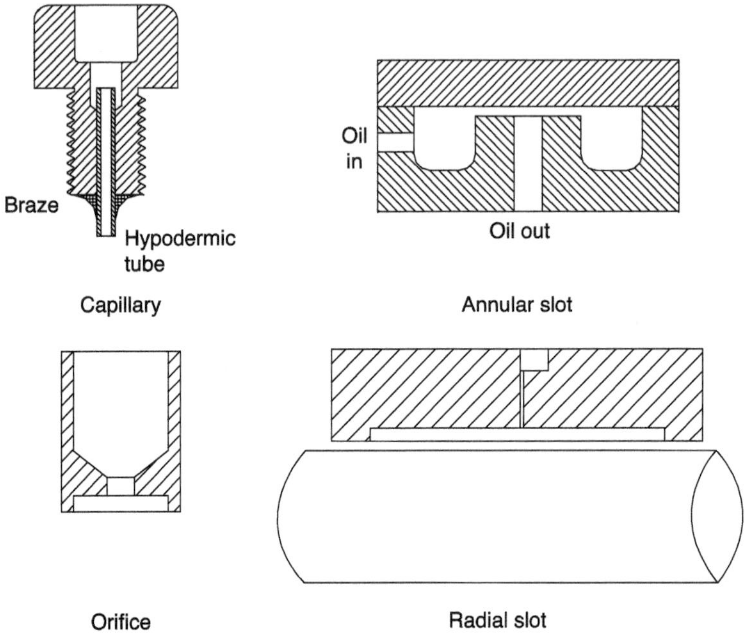

Figure 3.7 Examples of fixed compensation elements

through the device is proportional to the pressure difference across it and inversely proportional to oil viscosity. This is also the case for the bearing lands, so any change in oil viscosity has an equal effect on both components.

For an orifice, the mass flow rate of oil is proportional to the square root of the pressure difference across it and independent of oil viscosity. Thus any change in oil viscosity only affects the restriction of the bearing pad, altering its relationship to the orifice restriction. An advantage of the orifice, in comparison with capillary and annular pad restrictors, is that its non-linear mass flow rate–pressure difference characteristic generates a slightly higher static stiffness for a given bearing size and oil supply pressure.

A novel form of slot compensation, worthy of note, is described in [4], who used a back–back conical bearing arrangement similar to that shown in Figure 3.6(b). The conical bearing surfaces each contain 20 pockets, a relatively large number for a hydrostatic bearing, but they are fed with pressurised oil through a single cylindrical slot placed between the bearings, adjacent to the inboard end of the pockets. The pockets do not have a land around their inboard edge so oil flows into the pockets directly from the cylindrical slot. This results in a uniform pressure distribution over all pockets and very low motion errors.

3.6.2 Constant flow devices

By maintaining a constant flow rate of oil to each bearing pocket, any reduction in land gap produced by an increase in applied load also increases the pocket pressure to compensate. Constant flow rate may be achieved either by using a separate pump with each bearing pad or by using a single pump and a separate constant flow rate valve with each bearing pocket (Figure 3.8). Either way, the overall bearing system is complex and expensive. The potential advantage is a bearing with a higher static stiffness than can be achieved with fixed restriction elements. However, the benefit is not as great as may be achieved more economically with some pressure dependent compensation elements, so the use of this type of compensation in precision bearing arrangements is rare.

3.6.3 Pressure dependent restrictors

By making the restriction of a compensation element dependent on pocket pressure, the load–deflection relationship of a hydrostatic bearing

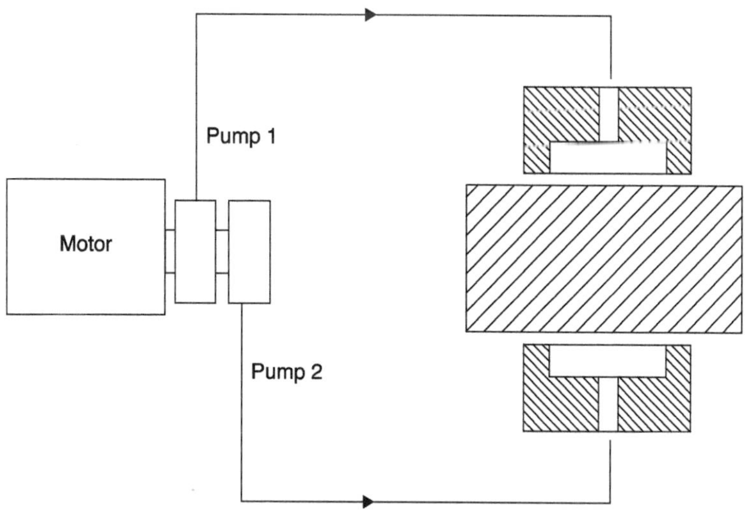

Figure 3.8 Example of constant flow rate compensation

can be substantially modified. The objective of such devices is usually to increase the bearing's overall static stiffness and the main challenge in their design is to achieve a good trade-off between high stiffness and additional system complexity. Pressure dependent compensation is an aspect of hydrostatic bearings subject to many innovations and patents. The devices may be sub-categorised as spool valve types, such as that developed by [5], or diaphragm types, such as those developed by [5,6]. Spool valve types are complex to manufacture, generate a temperature-dependent bearing performance and with moving parts are slow to respond and for these reasons are not considered here. In comparison, diaphragm-type compensation elements are simple in concept, relatively simple to manufacture, generate a bearing performance that is independent of temperature and have a rapid response to changes in bearing load.

The compensation element devised by [5] is shown in Figure 3.9. One element is required for every pocket of the bearing. Restriction to oil flow is created in an annular gap between a flexible diaphragm and a rigid land. Oil at supply pressure is fed to the bore of the annulus and allowed to leak out into a chamber of large area beneath the diaphragm. This chamber is connected directly to the bearing pocket, so the oil pressure in the chamber is the same as that in the pocket. An adjustable preloading spring contacts the diaphragm and is used to set the equilibrium gap

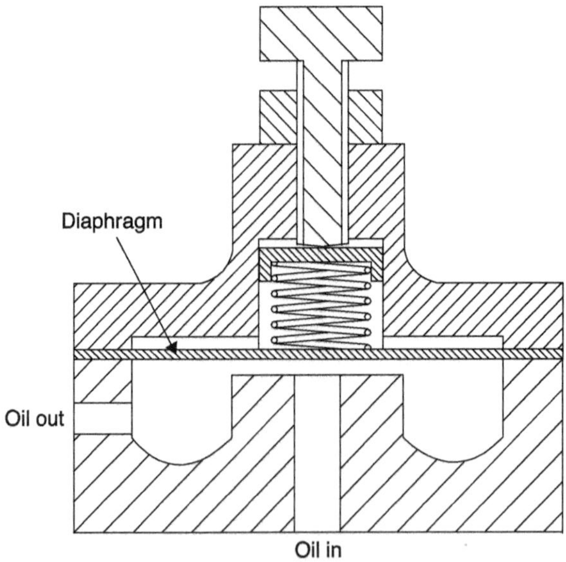

Figure 3.9 Diaphragm compensation element

between the diaphragm and land. The spring rate of this spring is not significant, but the flexibility of the diaphragm is important and is designed to increase bearing stiffness. Any reduction in the bearing land gap produced by an increase in applied load increases the bearings resistance to oil flow and hence increases oil pressure in the pocket. As this pressure acts over most of the area of the diaphragm, it produces a deflection that reduces the compensation element's resistance to oil flow, allowing pocket pressure to rise to a higher level than it would with a fixed compensation element. In effect, the diaphragm provides feedback controlling pocket pressure. The more flexible the diaphragm, the larger the increase in pocket pressure, and vice versa. In principle, the compensation element may be designed to generate infinite static stiffness over at least a part of the bearing's range of deflection. But too high a diaphragm flexibility may overcompensate for the applied load, resulting in an unstable bearing.

The fact that one element is required per pocket, the preload for each pocket must be set and the pocket pressures must be matched to centralise the moving part of the bearing, which adds significantly to the complexity of the bearing system. A somewhat simpler arrangement applicable to opposed pocket bearing arrangements was devised by Rowe [2].

Figure 3.10 shows a schematic arrangement of Rowe's compensation element. As for the Mohsin design, restriction to oil flow occurs in an annular gap between a rigid land and a flexible diaphragm but now there is

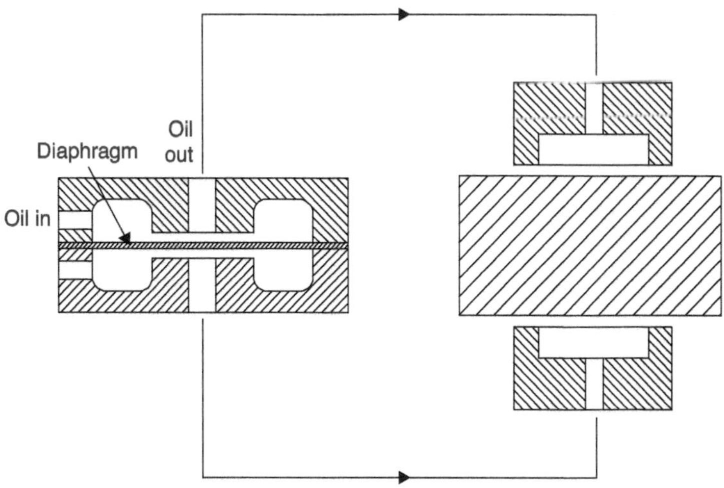

Figure 3.10 Diaphragm compensation element

a pair of annular gaps, one each side of the diaphragm and symmetrically disposed. Oil leaking out of the annular gaps fill the chambers either side of the diaphragm and as these are connected directly to the pockets of an opposed pair of bearing pockets, the oil pressure in the chambers is the same as that in the pockets. Any load applied to the bearing increases the land gap on one pocket and reduces it on the other, setting up an imbalance in pocket pressures. This pressure imbalance acts on the diaphragm in the compensation element and produces a deflection that increases the resistance to oil flow of one annular gap and reduces it for the other, allowing a greater pressure differential to occur in the bearing pockets than would otherwise occur if fixed compensation elements were used. As for the Mohsin compensation element, diaphragm flexibility is an important design parameter determining the overall static stiffness of the bearing arrangement.

Arranging the compensation elements in pairs for opposed bearing pockets eliminates the need for preloading the diaphragm and hence provides a useful level of simplification over and above that of the Mohsin arrangement, yet it offers similar bearing performance advantages.

Pressure dependent compensation elements can offer large improvements in the static stiffness of hydrostatic bearings compared to bearings with

Figure 3.11 Hydrostatic leadscrew with integral PM valves (courtesy of Hyprostatik)

elements of fixed geometry. But in many precision bearing applications, it can be difficult to integrate these devices into bearing housings without appreciably increasing their size and cost. Hyprostatik GmbH [7] have addressed this point by developing a compact design of diaphragm restrictor referred to as a PM Valve and by producing a range of hydrostatic products where such valves are fixed to convenient external surfaces, thus eliminating the need for external piping. Figure 3.11 is an example of a hydrostatic leadscrew with integral PM valves, where eight such valves are attached to the screw and another eight to the support bearing. Spindles are also available with integral PM valves and the valves may be sourced as separate items.

3.7 Approximate load capacity and static stiffness

It is often useful to have an estimate of the static stiffness and load capacity of a hydrostatic bearing prior to embarking upon its detailed design. This allows the machine designer to evaluate the bearing–shaft system's performance and find a near optimum bearing size and arrangement before undertaking detailed bearing design calculations. Load capacity can be simply expressed in terms of oil supply pressure, P_s and bearing area. For a double-sided annular thrust bearing, the load capacity is given by:

$$F_c = \xi . P_s . \pi (D_o^2 - D_i^2)/4$$

where D_o and D_i are the outer and inner diameters of the bearing respectively. ξ is referred to as the load coefficient and may be regarded as a measure of the effectiveness of the compensation system. In general, for a double-sided thrust bearing with a fixed compensation element $0.4 < \xi < 0.6$, 0.6 applies to a well-designed bearing for which the compensation system has been optimised. The bearing's static stiffness requires the 'no load' equilibrium gap, h_o, to be known and is given by:

$$K = \xi . P_s . \pi (D_o^2 - D_i^2)/(4h_o)$$

For a journal bearing:

$$F_c = \xi . P_s . L . D$$

$$K = \xi . P_s . L . D / h_o$$

where the projected bearing area is the product of shaft or journal diameter, D and pocket length, L and $0.3 < \xi < 0.6$. For a journal bearing, a load coefficient of 0.6, although possible is unusual, and a value of 0.4 is more typical. For a slide bearing with an opposing pair of rectangular pockets of length, L and width, W:

$$F_c = \xi.P_s.L.W$$

$$K = \xi.P_s.L.W/h_o$$

where $0.4 < \xi < 0.6$ as for the double-sided annular thrust bearing.

It should be noted that the dominant variables that determine the static load capacity of a hydrostatic bearing are its area and the oil supply pressure. Similarly, for static stiffness it is bearing area, oil supply pressure and working gap that are dominant. Whilst ξ has a significant effect, the difference between a bearing with a carefully optimised compensation system and one for which the compensation system has not been optimised, is relatively modest in comparison with the effect of area or pressure. Furthermore, oil viscosity has no influence on load capacity or static stiffness, but does have an important influence on power consumption.

3.8 Oil flow through a narrow gap

Fundamental to the design of hydrostatic bearings and an accurate calculation of load capacity, static stiffness and damping is the relationship between pressure gradient and mass flow rate for oil flowing through a narrow gap. The narrow gap may be a land surrounding a pocket or a hole or slot forming part of a compensation element. Knowing the pressure–flow rate relationship for each of the various elements that make up a hydrostatic bearing enables pocket pressures to be determined for different bearing gaps or deflections, and the detailed bearing design to be adjusted to ensure that it meets requirements. In a vast majority of hydrostatic bearings, the oil flow through the elements that determine bearing performance is one-dimensional (1-D) and so there is a relatively simple explicit relationship between pressure gradient and oil mass flow rate.

Fluid flow through a narrow gap was first considered by Reynolds [8] in 1886, who developed general equations describing the relationship between fluid properties, pressure gradients, gap size and surface velocities. The equations are based on the assumptions:

- Inertia forces in the fluid, due to acceleration, can be neglected compared to forces produced by viscous shearing.
- The fluid is Newtonian, for which its absolute viscosity is equal to shear stress divided by shear strain rate.
- Fluid density and viscosity are constant.
- Laminar flow conditions exist at all points in the fluid film.
- Pressure is constant over any section normal to the direction of fluid flow.
- The boundaries are solid and impervious.
- There is no slip at the boundaries between the fluid and the surfaces.

Here we simply consider a reduced form of the Navier–Stokes equation for the cases of 1-D flow between two rectangular plates, an annular slot and a capillary of constant bore.

3.8.1 Rectangular slot

Figures 3.12(a and b) show the velocity profile of oil flowing between two flat, parallel and stationary rectangular plates subject to a pressure that is constant in a plane normal to the surfaces of the plates. Under these conditions, oil flow is 1-D and viscous shearing within the oil creates a pressure drop in the direction of flow. The starting point for deriving the relationship between pressure and flow is the well-known Navier–Stokes equation [9], which for these conditions can be reduced to:

$$\frac{\partial^2 u}{\partial z^2} = \frac{1}{\eta}\frac{dp}{dx}$$

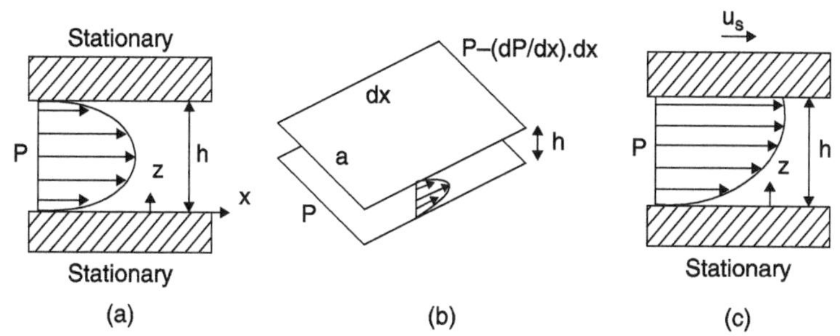

Figure 3.12 Oil flow between two rectangular plates

Integrating twice gives:

$$u = \frac{1}{\eta}\frac{dp}{dx}\frac{z^2}{2} + C_1 z + C_2 \qquad [3.1]$$

The volume flow rate of oil is:

$$dV/dt = a.\int_0^h u\,dz \qquad [3.2]$$

where a is the width of the plates. Integrating and putting in the boundary conditions, $u=0$ at $z=0$ and $u=0$ at $z=h$ enables the constants C_1 and C_2 to be determined and the equation simplified to:

$$dV/dt = \frac{-ah^3}{12\eta}\frac{dp}{dx} \qquad [3.3]$$

For the case where there is also relative movement between the plates (Figure 3.12(c)), the boundary conditions become $u=0$ at $z=0$ and $u=u_s$ at $z=h$, for which Eq. 3.2 can be integrated and simplified to give:

$$dV/dt = \frac{-ah^3}{12\eta}\frac{dp}{dx} + \frac{au_s h}{2}$$

$u_s/2$ may be considered as the average speed of oil in the gap due to relative motion between the surfaces, so that the above equation states that the oil flow rate generated by applied pressure and surface motion are linearly additive.

Integrating Eq. 3.3 with respect to x gives:

$$P_1 - P_2 = -(x_1 - x_2).12\eta.dV/dt/[h^3a] = 12\eta.LdV/dt/[h^3a] \qquad [3.4]$$

where P_1 and P_2 are the pressures at coordinates x_1 and x_2 respectively and $L=x_2 - x_1$. Equation 3.4 gives the relationship between pressure difference and volume flow rate of oil across a rectangular land. The pressure gradient is linear and the average pressure acting on the land is:

$$\underline{P} = [P_1 + P_2]/2 \qquad [3.5]$$

3.8.2 Annular slot

Oil flow rate through an annular slot subject to a constant pressure applied either to its bore or outside diameter can be found simply by applying Eq. 3.3 to a circular element of radius r and width dr

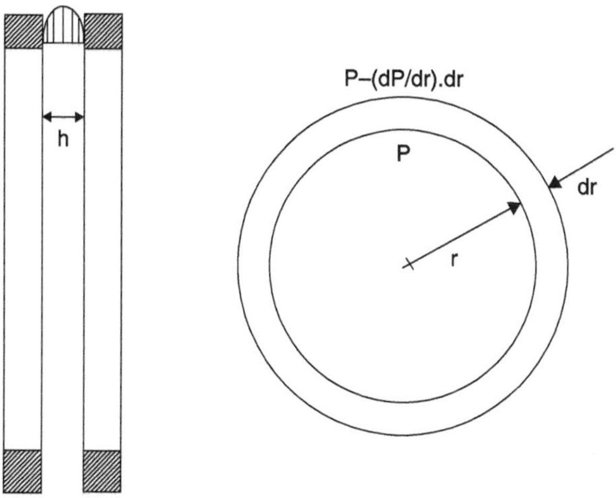

Figure 3.13 Oil flow between two annular plates

(Figure 3.13), where $x = r$ and $a = 2\pi r$. The assumptions and limitations 'a' to 'g' above still apply. This gives:

$$dV/dt = \frac{-2\pi r h^3}{12\eta}\frac{dp}{dr} \qquad [3.6]$$

Integrating from an inner radius of r_i to a radius of r and rearranging gives:

$$P_i - P = 6\eta.\ln(r/r_i).dV/dt/[\pi h^3] \qquad [3.7]$$

where P and P_i are the pressures at r and r_i, respectively. Equation 3.7 gives the pressure at any radius r in terms of the pressure at the bore and oil volume flow rate. Applying Eq. 3.7 to an annular slot of inner and outer radii of r_i and r_o, where the pressures are P_i and P_o, gives:

$$P_i - P_o = 6\eta.\ln(r_o/r_i).dV/dt/\ [\pi h^3] \qquad [3.8]$$

Eliminating volume flow rate from Eqs 3.7 and 3.8 gives:

$$[P_i - P]/[P_i - P_o] = \ln(r/r_i)/\ln(r_o/r_i) \qquad [3.9]$$

The average pressure on the land can be found by integrating Eq. 3.9 from the bore radius r_i to an outer radius r_o:

$$P_i - \underline{P} = [P_i - P_o][\ r_o/(r_o - r_i) - 1/\ln(r_o/r_i)] \qquad [3.10]$$

3.8.3 Capillary

The capillary is a long tube of constant bore diameter. In the context of hydrostatic bearings, long means a length/diameter ratio of 20 or more. This allows any fluid passing through the bore to develop laminar flow so that pressure losses are only due to viscous shearing. For the case of constant pressure applied to one end of the capillary, the relationship between volume flow rate and pressure gradient can be derived from the Navier-Stokes equation expressed in cylindrical polar coordinates [8], subject to the assumptions given above. The result is quoted as:

$$dV/dt = \frac{-\pi d^4}{128\eta}\frac{dp}{dz}$$

For a capillary of diameter d and length L, subject to a pressure difference of $P_2 - P_1$:

$$dV/dt = \frac{-\pi d^4}{128\eta L}(P_2 - P_1) \qquad [3.11]$$

3.9 Temperature dependence of oil viscosity

Operating temperature is normally an important consideration in precision bearing applications, as temperature rises of just a few degrees can create unwanted thermal distortions in machine structures. For hydrostatic bearings, oil is being pumped at high pressure and may also be subject to high shear rates when the bearings are set in motion. The heat generated not only affects the machine structure but increases the temperature of the oil and reduces its viscosity. The result is that higher oil mass flow rates are therefore required to maintain designed bearing pressures, in turn increasing the pumping power needed to supply the oil. Oils are available in a wide range of viscosity grades and the selection of the grade has a significant effect on bearing operating temperatures and overall power consumption. Figure 3.14 shows the viscosity–temperature relationship for a number of ISO grade mineral oils. In the ISO system, the oil grade is defined by the oil's kinematic viscosity in centistokes (cSt) at a temperature of 40 °C. For example, ISO VG 10 has a viscosity of 10 cSt at 40 °C. The system defines a wide range of grades where the difference in viscosity between one oil and the next is approximately 50% at 40 °C.

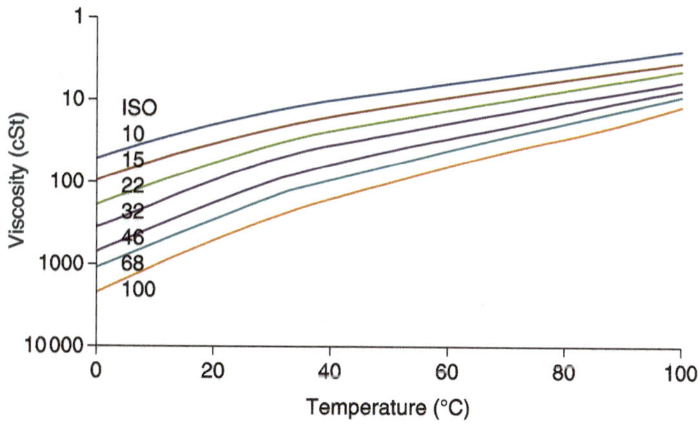

Figure 3.14 Oil viscosity–temperature relationships for standard mineral oils

Oil viscosity is strongly dependent on temperature (Figure 3.14). Roelands [10] used the following expression to relate oil viscosity to temperature:

$$\text{Log}\,(\log \eta + 1.2) = -S_o \log(1 + t_m/135) + \log G_o \qquad [3.12]$$

where η = absolute viscosity in cP, t_m = temperature in °C, G_o is a constant representing viscosity grade and S_o is a constant defining the slope of the viscosity–temperature relationship. In practical terms, oil viscosity approximately halves for every 15 °C rise in oil temperature.

3.10 Bearing load and deflection characteristics

If high load capacity and static stiffness are the only requirements of a hydrostatic bearing, then its design is relatively straightforward. However, in ultra precision bearing applications, heat generation is a particularly important consideration and as the hydrostatic bearing may have large areas of oil subject to high shear rates and require significant volumes of oil to be pumped at high pressure, then the influence of thermal effects on performance should be considered.

An accurate calculation of bearing load capacity and static stiffness and optimisation of the compensation system requires the pressure within the pocket to be determined at several different gaps produced by the action of an applied load. This is done by comparing the resistance to oil flow of the bearing pocket to that of the compensation element. As for the types of hydrostatic bearings considered here, pocket depths are many times larger than the gap under the surrounding lands, so the oil pressure in the pocket can be assumed to be uniform over its complete area and the resistance to oil flow of the bearing pad is essentially that of the lands. Furthermore, as the lands are also narrow, oil flow can be considered 1-D, normal to the land boundaries.

3.10.1 Pressure–flow relationship for rectangular pocket subject to central normal load

Figure 3.15 shows a rectangular pocket of overall length L and breadth B; in general it will have corner radii of external radius R and a constant land width W. W is typically several millimetres, whilst the gap over the land is normally between 0.01 and 0.1 mm. Oil flow across the land can therefore be considered as viscous and laminar and Reynold's equation for 1-D oil flow in a narrow gap separating two static, rectangular or annular plates (Eq. 3.2) can be used to determine the relationship between the pressure difference across the land and the volume flow rate of oil.

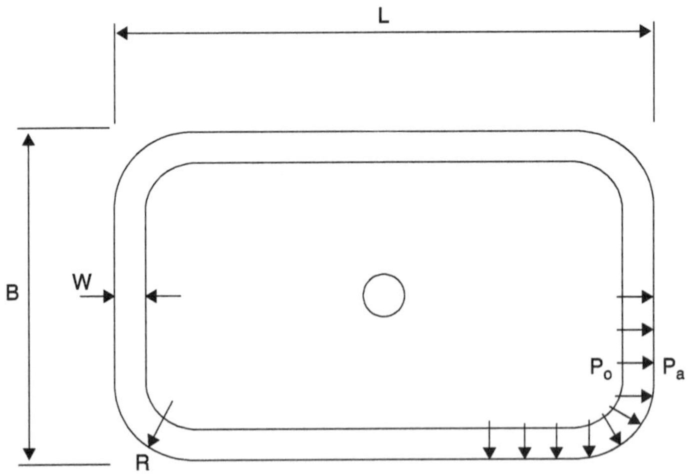

Figure 3.15 Oil flow across land of a rectangular pocket

Applying Eq. 3.4 to the rectangular lands of width W and lengths L – 2R and B – 2R for the bearing shown in Figure 3.8 gives:

$$dV_L/dt = (P_o - P_a)h^3(L - 2R)/(12\eta W) \qquad [3.13]$$

and:

$$dV_B/dt = (P_o - P_a)h^3(B - 2R)/(12\eta W) \qquad [3.14]$$

Applying Eq. 3.8 to the four corner lands of the bearing shown in Figure 3.15 gives:

$$dV_r/dt = \pi h^3(P_o - P_a)/[6\eta.\ln(R/(R - W))] \qquad [3.15]$$

Thus the total volume flow rate crossing the complete land is:

$$dV/dt = [h^3(P_o - P_a)/6\eta][(L - 2R)/W + (B - 2R)/W + \pi / \ln((R/(R - W))] \qquad [3.16]$$

This must balance the volume flow rate of oil passing through the compensation element. If the element is a capillary of diameter d and length l (Eq. 3.11) (Section 3.8.3):

$$dV/dt = \pi(P_s - P_o).d^4/(128.l.\eta) \qquad [3.17]$$

where P_s is the oil supply pressure and P_o is the oil pressure in the pocket. Equating Eqs 3.16 and 3.17 and rearranging gives an explicit expression for the pocket pressure P_o:

$$P_o = \{P_s.\ \pi d^4/(21.33l) + P_a.h^3[(L - 2R)/W + (B - 2R)/W + 1/\ln(R/(R - W)]\}/\ \{\pi d^4/(21.33l) + h^3[(L - 2R)/W + (B - 2R)/W + \pi/\ln(R/R - W)\} \qquad [3.18]$$

The load supported by the bearing can now be determined by integrating pressure in the pocket and across the lands over their respective areas and adding them together:

- The pocket area is $A_p = (L - 2W)(B - 2W) - (R - W)^2(4 - \pi)$;
- The total area of the rectangular lands is $A_r = 2(L - 2R) + 2(B - 2R)$;
- The area of the circular corner lands is $A_c = \pi(R^2 - [R - W)^2]$.

Thus the load supported by the bearing is:

$$W = (P_o - P_a).A_p + (P_o - P_a).A_r/2 + P_c.A_c \qquad [3.19]$$

where P_c is derived by applying Eq. 3.10 to the four corner lands:

$$P_c = [RP_a - (R - W)P_o]/W - [P_a - P_o]\ln(R/(R - W)) \qquad [3.20]$$

Bearing load capacity is defined at some practical minimum gap. On a geometrically perfect bearing, this is close to zero for which the pocket pressure P_0 approaches P_s, the supply pressure. In practice, defining load capacity at, for example, a gap of 5 μm, provides a margin of safety allowing for geometrically imperfect land surfaces. The load derived in this way is the load capacity of a single pocket. Usually pockets are arranged in opposing pairs (Figure 3.2), so the load capacity of the pair is defined at some maximum deflection from the zero load equilibrium position. The difference in the load supported by each pocket then gives the bearing's load capacity.

3.10.2 Design example for rectangular pocket

Consider a pocket for which (notation of Figure 3.15):

L = 100.0 mm; B = 50.0 mm; R = 10 mm and W = 5 mm supplied with oil of viscosity, 20 cSt at 150 Bar pressure.

For a capillary of 0.5 mm diameter and 10 mm in length, find the relationship between pocket pressure and land gap. Also find how the load capacity and static stiffness of an opposed pair of pockets depend on the designed 'Zero Load' pocket pressure. The total clearance of the pockets is 100 μm and the maximum permissible deflection is 45 μm.

The pocket pressure–gap relationship for a single pocket is shown in Figure 3.16. In this example, pocket pressure is almost equal to the supply

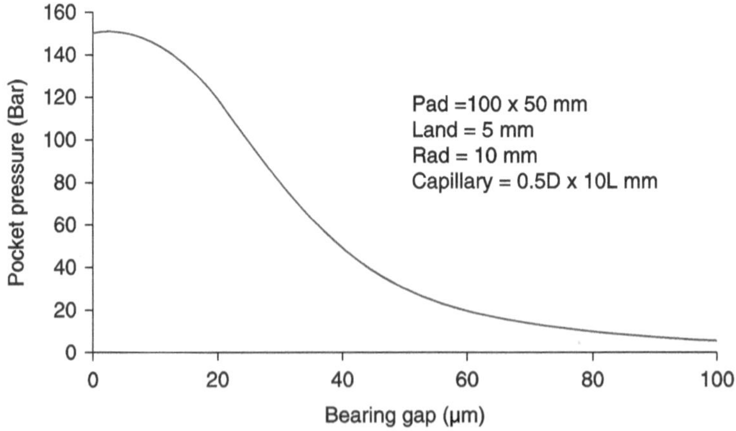

Figure 3.16 **Pocket pressure versus gap for a single rectangular pad**

pressure of 150 Bar for gaps of less than 10 μm. At such small gaps, oil flow and pocket pressure are governed by the land resistance and not that of the compensation element. As the gap continues to increase, pocket pressure is approximately, inversely proportional to gap cubed and the pressure–gap curve exhibits an inflection point at about 10 μm. As the gap continues to increase, oil flow becomes influenced by the compensation element and for gaps of 20 to 40 μm pocket pressure reduces almost linearly. At higher gaps, the resistance of the compensation element governs oil flow and pocket pressure decays in an exponential manner to about 5 Bar at a gap of 100 μm. Note that in this example a land gap of 32 μm gives a pocket pressure of half of the supply pressure and the highest rate of change of pressure with gap.

The load capacity of an opposed pair of pockets as a function of the 'Zero Load' pocket pressure is shown in Figure 3.17. In this example, the Zero Load pocket pressure is controlled by altering the resistance to oil flow of the compensation element. The specific load capacity is defined to be the ratio of actual load capacity to the product of bearing area and supply pressure, where actual load capacity is assumed to be the load supported by the pair of pockets at a deflection of 90% of the Zero Load gap. A maximum specific load capacity of 0.6 occurs for low pocket/supply pressure ratios, i.e. less than about 0.2, and load capacity reduces steadily as the pocket/supply pressure ratio increases. The reason for this is that at high pocket/supply pressure ratios, the pressure in the unloaded pocket is still well above ambient pressure at the maximum permissible bearing gap. Note that maximum load capacity occurs when the pressure

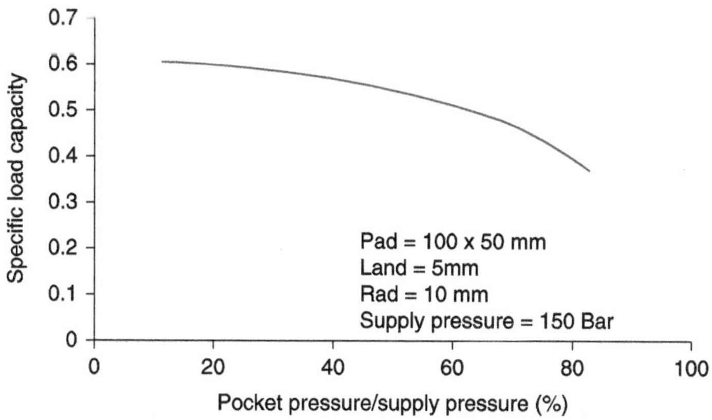

Figure 3.17 **Load capacity for an opposed pair of pockets**

in the loaded pocket approaches supply pressure and that in the unloaded pocket approaches ambient pressure.

The static stiffness as a function of Zero Load pocket/supply pressure ratio for the same pair of pockets is shown in Figure 3.18. Static stiffness is an important design criterion in many precision machines and is therefore frequently used as a criterion for hydrostatic bearing design. In Figure 3.18, static stiffness exhibits a clear maximum value at a pocket/supply pressure ratio of 0.5. This is where the resistance to fluid flow of the compensation element is the same as that of the land around the pocket at the Zero Load condition.

Figure 3.19 gives examples of load–deflection relationships for an opposed pair of hydrostatic pads. The three curves are for different design values of the Zero Load pocket pressure. For the low pressure of 22 Bar, the curve has poor linearity, exhibiting low stiffness for deflections below 10 μm, or more than 40 μm but a high load capacity, generated at deflections of more than 40 μm. The curve corresponding to a high Zero Load pocket pressure of 126 Bar exhibits good linearity, but both stiffness and load capacity are low. For a designed Zero Load pocket pressure of 75 Bar, half the supply pressure, the load–deflection relationship has good linearity for deflections of up to 25 μm and exhibits both high stiffness and load capacity.

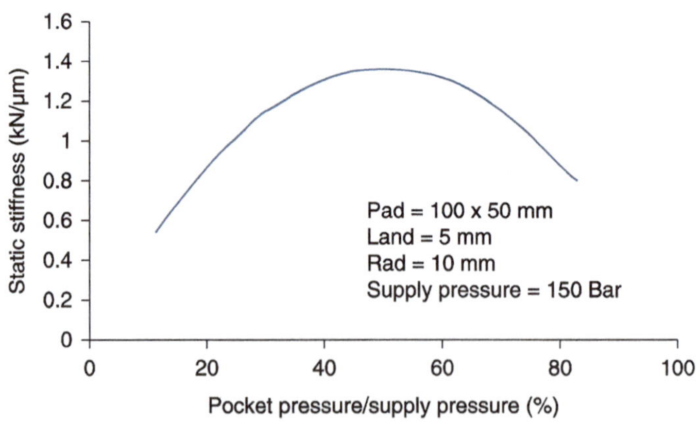

Figure 3.18 **Static stiffness versus Zero Load pocket pressure for an opposed pair of pockets**

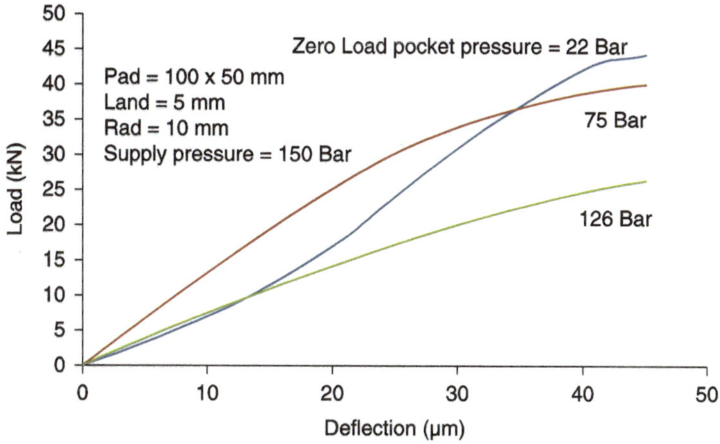

Figure 3.19 Load-deflection relationship for an opposed pair of pads

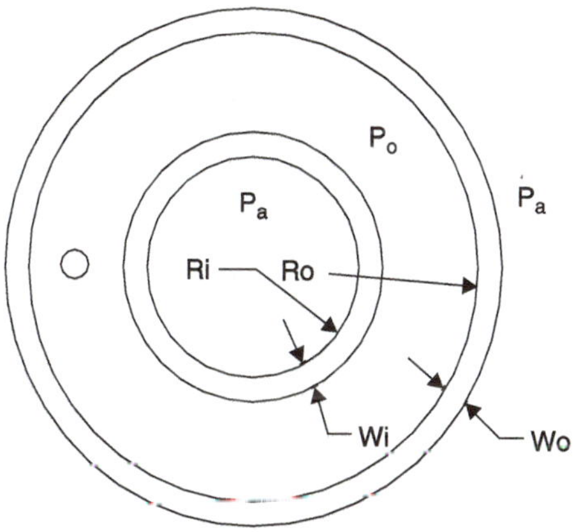

Figure 3.20 Annular hydrostatic bearing pocket with single oil inlet

3.10.3 Pressure–flow relationship for an annular pocket subject to a central axial load

Figure 3.20 shows an annular pocket of inner radius R_i and outer radius R_o with inner and outer land widths of W_i and W_o respectively. The bearing has a single oil inlet and therefore can only support a centrally applied normal load. For this situation, the volume flow rate of oil through the bearing can be found by applying Eq. 3.8 to the geometry defined in Figure 3.20. For both the inner and outer lands:

$$dV/dt = \pi h^3\ [P_o - P_a]\{1/\ln((R_i + W_i)/R_i) + 1/\ln(R_o/(R_o - W_o))\}/6\eta \qquad [3.21]$$

As for the rectangular pocket, the pocket pressure P_o is found by balancing the oil mass flow rate against that of the compensation element. For a capillary of diameter d and length l, the mass flow rate of oil is given by Eq. 3.11. Equating with Eq. 3.21 and simplifying gives:

$$\begin{aligned} P_o = \{&P_s d^4/(21.33l) + P_a h_o^3[1/\ln(R_i + W_i)/R_i) \\ &+ 1/\ln(R_o/(R_o - W_o))]\}/\{d^4/(21.33l) \\ &+ h_o^3[1/\ln(R_i + W_i)/R_i) + 1/\ln(R_o/(R_o - W_o))]\} \end{aligned} \qquad [3.22]$$

The load supported by the bearing is again found by integrating pressure over the bearing area:

The pocket area is $A_p = \pi[(R_o - W_o)^2 - (R_i + W_i)^2]$

The inner land area is $A_i = \pi[(R_i + W_i)^2 - R_i^2]$

and the outer land area is $A_o = \pi[R_o^{\ 2} - (R_o - W_o)^2]$

Thus the total load supported by the bearing is:

$$W = (P_o - P_a).A_p + P_{il}A_i + P_{ol}.A_o \qquad [3.23]$$

where now from Eq 3.10:

$$P_{il} = [P_o(R_i + W_i) - P_a R_i]/W_i - (P_o - P_a)\ln(R_i/(R_i - W_i)) \qquad [3.24]$$

and:

$$P_{ol} = [P_a R_o - P_o(R_o - W_o)]/W_o - (P_a - P_o)\ln(R_o - W_o)/R_o \qquad [3.25]$$

3.10.4 Pressure–flow relationship for an annular pocket with radial lands

The annular pocket with radial lands is a useful design, as it can take moment and offset normal loads and therefore provide tilt stiffness. The bearing shown in Figure 3.21 has an outer radius R_o, an inner radius R_i, outer and inner lands of width W_o and W_i respectively, but has in addition, radial lands of width W_d. The pressure difference across inner and outer lands is the pocket pressure less atmospheric pressure P_a, whereas the pressure difference across radial lands is the pressure difference between adjacent pockets. As the bearing is subject to tilt, the gap over the lands

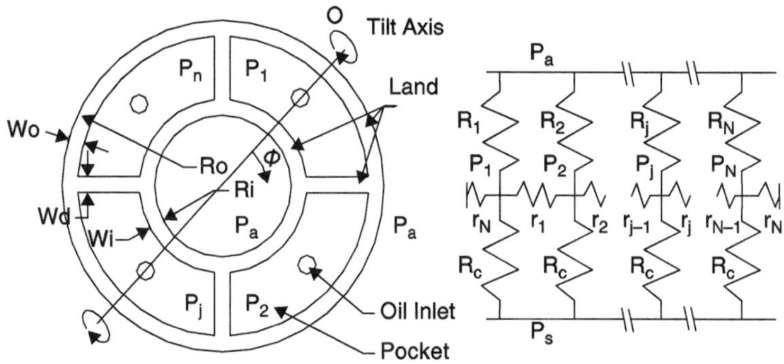

Figure 3.21 **Annular pocket with radial lands and fluid resistance model**

varies with angular position, so creating variation in pocket pressures. In this analysis, a local, mean effective gap is associated with each pocket and land and is calculated from the normal gap, the magnitude of the tilt and the angle that the centre of the land makes with the tilt axis.

A fluid resistance model of the bearing is also shown in Figure 3.21. In this analysis fluid resistance is defined by:

$$P_x - P_y = \acute{R}.dV/dt \qquad [3.26]$$

where $\acute{R}$ is the fluid resistance of a land, dV/dt is the volume flow rate of fluid flowing across the land and $P_x - P_y$ is the pressure difference across the land. From Eq. 3.3, it follows that the resistance of a rectangular land is:

$$\acute{R}_l = 12\eta.W/[h_o^3 L] \qquad [3.27]$$

where L is the length of the land normal to the direction of fluid flow, W is the width of the land (in the direction of fluid flow) and h_o is the gap over the land. For a circular land:

$$\acute{R}_c = 6\eta.\ln(r_o/r_i)/[\pi h_o^3] \qquad [3.28]$$

where r_o and r_i are the outer and inner radii of the land, h_o is the gap and the angular width of the land normal to fluid flow is 2π.

Each pocket is supplied with oil at a pressure P_s through a compensation element of resistance $\acute{R}_c$. Due to tilt varying the gap around the bearing, each pocket has a different resistance to fluid flow, $\acute{R}_1, \acute{R}_2 \ldots \acute{R}_j \ldots \acute{R}_N$, where N is the number of pockets in the bearing. These resistances are determined by the circular inner and outer land geometry and the local gap. Fluid flowing out of each pocket, across the circular lands, exhausts to a drain maintained at atmospheric pressure P_a. Thus the lands are subject to a pressure difference P_1-P_a, P_2-P_a . . . P_j-P_a . . . P_N-P_a for pockets 1, 2,. . .j. . .N, respectively. The pockets are separated by radial lands that also have different resistances to fluid flow, $\acute{r}_0$, $\acute{r}_1$ . . . $\acute{r}_j$. . .$\acute{r}_N$, depending on their location relative to the tilt axis. These lands are subject to a pressure difference P_2-P_1, P_j-P_{j-1}, P_N-P_{N-1} for radial land number 1. . .j. . .N, respectively.

For a given bearing gap and tilt, the normal and moment loads supported by the bearing are determined from the pocket pressures P_1. . .P_N and the bearing geometry. The pocket pressures are found by considering conservation of fluid flow through each pocket, i.e. for equilibrium, the mass flow rates of fluid into and out of a pocket must be equal. Thus for pocket number 1:

$$(P_s-P_1)/\acute{R}_c+(P_a-P_1)/\acute{R}_1+(P_N-P_1)/\acute{r}_N+(P_2-P_1)/\acute{r}_1=0 \qquad [3.29]$$

For pocket number 2:

$$(P_s-P_2)/\acute{R}_c+(P_a-P_2)/\acute{R}_1+(P_1-P_2)/\acute{r}_1+(P_3-P_2)/\acute{r}_2=0 \qquad [3.30]$$

and for pocket number j:

$$(P_s-P_j)/\acute{R}_c+(P_a-P_j)/\acute{R}_j+(P_{j-1}-P_j)/\acute{r}_{j-1}+(P_{j+1}-P_j)/\acute{r}_j=0 \qquad [3.31]$$

Thus for a bearing with N pockets, a set of N equations are obtained that can be solved to find the set of pressures P_1 to P_N. The above equations may be solved by any one of several numerical schemes. Relaxation techniques are widely used in the field of fluid bearings to iterate to a solution from an initial guess and is a viable approach here. But as hydrostatic bearings generally have a low number of pockets, usually less than 10, a direct solution scheme based on matrix arithmetic is preferred. With this approach, it is convenient to work in terms of the inverse of fluid resistance, $\acute{A}$, where:

$$\acute{A}=1/\acute{R}$$

and $\acute{A}$ is referred to as the fluid admittance. Expressing the set of Eq. 3.31 in matrix form:

$$\begin{bmatrix} (\acute{A}_c+\acute{A}_1+\acute{a}_N+\acute{a}_1)\ (-\acute{a}_1)\ldots 0 \ldots 0 \ldots (-\acute{a}_N) \\ (-\acute{a}_1)\ (\acute{A}_c+\acute{A}_2+\acute{a}_1+\acute{a}_1)\ (-\acute{a}_2)\ldots 0 \ldots 0 \ldots \\ 0 \quad (-\acute{a}_{j-1})\ (\acute{A}_c+\acute{A}_j+\acute{a}_{j-1}+\acute{a}_j)\ (-\acute{a}_{j+1})\ldots \\ (-\acute{a}_N)\ldots 0 \ldots 0 \ldots (-\acute{a}_{N-1})\ (\acute{A}_c+\acute{A}_N+\acute{a}_{N-1}+\acute{a}_N) \end{bmatrix} \begin{bmatrix} P_1 \\ P_2 \\ P_j \\ P_N \end{bmatrix} = \begin{bmatrix} P_s\acute{A}_c+P_a\acute{A}_1 \\ P_s\acute{A}_c+P_a\acute{A}_2 \\ P_s\acute{A}_c+P_a\acute{A}_j \\ P_s\acute{A}_c+P_a\acute{A}_N \end{bmatrix} \qquad [3.32]$$

The set of pocket pressures P_1 to P_N are found by inverting the admittance matrix and multiplying by the right-hand column matrix:

$$\begin{matrix} P_1 \\ P_2 \\ \\ P_j \\ P_N \end{matrix} = \begin{bmatrix} & & \\ & Á & \\ & & \end{bmatrix}^{-1} \begin{bmatrix} P_sÁ_c + P_aÁ_1 \\ P_sÁ_c + P_aÁ_2 \\ \\ P_sÁ_c + P_aÁ_j \\ P_sÁ_c + P_aÁ_N \end{bmatrix} \qquad [3.33]$$

Expressions for the admittances are found using Eqs 3.17 or 3.21, and the notation given in Figure 3.21. If the designed bearing gap is h_o then for a tilt of θ, the gap at any point in the bearing is given by:

$$h = h_o + \theta.r.\sin(\phi)$$

where r is the radius of the point and ϕ is its angular position with respect to the tilt axis. Thus if the bearing has N pockets and capillary compensation, the admittances for pocket j are:

$$\begin{aligned} &Á_c = \pi d^4/(128\eta l) \\ &Á_j = \pi h_{oj}^{\ 3}/[6\eta \ln(R_o/(R_o - W_o))]/N + \pi h_{ij}^{\ 3}/[6\eta.\ln((R_i + W_i)/R_i)]/N \\ &á_{j-1} = h_{j-1}^{\ 3}(R_o - R_i - W_o - W_i)/(12\eta W_d) \\ &á_j = h_j^{\ 3}(R_o - R_i - W_o - W_i)/(12\eta W_d) \end{aligned} \qquad [3.34]$$

where to a first approximation:

$$\begin{aligned} &h_{oj} = h_o + \theta.(R_o - W_o/2).\sin(\phi_j) \\ &h_{ij} = h_o + \theta.(R_i + W_i/2).\sin(\phi_j) \\ &h_{j-1} = h_o + 0.5.\theta.(R_o - W_o + R_i + W_i).\sin(\phi_j - \pi/N) \\ &h_j = h_o + 0.5.\theta.(R_o - W_o + R_i + W_i).\sin(\phi_j + \pi/N) \end{aligned} \qquad [3.35]$$

and $1 \le j \le N$.

Equations 3.33 to 3.35 define the set of pocket pressures P_1 to P_N. These in turn enable the load supported by the bearing to be determined. Each pocket supports a load:

$$\begin{aligned} Q_j = {} & (P_j - P_a)\ A_p + (P_{oj} - P_a)A_{ol} + (P_{ij} - P_a)A_{il} \\ & + (P_{j+1} + 2P_j + P_{j-1} - 2P_a).A_d/4 \end{aligned} \qquad [3.36]$$

where P_{oj} and P_{ij} are the average pressures acting on outer and inner lands as given by Eq. 3.5, A_{ol} and A_{il} are the land areas local to the pocket and

A_d is the area of the radial land. The total axial load supported by the bearing is given by:

$$Q_a = \Sigma_1^N Q_j \qquad [3.37]$$

and the moment load is:

$$M_{xx} = \Sigma_1^N Q_j\ r.\theta.\cos(\phi) \qquad [3.38]$$

$$M_{yy} = \Sigma_1^N Q_j\ r.\theta.\sin(\phi) \qquad [3.39]$$

where $r = (R_i + W_i + R_o - W_o)/2$

3.10.5 Design example for annular pocket with radial lands

Consider an annular pocket of outside diameter 150 mm and bore diameter of 70 mm with an outer circular land of width 7.5 mm, an inner circular land width of 3.5 mm and 4 radial lands of width 10 mm supplied with oil at a pressure of 150 Bar. Determine the Zero Load pocket pressure that gives maximum axial stiffness and its corresponding tilt stiffness. Find the effect of the number of radial lands on tilt stiffness.

The axial load–deflection characteristics of the annular thrust bearing are similar to the normal load–deflection characteristics of the rectangular pad bearing, in as much as the maximum static stiffness of an opposed pair of bearings occurs for a designed Zero Load pocket pressure of half the supply pressure (Figure 3.22) and the axial load capacity reduces steadily as the designed Zero Load pocket pressure is increased or decreased.

With more than three pockets, this bearing design has the capacity to support moment loads in addition to axial loads. Moment load capacity of the bearing is defined at a tilt that generates a minimum permissible gap at the outside diameter of the bearing. Figure 3.23 shows the moment load–tilt deflection characteristics of the bearing of Figure 3.21 for different Zero Load pocket pressures. The optimum Zero Load pocket pressure is half the supply pressure for which the moment–tilt relationship is very linear.

Increasing the number of pockets has a negligible effect on axial load carrying capacity but a significant effect on moment load capacity (Figure 3.24). In this bearing, the radial lands are 10 mm wide and increasing the number of pockets from 3 to 6 increases the total land width, providing

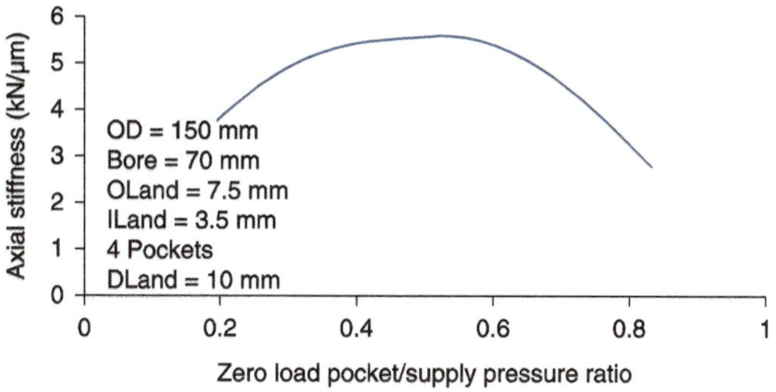

Figure 3.22 Static stiffness versus pocket pressure for double-sided annular thrust bearing

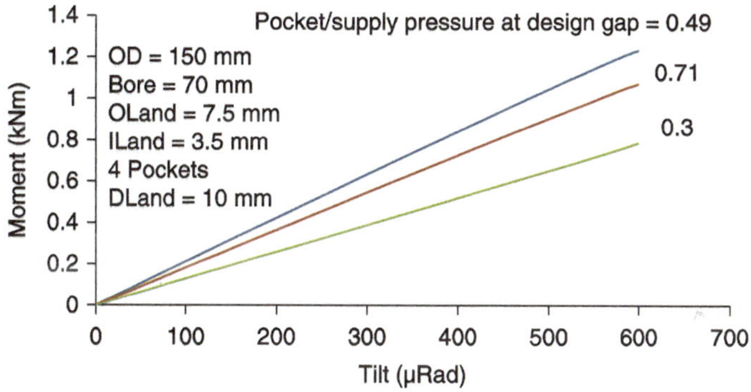

Figure 3.23 Moment–tilt relationship for double-sided annular thrust bearing

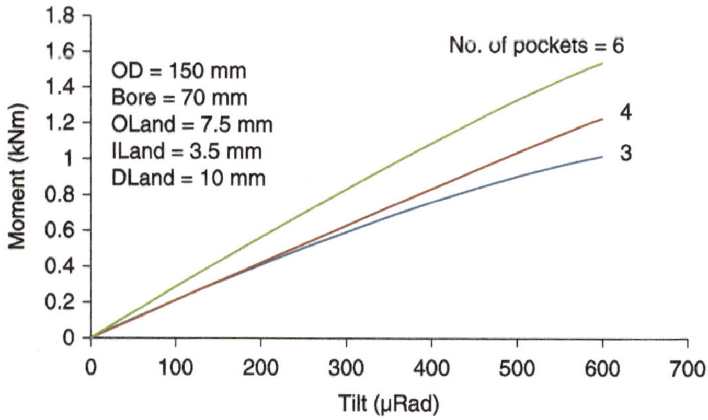

Figure 3.24 Effect of number of pockets on moment load capacity

resistance to cross flow of oil around the bearing. For the bearing design shown, 6 pockets give 50% more moment load capacity than 3.

3.10.6 Pressure–flow relationship for a cylindrical journal bearing

Figure 3.25 shows a journal bearing of diameter D and overall length L, with end lands of width W and axial lands of width W_d. Note that the axial lands are essential for the bearing to support any load at all. The bearing has a designed clearance of h_o and is subject to a radial load that acts through the centre of pocket number 1. Single journal bearings of this design cannot support any tilt or moment loading, so the load vector must act in a central radial plane. The load produces an eccentricity, e, so that the bearing gap is a function of angular position:

$$h = h_o(1 - e.\cos(\phi)) \qquad [3.40]$$

where e is defined to be equal to h/h_o and therefore $0 \leq e \leq 1$.

A fluid resistance model of the bearing is also shown in Figure 3.25. As for the annular thrust bearing with radial lands, the compensation element is a capillary or resistance, $\acute{R}_c$; the two end lands of each pocket have a combined resistance, $\acute{R}_1$, $\acute{R}_2$ . . . $\acute{R}_j$ . . . $\acute{R}_N$, where N is the number of pockets and the axial lands have resistances, $\acute{r}_1$, $\acute{r}_2$, . . . $\acute{r}_j$ . . . $\acute{r}_N$. The oil pressure supplied to the compensation element is P_s and it exhausts from the bearing into a drain at ambient pressure, P_a.

The fluid resistance model is identical to that for the thrust bearing with radial lands and the pocket pressures may be found using the same analytical procedure and numerical techniques. For a given bearing gap and eccentricity, the radial load supported by the bearing is determined from the pocket pressures P_1 . . . P_N and the bearing geometry. Conservation of fluid flow through each pocket, gives:

For pocket number 1:

$$(P_s - P_1)/\acute{R}_c + (P_a - P_1)/\acute{R}_1 + (P_N - P_1)/\acute{r}_N + (P_2 - P_1)/\acute{r}_1 = 0 \qquad [3.41]$$

For pocket number 2:

$$(P_s - P_2)/\acute{R}_c + (P_a - P_2)/\acute{R}_1 + (P_1 - P_2)/\acute{r}_1 + (P_3 - P_2)/\acute{r}_2 = 0 \qquad [3.42]$$

and for pocket number j:

$$(P_s - P_j)/\acute{R}_c + (P_a - P_j)/\acute{R}_j + (P_{j-1} - P_j)/\acute{r}_{j-1} + (P_{j+1} - P_j)/\acute{r}_j = 0 \qquad [3.43]$$

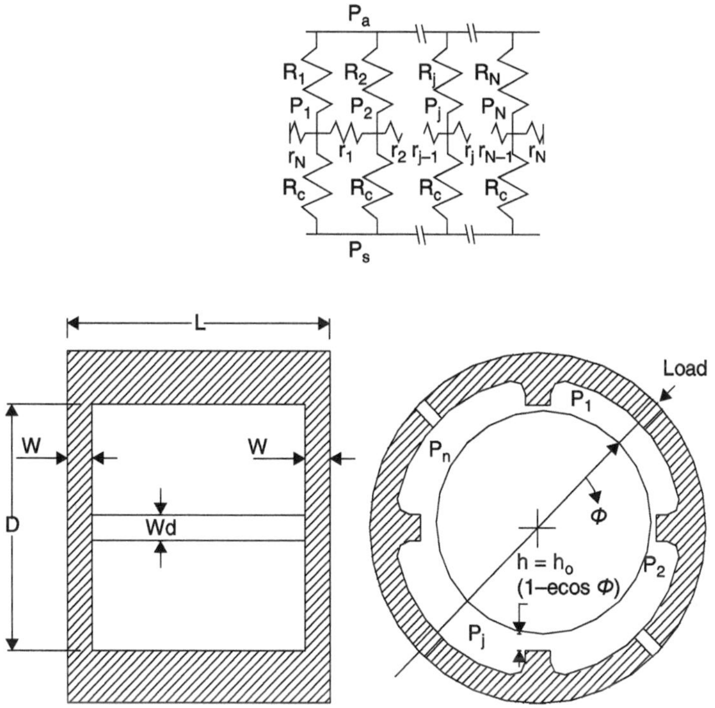

Figure 3.25 **Journal bearing with radial lands and fluid flow model**

Thus for a bearing with N pockets, a set of N equations are obtained that can be solved to find the set of pressures, P_1 to P_N. Using matrix arithmetic and working in terms of the inverse of fluid resistance, Á, where:

$$\acute{A} = 1/\acute{R}$$

the set of Eqs 3.41 to 3.43 may be expressed as:

$$\begin{bmatrix} (\acute{A}_c+\acute{A}_1+\acute{a}_N+\acute{a}_1)\ (-\acute{a}_1)\ldots\ldots 0\ldots\ldots 0\ldots\ldots (-\acute{a}_N) \\ (-\acute{a}_1)\ (\acute{A}_c+\acute{A}_2+\acute{a}_1+\acute{a}_2)\ (-\acute{a}_2)\ldots\ldots 0\ldots\ldots 0\ldots \\ 0\ldots\ldots(-\acute{a}_{j-1})\ (\acute{A}_c+\acute{A}_j+\acute{a}_{j-1}+\acute{a}_j)\ (-\acute{a}_j)\ldots \\ (-\acute{a}_1)\ldots\ldots 0\ldots\ldots 0\ldots\ldots (-\acute{a}_{N-1})\ (\acute{A}_c+\acute{A}_N+\acute{a}_{N-1}+\acute{a}_1) \end{bmatrix} \begin{bmatrix} P_1 \\ P_2 \\ P_j \\ P_N \end{bmatrix} = \begin{bmatrix} P_s\acute{A}_c+P_a\acute{A}_1 \\ P_s\acute{A}_c+P_a\acute{A}_2 \\ P_s\acute{A}_c+P_a\acute{A}_j \\ P_s\acute{A}_c+P_a\acute{A}_N \end{bmatrix} \quad [3.44]$$

The set of pocket pressures, P_1 to P_N, are found by inverting the admittance matrix and multiplying by the right-hand column matrix:

$$\begin{bmatrix} P_1 \\ P_2 \\ P_j \\ P_N \end{bmatrix} = \begin{bmatrix} & & \\ & Á & \\ & & \end{bmatrix}^{-1} \begin{bmatrix} P_sÁ_c + P_aÁ_1 \\ P_sÁ_c + P_aÁ_2 \\ P_sÁ_c + P_aÁ_j \\ P_sÁ_c + P_aÁ_N \end{bmatrix} \qquad [3.45]$$

Expressions for the admittances are found using Eqs 3.4 or 3.11, and the notation given in Figure 3.25. For a bearing with n pockets, a gap given by Eq. 3.40 and capillary compensation the admittances are given by:

$$\begin{aligned} &Á_c = \pi d^4/(128\eta l) \\ &Á_j = \pi h_{oj}^{\,3}/[6\eta.\ln(R_o/(R_o - W_o))]/N + \pi h_{ij}^{\,3}/[6\eta.\ln((R_i + W_i)/R_i)]/N \\ &á_{j-1} = h_{j-1}^{\,3}(R_o - R_i - W_o - W_i)/(12\eta W_d) \\ &á_j = h_j^{\,3}(R_o - R_i - W_o - W_i)/(12\eta W_d) \end{aligned} \qquad [3.46]$$

where to a first approximation:

$$\begin{aligned} &h_{oj} = h_o + \theta.(R_o - W_o/2).\sin(\phi_j) \\ &h_{ij} = h_o + \theta.(R_i + W_i/2).\sin(\phi_j) \\ &h_{j-1} = h_o + 0.5.\theta.(R_o - W_o + R_i + W_i).\sin(\phi_j - \pi/N) \\ &h_j = h_o + 0.5.\theta.(R_o - W_o + R_i + W_i).\sin(\phi_j + \pi/N) \end{aligned} \qquad [3.47]$$

and $1 \le j \le N$

Knowing the pocket pressures, P_1 to P_N, enables the load supported by the bearing to be determined. Each pocket supports a load given by:

$$Q_j = (P_j - P_a)A_p + (P_j - P_a)./2A_l + (P_{j+1} + 2P_j + P_{j-1} - 2P_a).A_d/4 \qquad [3.48]$$

where A_p is the projected area of the pocket, A_l is the combined area of the two circular end lands and A_d is the area of the radial lands. Q_j acts through the centre of the pocket, so that the total bearing load is the sum of the pocket loads:

$$Q_{rx} = \Sigma_1^{\,N}\, Q_j \cos(\phi) \qquad [3.49]$$

$$Q_{ry} = \Sigma_1^{\,N}\, Q_j \sin(\phi) \qquad [3.50]$$

3.10.7 Design example for a cylindrical journal bearing

Consider a cylindrical journal bearing of diameter 60 mm and length 100 mm, with circular end lands of width 10 mm and 4 axial lands of

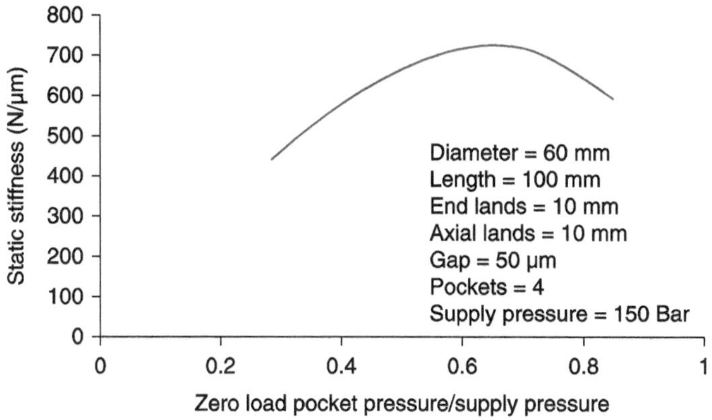

Figure 3.26 **Optimum pocket pressure for journal bearing**

width 10 mm supplied with oil at 150 Bar. Determine the Zero Load pocket pressure that gives maximum stiffness and the effect of axial land width and the number of axial lands on stiffness.

Figure 3.26 shows how the Zero Load pocket pressure affects the static stiffness of the bearing. The bearing has four pockets and has capillary compensation. The Zero Load pocket pressure that gives maximum static stiffness is higher than for the flat rectangular and annular pad bearings, being 0.64 times the supply pressure compared to a value of 0.5 for these other bearing types. At the optimum pocket pressure, the load coefficient of the bearing is 0.4.

In journal bearings, reducing the cross flow of oil around the bearing is reasonably expected to have a significant effect on both static stiffness and load carrying capacity. Figure 3.27 shows how increasing the number of pockets and hence the number of axial lands improves static stiffness. However, there is a law of diminishing returns where the gain in performance for bearings with more than five pockets is very low and may well not be worth the additional complexity. The load coefficient for a bearing with 10 pockets is 0.59.

Static stiffness may also be expected to improve by keeping to a low number of pockets and simply increasing the width of the axial lands. Figure 3.28 shows how static stiffness depends on axial land width for a bearing with four pockets. Widths of 7.5, 10, 15, 20 and 25 mm are equivalent to 3, 4, 6, 8 and 10 pockets for the bearing of Figure 3.27, the total axial land widths being the same. The load coefficient for 4 pockets and an axial land width of 25 mm is 0.52 and compares with a value of 0.59 for a bearing with 10 pockets and an axial land width of 10 mm.

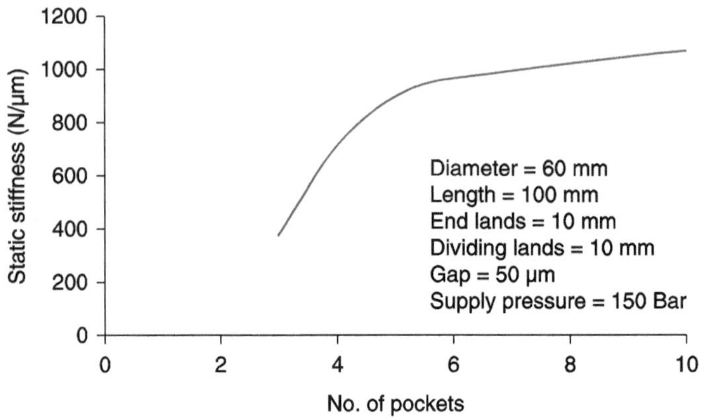

Figure 3.27 Effect of number of pockets on static stiffness

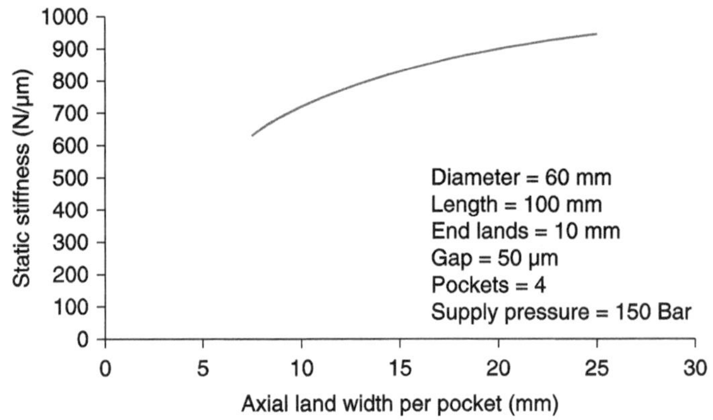

Figure 3.28 The effect of axial land width on static stiffness

Increasing the number of pockets gives a modest improvement in bearing performance compared to simply increasing land width, but at the expense of additional complexity. However, there is another benefit in using more pockets and that is less variation in stiffness and capacity with angular position. With just four pockets there is a significant variation in these parameters, depending on whether the load acts through the centre of a pocket or midway between two adjacent pockets. This variation reduces rapidly with increasing number of pockets, enabling the bearing to maintain a consistent deflection in the presence of rotating loads.

The gains in load capacity and stiffness by increasing the total width of axial lands is at the expense of power consumption and bearing running temperature, as the total area of oil subject to a high shear rate has been increased. This aspect of hydrostatic bearing performance is considered further in Section 3.13.

3.11 Damping

3.11.1 Dynamic model of a hydrostatic bearing

Of all precision bearing types, it is the hydrostatic bearing that exhibits outstanding damping characteristics. Damping coefficients are not only high but can be designed to meet application requirements. In this section, a dynamic model of a hydrostatic bearing is described and a means of calculating its damping coefficients presented.

Figure 3.29 shows a hydrostatic bearing with a single pocket, where the housing is subject to a dynamic displacement dh.sin(ωt), and the opposing surface is fixed in position. As a result, the gap under the bearing lands is given by h_o + dh.sin(ωt), where h_o is the static equilibrium gap. The bearing is supplied with oil at pressure P_s through a capillary restrictor with fluid resistance R_c. The bearing's static equilibrium resistance to oil flow is R_p and the oil pressure in the pocket is P_p, where $P_p = P_o + p\sin(\omega t - \phi)$. P_o is the static equilibrium value and dP is a dynamic

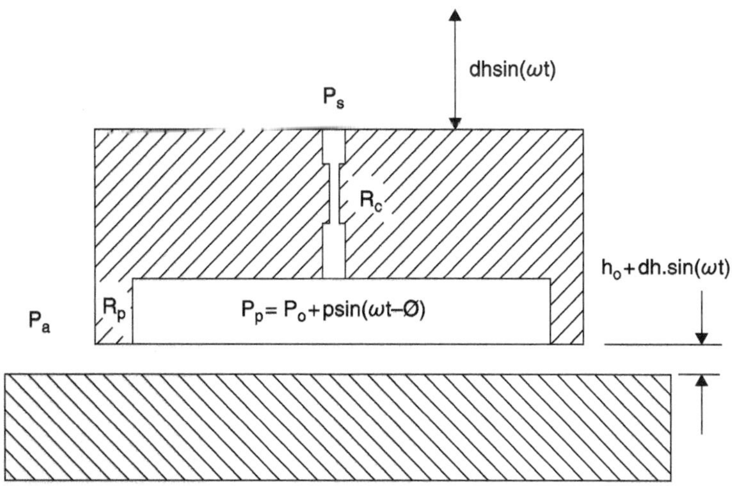

Figure 3.29 Dynamic model of hydrostatic bearing

component resulting from the applied dynamic displacement, dh. In general, there will be a phase lag ϕ between dh and p.

In this analysis, the dynamic deflection of the bearing housing, dh. sin(ωt), is assumed to be small in comparison with the static equilibrium gap, h_o, and that the inertia of the bearing and the oil that it contains is negligible. Damping is assumed to be due to two mechanisms:

1. a squeeze film action occurring in the bearing lands, where the damping force is dependent upon the velocity of approach of opposing surfaces separated by a narrow gap; and
2. displacement of oil in and out of the pocket, creating a dynamic pressure fluctuation that depends on both pocket volume and rate of change of volume.

The pocket volume is dependent on the instantaneous gap $- h_o + dh.\sin(\omega t)$ and the bearing area, but the rate of change of volume creates a dynamic pressure that has a component proportional to the velocity of approach, $\omega dh\cos(\omega t)$, of the bearing surfaces, which produces a phase lag in the dynamic component of pocket pressure, i.e.:

$$P_p = P_o + p\sin(\omega t - \phi)$$

Damping forces are determined by considering the dynamic equilibrium of the bearing. The starting point is to determine the relationship between oil pressure distribution and rate of change of gap size for rectangular and circular land geometries.

3.11.2 Rectangular land subject to a squeeze action

Oil flow under a rectangular land subject to a squeeze action is driven by the pocket pressure, P_p, and the rate of change of gap, dh/dt. The velocity profile of the oil therefore changes with distance across the land (Figure 3.30). For 1-D fluid flow through a rectangular gap subject to squeeze, the relevant reduced form of the Navier-Stokes equation is [9]:

$$\frac{\partial^2 p}{\partial x^2} = -\frac{12\eta}{h^3}\frac{dh}{dt}$$

Integrating twice with respect to x and applying the boundary conditions $p = P_p$ at $x = 0$ and $p = P_a$ at $x = W$, gives the pressure distribution across the land:

$$p - P_p = \frac{6\eta}{h^3}\frac{dh}{dt}(Wx - x^2) + \frac{(P_a - P_p)}{W}x \qquad [3.51]$$

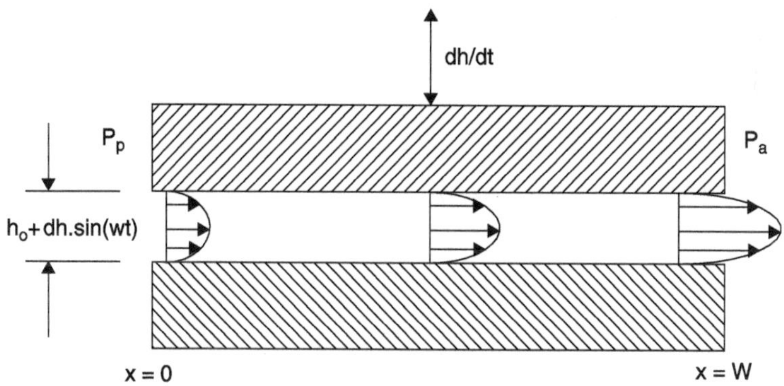

Figure 3.30 **Rectangular land subject to squeeze**

where P_p is the pressure in the bearing pocket. For the model of Figure 3.30:

$$dh/dt = \omega.dh.\cos(\omega t) \qquad [3.52]$$

and:

$$h = h_o + dh.\sin(\omega t) \qquad [3.53]$$

The volume flow rate of oil is given by Eq. 3.3, where dp/dx is a function of x, and time as can be derived from Eq. 3.51:

$$\frac{dp}{dx} = \frac{6\eta}{h^3}\frac{dh}{dt}(W - 2x) + \frac{(P_a - P_p)}{W} \qquad [3.54]$$

and therefore:

$$\frac{dV}{dt} = \frac{-a}{2}(W - 2x)\frac{dh}{dt} - \frac{-ah^3}{12\eta W}(P_a - P_p) \qquad [3.55]$$

Where a is the length of the land normal to the direction of the fluid flow. Of particular interest is the flow rate at entry to the land, i.e. when $x = 0$:

$$dV/dt = -\frac{a}{2}\frac{Wdh}{dt} - \frac{ah^3}{12\eta W}(P_a - P_p) \qquad [3.56]$$

Approximating h^3 to $h_o^3(1 + 3dh\sin(\omega t)/h_o)$ and P_p to $P_o + p\sin(\omega t - \phi)$ and then using Eqs 3.52 and 3.53 to replace dh/dt and h, gives to a first approximation:

$$dV/dt = \frac{ah_o^3}{12\eta W} p\sin(\omega t - \phi) + \frac{ah_o^2}{4\eta W}(P_o - P_a) dh\sin(\omega t) - \frac{a}{2} W dh.\omega\cos(\omega t) + \frac{ah_o^3(P_o - P_a)}{12\eta W} \quad [3.57]$$

Thus the volume flow rate of oil entering the land has a static component and first-order dynamic components describing the effect of squeeze motion, the change in resistance to oil flow rate due to a change in gap size and the change in pressure gradient across the land.

The average pressure under the land is found by integrating Eq. 3.51 with respect to x over the interval 0 to W and dividing the result by W. Putting $dh/dt = dh\omega\cos\omega t$ and rearranging gives to a first approximation:

$$P_{av} = \frac{\eta W^2}{h_o^3} dh\omega\cos(\omega t) + \frac{(P_a + P_p)}{2} \quad [3.58]$$

Note that the average pressure comprises two terms, one of which is equal to the average of the external pressures present at the two edges of the land, the other, a squeeze term, dependent on the velocity of approach of the land surfaces, the land width, gap and the oil viscosity. The land is part of a bearing subject to a dynamic variation in gap and the pocket pressure P_p is dynamic and given by:

$$P_p = P_o + p\sin(\omega t - \phi) \quad [3.59]$$

where P_o is the mean static pressure in the pocket and p is a small dynamic variation from the mean. Thus the average dynamic pressure under the land is:

$$P_{avd} = \frac{\eta W^2}{h_o^3} dh\omega\cos(\omega t) + \frac{p}{2}\sin(\omega t - \phi) \quad [3.60]$$

For a land considered in isolation, the dynamic pressure p has to be known to determine P_{avd}, but for a land that is part of a bearing, it is determined by considering equilibrium of oil flow rate between the pocket and the land.

3.11.3 Annular land subject to a squeeze action

As for the rectangular land, the starting point for determining the pressure distribution in and mass flow rate through an annular gap subject to squeeze is the reduced form of the Navier-Stokes equation but this time expressed in polar coordinates [9]:

$$\frac{\partial}{\partial r}\left(rh^3\frac{\partial p}{\partial r}\right)=-12\eta r\frac{\partial h}{\partial t}$$

Integrating with respect to r:

$$\frac{\partial p}{\partial r}=\frac{-6\eta r}{h^3}\cdot\frac{\partial h}{\partial t}+\frac{C_1}{r}$$

and:

$$p=\frac{-3\eta r^2}{h^3}\cdot\frac{\partial h}{\partial t}+C_1.\ln(r)+C_2$$

where C_1 and C_2 are constants of integration determined from the boundary conditions $p=P_p$ when $r=r_i$ and $p=P_a$ when $r=r_o$. The pressure at any point in the land may be expressed as:

$$p-P_a=\frac{-3\eta}{h^3}\frac{dh}{dt}[r^2-r_o^2+(r_o^2-r_i^2)\ln(r/r_o)/\ln(r_i/r_o)]+[P_p-P_a]\ln(r/r_o)/\ln(r_i/r_o) \quad [3.61]$$

The average pressure under the land is found by integrating Eq. 3.61 with respect to r from r_i to r_o and dividing by the width of the land, r_o-r_i:

$$P_{av}=\frac{3\eta}{h^3}\frac{dh}{dt}\left[\frac{2}{3}(r_i^2+r_or_i+r_o^2)+(r_o^2+r_i^2)/\ln(r_i/r_o)\right] \quad [3.62]$$
$$+\left[P_p-P_a\right]\left[r_i-r_o-r_i\ln(r_i/r_o)\right]/\left[(r_o-r_i)\ln(r_i/r_o)\right]$$

Again, the average pressure acting on the land contains a dynamic term due to the effect of the squeeze film and a term giving the average effect of pressures present at the edges of the land. The dynamic component of average pressure is:

$$P_{avd}=\frac{3\eta}{h_o^3}\frac{dh}{dt}\left[\frac{2}{3}(r_i^2+r_or_i+r_o^2)+(r_o^2-r_i^2)/\ln(r_i/r_o)\right] \quad [3.63]$$
$$+p\sin(\omega t-\phi)\left[r_i-r_o-r_i\ln(r_i/r_o)\right]/\left[(r_o-r_i)\ln(r_i/r_o)\right]$$

The volume flow rate of oil is derived from:

$$dV/dt=\frac{-\pi rh^3}{6\eta}\frac{dp}{dr} \quad [3.64]$$

where dp/dr is found from Eq. 3.61:

$$\frac{dp}{dr}=\frac{-3\eta}{h^3}\frac{dh}{dt}\left[2r+\frac{(r_o^2-r_i^2)}{\ln(r_i/r_o)}\right]+\frac{[P_p-P_a]}{\ln(r_i/r_o)} \quad [3.65]$$

Of particular interest is the volume flow rate at entry to the land, i.e. when $r=r_i$:

$$\frac{dV}{dt}=\frac{\pi}{2}\frac{dh}{dt}\left[2r_i^2+\frac{(r_o^2-r_i^2)}{\ln(r_i/r_o)}\right]-\frac{\pi h^3}{6\eta}\frac{[P_p-P_a]}{\ln(r_i/r_o)} \quad [3.66]$$

Approximating h^3 to $h_o^3(1+3dh\sin(\omega t)/h_o)$ and P_p to $P_o+p\sin(\omega t-\phi)$ and then using Eqs 3.52 and 3.53 to replace dh/dt and h gives a first approximation:

$$\frac{dV}{dt}=\frac{\pi}{2}\left[2r_i^2+\frac{(r_o^2-r_i^2)}{\ln(r_i/r_o)}\right]dh\omega\cos\omega t-\frac{\pi h_o^2}{2\eta}\frac{(P_o-P_a)}{\ln(r_i/r_o)}dh\sin(\omega t)$$
$$-\frac{\pi h_o^3 p\sin(\omega t-\phi)}{6\eta\ln(r_i/r_o)}-\frac{\pi h_o^3(P_o-P_a)}{6\eta\ln r_i/r_o} \quad [3.67]$$

As for the rectangular land, the volume flow rate has a static component and dynamic components describing the squeeze, dynamic pressure and flow resistance effects.

3.11.4 Damping coefficient for a rectangular pad

For the model shown in Figure 3.29, damping forces and coefficients are determined by considering the equilibrium of oil flow rate leaving the pocket and entering the lands. If the compensation element is a capillary of flow resistance, R_c, the volume flow rate entering the pocket is:

$$dV/dt=(P_s-P_p)/R_c \quad [3.68]$$

and the volume flow rate leaving the pocket is that entering the pocket plus the rate of change of pocket volume:

$$dV/dt=(P_s-P_p)/R_c+A_p dh/dt \quad [3.69]$$

where A_p is the cross-sectional area of the pocket and P_p is the dynamic pressure given by Eq. 3.59. Volume flow rate may therefore be expressed as:

$$dV/dt = (P_s - P_o)/R_c - (p/R_c)\sin(\omega t - \phi) + A_p dh\omega\cos(\omega t) \qquad [3.70]$$

For equilibrium, this must equal the flow rate entering the lands. If the pocket has four rectangular lands of width W and total length L_T and four corner radii of internal radius r_i and external radius r_o, where $r_o - r_i = W$, the total flow rate entering the lands is found by adding Eqs 3.57 and 3.67, first putting a = L:

$$\begin{aligned} dV/dt = {} & \frac{L_T h_o^3}{12\eta W} p\sin(\omega t - \phi) + \frac{L_T h_o^2}{4\eta W}(P_o - P_a) dh\sin(\omega t) \\ & - \frac{L_T}{2} W dh.\omega\cos(\omega t) + \frac{\pi h_o^3 p\sin(\omega t - \phi)}{6\eta \ln(r_i/r_o)} \\ & + \frac{\pi h_o^2}{2\eta}\frac{[P_o - P_a]}{\ln(r_i/r_o)} dh\sin(\omega t) + \frac{\pi}{2}\left[2r_i^2 + \frac{(r_o^2 - r_i^2)}{\ln(r_i/r_o)}\right] dh\omega\cos(\omega t) \\ & + \frac{L_T h_o^3 (P_o - P_a)}{12\eta W} + \frac{\pi h_o^3}{6\eta}\frac{(P_o - P_a)}{\ln(r_i/r_o)} \end{aligned} \qquad [3.71]$$

Equations 3.70 and 3.71 have static, dynamic in-phase and quadrature terms. Considering just dynamic equilibrium and equating in-phase and quadrature components of these two equations gives

In phase:

$$\frac{p}{R_c}\cos\phi = \frac{L_T h_o^2}{4\eta W} dh\,(P_o - P_a)\,dh + \frac{\pi h_o^2}{2\eta}\frac{[P_o - P_a]}{\ln(r_i/r_o)} dh + \frac{L_T h_o^3}{12\eta W} p\cos\phi + \frac{\pi h_o^3}{6\eta \ln(r_i/r_o)} p\cos\phi$$

$$p\cos\phi = (P_o - P_a)\frac{h_o^2}{4\eta}\left[\frac{L_T}{W} + \frac{2\pi}{\ln(r_i/r_o)}\right] \Bigg/ \left[\frac{1}{R_c} - \frac{L_T h_o^3}{12\eta W} - \frac{\pi h_o^3}{6\eta \ln(r_i/r_o)}\right] dh \qquad [3.72]$$

and quadrature:

$$\begin{aligned} \frac{-p}{R_c}\sin\phi + A_p dh\omega = {} & \frac{-L_T}{2} W dh.\omega - \frac{L_T h_o^3}{12\eta W} p\sin\phi \\ & + \pi\left[2r_i^2 + \frac{(r_o^2 - r_i^2)}{2\ln(r_i/r_o)}\right] dh\omega - \frac{\pi h_o^3 p\sin\phi}{6\eta \ln(r_i/r_o)} p\sin\phi \end{aligned} \qquad [3.73]$$

$$p\sin\phi = w dh\left[\frac{\pi}{2}\left(2r_i^2 + \frac{(r_o^2 - r_i^2)}{\ln(r_i/r_o)}\right) - \frac{L_T W}{2} + A_p\right] \Bigg/ \left[\frac{L_T h_o^3}{12\eta W} + \frac{\pi h_o^3}{6\eta \ln(r_i/r_o)} + \frac{1}{R_c}\right]$$

$p\sin\phi$ is the quadrature component of dynamic pressure that gives rise to damping. The damping force is found by summing the products of dynamic pressure and area for the pocket, rectangular and corner lands. Equations 3.60 and 3.63 give the dynamic pressures for the rectangular and corner lands respectively. Summing gives:

$$F_d = \left[\frac{\eta\, W^2}{h_o^3} dh\omega + \frac{p}{2}\sin\phi\right] L_T\, W + A_p\, p\sin\phi$$
$$+ \frac{3\eta}{h_o^3}\frac{dh}{dt}\left[2(r_o^2 + r_i r_o + r_i^2)/3 + (r_o^2 - r_i^2)/\ln(r_i/r_o)\right] \quad [3.74]$$
$$\left. + p\sin\phi[r_i - r_o - r_i \ln(r_i/r_o)]/(r_o - r_i)\ln(r_i/r_o)\right\}\pi(r_o^2 - r_i^2)$$

The damping coefficient is:

$$C = F_d/(\omega dh) \quad [3.75]$$

3.11.5 Damping coefficient for an annular thrust pad

An annular thrust pad with a single pocket, i.e. no radial lands, is considered first and damping is determined by considering equilibrium of oil flow rate between the pocket and the inner and outer circular lands when the bearing is subjected to a small dynamic deflection. For a deflection $dh\sin(\omega t)$ the bearing gap is:

$$h = h_o + dh\sin(\omega t)$$

and the pocket pressure is:

$$P_p = P_o - p\sin(\omega t - \phi)$$

The flow rate out of the pocket is given by the flow rate into the pocket plus the rate of change of pocket volume:

$$dV/dt = (P_s - P_o)/R_c - p/R_c \sin(\omega t - \phi) + A_p dh\omega\cos(\omega t) \quad [3.76]$$

The volume flow rate of oil entering the inner and outer annular lands is derived from Eq. 3.67:

$$\frac{dV}{dt}=\frac{-\pi h_o^3}{6\eta}[P_o-P_a][1/\ln((R_o-W_o)/R_o)+1/\ln(R_i/(R_i+W_i))]$$
$$-dh\frac{-\pi h_o^3}{6\eta}[1/\ln((R_o-W_o)/R_o)+1/\ln(R_i/(R_i+W_i))]p\sin(\omega t-\phi)$$
$$+\frac{\pi h_o^2}{2\eta}[P_o-P_a][1/\ln((R_o-W_o)/R_o)+1/\ln(R_i/(R_i+W_i))] \qquad [3.77]$$
$$+\frac{\pi}{2}\left[2((R_o-W_o)^2+(R_i+W_i)^2)+\frac{(R_o^2-(R_o-W_o)^2)}{\ln(R_o-W_o)/R_o)}\right.$$
$$\left.+\frac{((R_i+W_i)^2-R_i^2)}{\ln(R_i/(R_i+W_i))}\right]dh\omega\cos(\omega t)$$

where the notation given in Figure 3.20 is used for land radii and widths. Equating the sin(ωt) terms from Eqs 3.76 and 3.77 gives the in phase equilibrium:

$$\frac{p}{R_c}\cos\phi=\frac{-\pi h_o^2}{2\eta}[P_o-P_a][1/\ln((R_o-W_o)/R_o)+1/\ln(R_i/(R_i+W_i))]dh$$
$$-\frac{\pi h_o^3}{6\eta}[1/\ln((R_o-W_o)/R_o)+1/\ln(R_i/(R_i+W_i))]p\cos\phi \qquad [3.78]$$

Rearranging:

$$p\cos\phi=\left\{\frac{-\pi h_o^2}{2}[P_o-P_a][1/\ln((R_o-W_o)/R_o)+1/\ln(R_i/(R_i+W_i))]dh\right\}\Big/$$
$$\left\{\frac{\pi h_o^3}{6\eta}[1/\ln((R_o-W_o)/R_o)+1/\ln(R_i/(R_i+W_i))]+\frac{1}{R_c}\right\} \qquad [3.79]$$

Similarly for the quadrature terms:

$$\frac{p}{R_c}\sin\phi=\left\{A_p+\frac{\pi}{2}\left[2((R_o-W_o)^2+(R_i+W_i)^2)\right.\right.$$
$$\left.\left.+\frac{(R_o^2-(R_o-W_o)^2)}{\ln((R_o-W_o)/R_o)}+\frac{((R_i+W_i)^2-R_i^2)}{\ln(R_i/(R_i+W_i))}\right]\right\}\omega dh$$
$$+\frac{\pi h_o^3}{6\eta}[1/\ln((R_o-W_o)/R_o)+1/\ln(R_i/(R_i+W_i))]p\sin\phi$$

Rearranging:

$$\begin{aligned}\mathrm{psin}\phi = & \left\{ A_p + \frac{\pi}{2}\left[2((R_o - W_o)^2 + (R_i + W_i)^2) \right.\right. \\ & \left.\left. + \frac{(R_o^2 - (R_o - W_o)^2)}{\ln((R_o - W_o)/R_o)} + \frac{((R_i + W_i)^2 - R_i^2)}{n(R_i/(R_i + W_i))} \right] \right\} \omega dh / \\ & \left\{ \frac{1}{R_c} - \frac{\pi h_o^3}{6\eta} [1/\ln((R_o - W_o)/R_o) + 1/\ln(R_i/(R_i + W_i))] \right\}\end{aligned} \quad [3.80]$$

$\mathrm{psin}\phi$ is the component of dynamic pressure that gives rise to damping. The total damping force is the sum of the products of average dynamic pressure and area for the pocket and lands. Using Eq. 3.63 to determine the damping in the circular lands and Eq. 3.80 for the pocket gives:

$$\begin{aligned}F_d = & \pi\left[(R_o - W_o)^2 - (R_i + W_i)^2\right] \mathrm{p}\sin\phi \\ & + \pi\left[R_o^2 - (R_o - W_o)^2\right] P_{dyn}(\text{outer}) \\ & + \pi\left[(R_i - W_i)^2 - R_i^2\right] P_{dyn}(\text{inner})\end{aligned} \quad [3.81]$$

The damping coefficient is given by Eq. 3.75. For an annular thrust pad with radial lands subject to a concentric axial load, the total damping force is assumed to be the sum of that due to squeeze in the radial lands and that given by Eq. 3.81. Under a concentric load, the pressures in the pockets are the same and there is no pressure gradient across these lands. For the damping calculation the dynamic pressure (Eq. 3.76) and damping force is calculated for a single annular pocket fed by N capillaries, where N is the actual number of pockets in the bearing and added to the squeeze film forces for the radial lands. The average squeeze film pressure in the radial lands is the same as for a rectangular land and is given by Eq. 3.60:

$$P_{avd} = \frac{\eta W_r^2}{h_o^3} dh\omega\cos(\omega t) \quad [3.82]$$

For N lands of width W_r and length $R_o - W_o - R_i - W_i$ the total damping force generated by the radial lands is:

$$F_r = \frac{\eta N}{h_o^3}(R_o - W_o - R_i - W_i)W_r^3 dh\omega \quad [3.83]$$

The total axial damping force for the annular thrust bearing with radial lands is therefore the sum of Eqs 3.81 and 3.83.

For an annular thrust pad with radial lands subject to a dynamic moment load an approximate calculation of the tilt damping coefficient may be made by considering the damping in each pocket separately and summing the results for all pockets. Average values for the gap and rate of change of gap over the radial and circular lands local to the pocket are used to estimate the local damping force. For the bearing of Figure 3.21, subject only to a small dynamic angular tilt of $d\theta$, the gap at any radius r and angular position ϕ is given as:

$$h = h_o + d\theta.r.\sin\phi \quad [3.84]$$

Assuming the dynamic component of tilt is given by:

$$d\theta = d\theta.\sin(\omega t) \quad [3.85]$$

then the dynamic variation in gap is:

$$dh = d\theta.r.\sin\phi\ \sin(\omega t) \quad [3.86]$$

and the velocity of approach given by:

$$dh/dt = d\theta r\omega.\sin(\phi).\cos(\omega t) \quad [3.87]$$

In the approximate solution scheme, Eqs 3.84 and 3.87 are used to find the gap and velocity of approach of the centre of each pocket. These values are then used in Eqs 3.80 and 3.81 to estimate the dynamic pressure and damping force associated with each pocket. Pocket area and perimeter lengths are divided by the number of pockets and the capillary resistance is that of a single capillary. The damping forces in the radial lands are determined using Eq. 3.82 and the local clearance. Half of the damping force in each radial land is added to adjacent pockets. If F_{dj} is the damping force associated with pocket number, j, then the damping moment is:

$$M_{dxx} = \Sigma_1^N F_{dj}\ r.\cos\phi_j \quad [3.88]$$

$$M_{dyx} = \Sigma_1^N F_{dj}\ r.\sin\phi_j \quad [3.89]$$

where:

$$r = (R_i + R_o)/2$$

The corresponding damping coefficients are:

$$C_{yx} = \Sigma_1^N F_{dj}\ r.d\theta.\cos\phi_j/\omega dh_j \qquad [3.90]$$

$$C_{yy} = {}_1^N F_{dj}\ r.d\theta.\sin\phi_j/\omega dh_j \qquad [3.91]$$

3.11.6 Damping coefficient for a cylindrical journal bearing

For a cylindrical journal bearing subject to a small dynamic deflection about the zero load static equilibrium position, the overall damping coefficient is the vector sum of the coefficients associated with each pocket and its surrounding lands.

Each pocket is assumed to be rectangular in shape and surrounded by four rectangular lands, so the analysis follows that of Section 3.11.4 for the rectangular pad; however, now the deflection over the lands is a function of angular position and the pressure gradients across the axial lands are dependent on the pressures in adjacent pockets. Figure 3.31 shows a journal bearing subject to a small radial dynamic deflection, $dh\sin\omega t$. The clearance at any point around the bearing is:

$$h = h_o + dh\sin\omega t.\cos\phi \qquad [3.92]$$

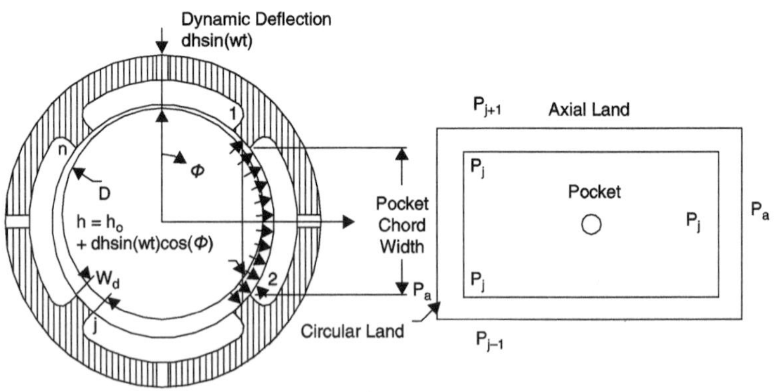

Figure 3.31 Journal bearing subject to dynamic deflection

The pressure in pocket number j is:

$$P_j = P_o - p_j \sin(\omega t - \phi) \tag{3.93}$$

where P_o is the static equilibrium pressure and p_j is the dynamic pressure produced by the deflection dh. The dynamic pressure in adjacent pockets is P_{j+1} and P_{j-1}. The axial lands are of width W_d and effective length L_d and the circular lands at the ends of the pocket are of width W_e and effective length L_e, where $L_d = L - W_e$ and $L_e = \pi D/N$. L is the length of the bearing, D is its diameter and N is the number of pockets. The length of the pocket is $L_p = L - 2W_e$ and the effective width of the pocket is its chord width, $W_p = D\sin(\pi/N)$. As for the rectangular pad, the damping force can be found by considering dynamic equilibrium of oil flow rates into and out of the pocket. If ϕ_j is the angular position of the centre of pocket, then flow rate into pocket number j is given by Eq. 3.70:

$$dV/dt = (P_s - P_o)/R_c - (p_j/R_c)\sin(\omega t - \psi) + A_p dh\omega\cos(\omega t)\cos\phi_j \tag{3.94}$$

and flow out of the pocket across the lands is derived from Eq. 3.57. For each of the circular end lands:

$$\begin{aligned} dV/dt = {} & \frac{\pi D h_o^3}{12N\eta W_e} p_j \sin(\omega t - \psi) + \frac{\pi D h_o^2}{4N\eta W_e}(P_o - P_a)\, dh\sin(\omega t)\cos\phi_j \\ & \frac{-\pi D}{2N} W_e dh\omega\cos(\omega t)\cos\phi_j + \frac{\pi D h_o^3 (P_o - P_a)}{12N\eta W_e} \end{aligned} \tag{3.95}$$

and from Eq. 3.56 for the axial land at position $\phi_j + \pi/N$:

$$dV/dt = \frac{L_d h_o^3}{12\eta W_d}(p_j - p_{j+1})\sin(\omega t - \psi) - \frac{L_d}{2} W_d dh\omega\cos(\omega t)\cos(\phi_j + \pi/N) \tag{3.96}$$

Note that for small deflections about the zero load position, the static pressure difference across the axial lands is zero. Equation 3.96 also applies to the axial land at angular position $\phi_j - \pi/N$, where p_{j+1} becomes p_{j-1}. Equating Eqs 3.94 to 3.96 for the in phase components, and taking into consideration two circular and two axial lands gives:

$$\begin{aligned} p_j\cos\psi = {} & -\left[\frac{\pi D h_o^2}{2N\eta W_e}(P_o - P_a) dh\cos\phi_j + \frac{L_d h_o^3}{12\eta W_d}(p_{j+1} + p_{j-1})\cos\psi\right] \Bigg/ \\ & \left[\frac{1}{R_c} + \frac{\pi D h_o^3}{6N\eta W_e} + \frac{L_d h_o^3}{6\eta W_d}\right] \end{aligned} \tag{3.97}$$

and for the quadrature component:

$$p_j \sin\psi = \left[\frac{\pi D W_e \omega}{N} dh\cos\phi_j + L_d W_d \omega dh \cos\phi_j \cos(\pi/N) + \frac{L_d h_o^3}{6\eta W_d}(p_{j+1} + p_{j-1})\sin\psi + A_p \omega dh \cos\phi_j\right] \Bigg/ \left[\frac{1}{R_c} + \frac{\pi D h_o^3}{6N\eta W_e} + \frac{L_d h_o^3}{6\eta W_d}\right] \quad [3.98]$$

Equations 3.97 and 3.98 give the in-phase and quadrature components of dynamic pressure for pocket number j in terms of the bearing geometry, oil viscosity and the pressures in adjacent pockets. Applying these equations to all pockets gives two sets of equations that can each be solved by the matrix inversion technique described in Section 3.10.6 and used to find the static pocket pressures. Equations 3.97 and 3.98 are each put in the form:

$$\begin{bmatrix} a_1 & -a_2 & 0 & 0 & \dots & 0 & -a_N \\ -a_1 & a_2 & -a_3 & 0 & \dots & 0 & 0 \\ 0 & \dots\dots & -a_{j-1} & a_j & \dots & -a_{j+1} & 0 \\ -a_1 & \dots\dots & 0 & 0 & \dots & -a_{n-1} & a_N \end{bmatrix} \begin{bmatrix} p_1 \\ p_2 \\ p_j \\ p_N \end{bmatrix} = \begin{bmatrix} b_1 \\ b_2 \\ b_j \\ b_N \end{bmatrix} \quad [3.99]$$

where:

$$\begin{bmatrix} p_1 \\ p_2 \\ p_j \\ p_N \end{bmatrix} = \begin{bmatrix} & & \\ & A & \\ & & \end{bmatrix}^{-1} \begin{bmatrix} b_1 \\ b_2 \\ b_j \\ b_N \end{bmatrix} \quad [3.100]$$

The dynamic pressure p_j is assumed to act over the projected width of the pocket and circular lands. For damping, the force associated with pocket number j is:

$$F_{dj} = [p_j(L_d - 2W_e) + 2\eta W_e^3 dh\omega/h_o^3] D\sin(\pi/N) + \eta W_d^3 L_d dh\omega/h_o^3 \quad [3.101]$$

where the second term is the squeeze film component acting in the circular land over the angular width of the pocket and is determined from Eq. 3.60. The third term is the squeeze term for the axial lands and is half of the force in each of the lands either side of the pocket. The total damping force acting on the bearing is:

$$F_{dx} = \Sigma_1^N F_{dj} \cos\phi_j \quad \text{and} \quad F_{dy} = \Sigma_1^N F_{dj} \sin\phi_j \qquad [3.102]$$

and the damping coefficients are:

$$C_{dx} = F_{dx}/(\omega dh) \quad \text{and} \quad C_{dy} = F_{dy}/(\omega dh) \qquad [3.103]$$

3.11.7 Example of damping coefficients for an annular thrust pad

For the annular thrust bearing of Section 3.10.5, Figure 3.32 shows how the axial damping coefficient depends on clearance and viscosity. This bearing is 150 mm outside diameter, 70 mm bore with outer and inner land widths of 7.5 and 3.5 mm and 4 pockets separated by 10 mm wide radial lands. At more than 0.2 MNs/m, the damping coefficient is very high compared to the other precision bearing types considered in this book. It is inversely proportional to clearance cubed and proportional to inner and outer land width cubed and to viscosity. Thus small changes in clearance and inner and outer land width have a large effect on damping levels.

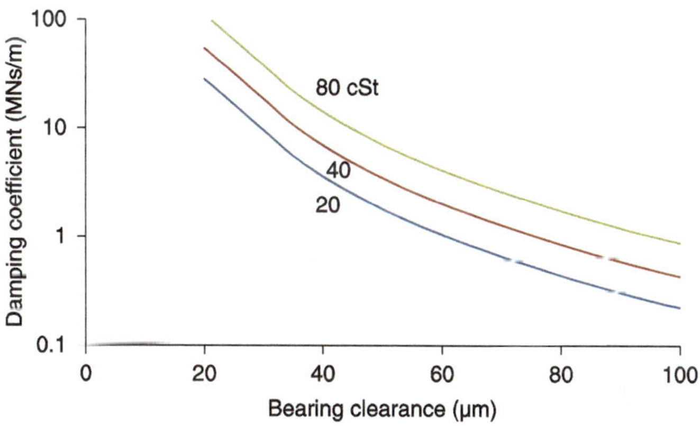

Figure 3.32 Example of damping coefficient versus bearing clearance

3.12 Motion error

3.12.1 Sources of motion error

For hydrostatic bearings, the primary sources of motion error are geometrical errors present on the bearing surfaces and to a lesser extent pressure fluctuations created within the oil supply. The former create motion errors synchronous with speed of rotation, whilst the latter create asynchronous errors.

Out of round or out of flat bearing surfaces create pocket-to-pocket pressure variations when the bearing is in motion and these in turn create relative oscillatory motion between the bearing surfaces. Whereas other aspects of bearing performance such as static stiffness, speed and power loss do not depend greatly on manufacturing precision, motion error does. Oil films within the bearing average out the effect of geometrical errors on the surfaces and in precision bearing applications the degree to which this happens is an important aspect of performance. In this section a relationship between the amplitude and wavelength of geometrical errors on the bearing surfaces and the amplitude and frequency of motion errors is derived for annular thrust and journal bearing types.

As for the ball bearing (Chapter 2, Section 2.11.1), the hydrostatic bearing is modelled as a source of excitation in series with a spring of linear stiffness K (Figure 3.33). As part of a mechanical system, the motion error generated by the excitation will depend upon the transfer function of the system. Two limiting cases are considered, one in which both housing and shaft are rigidly fixed. The other case is for the housing and shaft completely unrestrained and free to move in response to the surface imperfections.

When housing and shaft are fixed in position, then any geometrical errors of the bearing surfaces will generate fluctuations in oil pressure that create a net excitation force, $F_c \sin\omega t$, and when the housing and

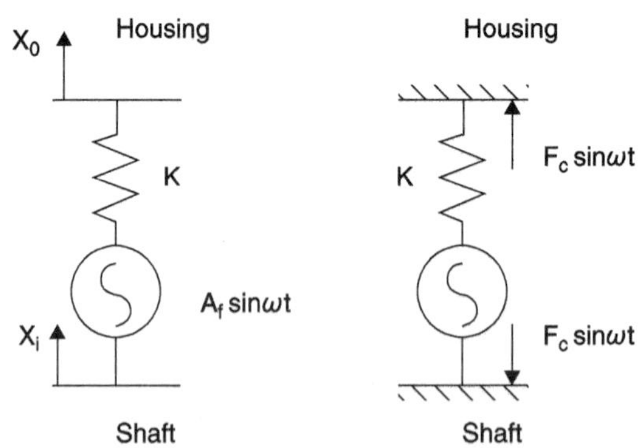

Figure 3.33 Relationship between dynamic force and displacement

shaft are completely free to move, the excitation is displacement, $A_f \sin\omega t$. The two are assumed to be related by:

$$F_c = K\,A_f$$

It is convenient to consider the case for which housing and shaft are rigidly fixed and to determine the net excitation force created by bearing surface geometrical error.

3.12.2 Synchronous motion error of annular thrust bearings

A model of an annular thrust bearing containing geometrical errors of its surfaces is shown in Figure 3.34. The bearing comprises a static housing and a disc rotating at a constant speed of ω. The housing has N pockets separated by lands of width W_d and inner and outer annular lands of width W_i and W_o respectively. Housing and disc are separated by an average gap h_o, which is constrained to be fixed in magnitude. The bearing considered has capillary restriction and it is assumed that without surface geometry errors and with constant gap, pocket pressures are all equal.

The significant geometrical errors that influence the bearing's motion error occur on the surface of the rotating disc and the inner and outer lands of the housing. With the bearing gap fixed in value, then the effect

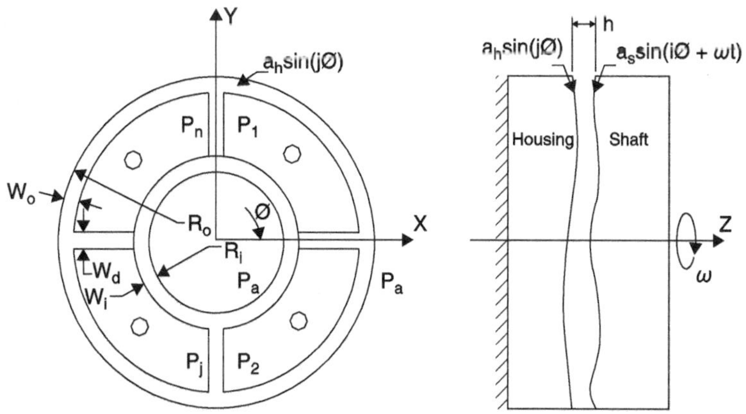

Figure 3.34 Hydrostatic bearing with geometrical errors on land surfaces

of these errors is to create a variation in pocket pressures. The dominant mechanism by which this happens is that surface errors on the annular lands or under the lands on the disc surface create local variations in the resistance to oil flow. As the bearing is compensated by a fixed restrictor, the local resistance variations integrated over the arc length of a pocket produce small variations in pocket pressure proportional to the resistance changes.

Any surface geometry error occurring on a flat annular surface can be described by a Fourier Series:

$$X = x_o + \Sigma_1^n a_i \sin(i\phi) + \Sigma_1^n b_i \cos(i\phi) \quad [3.104]$$

By considering surface errors to be simple harmonic with constant amplitude and fixed wavelength, enables their effect on pocket pressure and motion error to be determined analytically. The effect of more complex surface errors can also be determined analytically by considering them to consist of two or more simple harmonic terms and summing the results based on a Fourier analysis.

In the model of Figure 3.34, surface geometry errors on housing lands are described by:

$$a = a_h \sin(j\phi)$$

and on the disc:

$$a = a_d \sin i(\omega t + \phi')$$

where i and j is the number of waves per circumference, a_h and a_d are amplitudes, ϕ' is a phase angle and ω is the speed of rotation of the disc. In this analysis, a_h and a_d are assumed small compared to the equilibrium bearing gap, h_o. With the surface geometry errors, the bearing gap under the lands is a function of position and time:

$$h'(\phi,t) = h_o + a_h \sin(j\phi) + a_d \sin i(\omega t + \phi') = h_o + \Delta h(\phi,t) \quad [3.105]$$

For h_o constant in value, the bearing's resistance to oil flow is that due to inner and outer annular lands. Equation 3.8 relates the pressure difference across a circular land to volume flow rate through it. The resistance to oil flow is the ratio of pressure difference to volume flow rate, thus from Eq. 3.8:

$$\acute{R} = (P_i - P_o)/(dV/dt) = 6\eta \ln((r_i)/r_o)/(\pi h^3) \quad [3.106]$$

Applying Eq. 3.106 to the inner and outer lands of one pocket of the bearing shown in Figure 3.34 gives:

$$\acute{R}_i = 6\eta \ln((R_i + W_i)/R_i)/(N\pi h'^3) \quad [3.107]$$

and:

$$\acute{R}_o = 6\eta \ln((R_o/(R_o - W_o))/(N\pi h'^3) \quad [3.108]$$

where the surface errors are assumed to have the same amplitude and number of waves/circumference on inner and outer lands. The two lands act in parallel so the total resistance is:

$$\acute{R}_{to} = \frac{6\eta c_o c_i}{N\pi h'^3 (c_o + c_i)} \quad [3.109]$$

$$c_o = N((R_o/(R_o - W_o) \qquad \text{and} \qquad c_i = \ln((R_i + W_i)/R_i) \quad [3.110]$$

Using the Binomial Theorem to approximate h'^3:

$$h'^3 = (h_o + \Delta h)^3 = h_o(1 + 3\Delta h/h_o + 3(\Delta h/h_o)^2 \ldots) \quad [3.111]$$

A first-order approximation effectively linearises and hence considerably simplifies the analysis, but this is only reasonable for $\Delta h/h_o < 0.1$ (Figure 3.35). A second-order approximation gives a much improved accuracy generating an error of less than 0.5% for $\Delta h/h < 0.2$ and is more appropriate to the magnitude of surface geometry errors and gaps typical of hydrostatic bearings.

Substituting for h′ in Eq. 3.109 using Eq. 3.105 and then using the Binomial Theorem to simplify the resulting expression gives:

$$\acute{R}_t = \acute{R}_{to} \,(1 - 3\Delta h/h_o + 3(\Delta h/h_o)^2) \quad [3.112]$$

where $\acute{R}_t$ is the resistance of a pocket with geometrically perfect lands.

The compensation element has resistance $\acute{R}_c$ so the static pocket pressure is given by the supply pressure multiplied by the ratio of the pocket resistance to the total resistance:

$$P_k - P_a = (P_s - P_a)\, \acute{R}_t/(\acute{R}_t + \acute{R}_c) \quad [3.113]$$

where k denotes pocket number k. Substituting for $\acute{R}_t$ using Eq. 3.112 gives:

$$P_k - P_a = \frac{(P_s - P_a)\acute{R}_{to}}{(\acute{R}_{to} + \acute{R}_c)} [1 - 3(\Delta h_k/h_o + (\Delta h_k/h_o)^2)\acute{R}_c/(\acute{R}_{to} + \acute{R}_c)] \quad [3.114]$$

where Δh_k is the local average variation in gap associated with pocket number k. The term $(P_s - P_a)\, \acute{R}_{to}/(\acute{R}_{to} + \acute{R}_c)$ is put equal to $P_o - P_a$, where P_o is the designed level of static pocket pressure for a bearing with

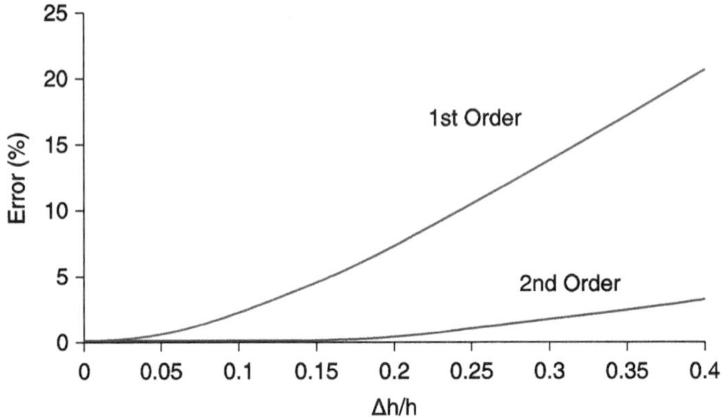

Figure 3.35 Error in binomial expansion

geometrically perfect lands operating at a defined gap of h_o. Replacing $\acute{R}_{to}/(\acute{R}_{to}+\acute{R}_c)$ gives:

$$P_k - P_a = (P_o - P_a)[1 - 3(\Delta h_k/h_o + (\Delta h_k/h_o)^2)(P_o - P_a)/(P_s - P_a)] \qquad [3.115]$$

The effect of Δh_k is to create a variation in pocket pressures. Surface geometry errors on the stationary housing produce a static imbalance between pocket pressures, whilst those on the rotating disc create a time dependent fluctuation in pocket pressures. Of interest here is the fluctuation in pressure produced by $\Delta h_k/h_o$ that may now be expressed as:

$$\Delta P'_k = -3\frac{P'^2_o}{P'_s}[\Delta h_k/h_o + (\Delta h_k/h_o)^2)] \qquad [3.116]$$

where ′ denotes gauge pressure. For surface geometry errors on the housing expressed as a proportion of the static gap h_o:

$$\Delta h/h_o = (a_h/h_o)\sin(j\phi) \qquad [3.117]$$

The effect of this on the average pocket gap is:

$$\underline{\underline{\Delta h/h}}_o = \frac{a_h}{(\phi_2 - \phi_1)h_o}\int_{\phi^1}^{\phi^2}\sin(j\phi)\,d\phi$$

where $\phi_2 - \phi_1$ is the arc length of the pocket $= 2\pi/N$. Integrating and simplifying gives:

$$\underline{\underline{\Delta h/h}}_o = \frac{-Na_h}{j\pi h_o}\sin(j\pi/N).\sin(j\phi_1 + j\pi/N) \qquad 1 \le j \le \infty \qquad [3.118]$$

For pocket number k, $\phi_1 = (k-1)2\pi/N$ and the pocket–pocket gap variation is:

$$\Delta h_k/h_o = \frac{-Na_h}{j\pi h_o}\sin(j\pi/N).\sin((2k-1)j\pi/N) \quad 1 \le k \le N \quad 1 \le j \le \infty \quad [3.119]$$

Substituting for $\Delta h_k/h_o$ into Eq. 3.116 enables pocket pressures to be determined. Of interest is that a single pocket enables surface geometry errors to be integrated over an interval of 2π resulting in a theoretical average gap variation $\Delta h_k/h_o$ of zero and hence no pressure fluctuation. However, for two or more pockets, $\Delta h_k/h_o$ has a finite value that depends on both the wavelength of the surface error as well as its amplitude. If the number of waves per circumference is an integral times the number of pockets, the average $\Delta h_k/h_o$ again theoretically integrates to zero having no effect on pressure, whereas for all other wavelengths a pressure imbalance between pockets is set up. In general, the shorter the wavelength of the surface error, i.e. the higher the value of j, the greater the averaging effect and hence the less the effect on pocket pressure.

Examples of the effect of surface geometry errors on pocket pressure variation are given in Tables 3.1 to 3.3.

Table 3.1 Annular thrust bearing with three pockets

$P_o/P_s = 0.5$ and $a_h/h_o = 0.1$

Type of geometry error	Pocket pressure variation (%)		
	Pocket No. 1	*Pocket No. 2*	*Pocket No. 3*
1 point lobing	11.5	0	−9.97
2 point lobing	5.56	0	−5.18
3 point lobing	0	0	0

Table 3.2 Annular thrust bearing with four pockets

$P_o/P_s = 0.5$ and $a_h/h_o = 0.1$

Type of geometry error	Pocket Pressure Variation (%)			
	Pocket No. 1	*Pocket No. 2*	*Pocket No. 3*	*Pocket No. 4*
1 point lobing	10.2	10.2	−.94	−8.94
2 point lobing	10.2	−8.94	10.2	−8.94
3 point lobing	3.25	3.25	−3.12	−3.12

Table 3.3 Annular thrust bearing with six pockets

$P_o/P_s = 0.5$ and $a_h/h_o = 0.1$						
Type of geometry error	Pocket Pressure Variation (%)					
	Pocket 1	*Pocket 2*	*Pocket 3*	*Pocket 4*	*Pocket 5*	*Pocket 6*
1 point lobing	7.5	15.7	7.5	−6.82	−13.0	−6.82
2 point lobing	11.5	0	−9.97	11.5	0	−9.97
3 point lobing	10.2	−8.94	10.2	−8.94	10.2	−8.94

In the above tables, only the longer wavelength surface errors having 1, 2 or 3 lobes/circumference are considered, as generally they are higher in amplitude, 1-point lobing is in effect misalignment. The pressure variations are for a surface error amplitude of 10% of the operating gap of the bearing and may be as high as 13% of the designed operating pocket pressure. Note that the sum of the pocket pressures is in all cases small compared to the fluctuation in pressure from one pocket to another. In fact, to a first approximation, i.e. neglecting the second-order term in Eq. 3.116, the pocket pressures sum to zero. The small departure from zero in summing the pocket pressures in the above tables is due to the non-linear dependence of pressure on gap represented by the second-order term. The significance of a near-zero summation is that the bearing is very effective in averaging out the effect of surface geometry errors on bearing load.

For the rotating disc, the analysis follows the same steps described above but with ϕ replaced by $\omega t + \phi'$. For this it can be shown that:

$$\underline{\Delta h}_k / h_o = \frac{Na_d}{j\pi h_o} \sin(j\pi/N).\sin(j\omega t + (2k-1)j\pi/N) \quad 1 \le k \le N \quad 1 \le j \le \infty \qquad [3.120]$$

Because the disc is rotating each pocket experiences the same average gap variation of amplitude:

$$\underline{\Delta h}_k / h_o = \frac{Na_d}{j\pi h_o} \sin(j\pi/N) \qquad 1 \le j \le \infty \qquad [3.121]$$

but displaced in time. The frequency of the gap variation is $j\omega$.

The total dynamic force and moments generated by the bearing is found by summing the product of pocket pressures and pocket areas and the moments of these products. Assuming the designed level of static pocket pressure P'_o is the same for all pockets, then $\Delta h_k/h_o$ may be summed over all pockets. Thus axial force may be written as:

$$F_a = \Sigma_1^N \Delta P'_k A' = \frac{-3P'^2_o}{P'_s} A'[\Sigma_1^N \Delta h_k/h_o + \Sigma_1^N (\Delta h_k/h_o)^2)] \qquad 1 \le k \le N \qquad [3.122]$$

where A′ is the effective area of each pocket. The moments are given by:

$$M_{xx} = \Sigma_1{}^N \Delta P_k A' R_p \cos((k-1)2\pi/N + \pi/N)$$
$$M_{yy} = \Sigma_1{}^N \Delta P_k A' R_p \sin((k-1)2\pi/N + \pi/N) \qquad 1 \le k \le N \qquad [3.123]$$

Expressed in terms of $\Delta h_k/h_o$:

$$M_{xx} = \frac{-3P'^2_o}{P'_s} A' R_p [\Sigma_1^N \Delta h_k/h_o + \Sigma_1^N (\Delta h_k/h_o)^2)] \cos((k-1)2\pi/N + \pi/N)$$
$$M_{yy} = \frac{-3P'^2_o}{P'_s} A' R_p [\Sigma_1^N \Delta h_k/h_o + \Sigma_1^N (\Delta h_k/h_o)^2)] \sin((k-1)2\pi/N + \pi/N) \qquad [3.124]$$
$$1 \le k \le N$$

where R_p is the pitch radius of the pockets. The terms $\Sigma_1{}^N \Delta h_k/h_o$ can be expressed in terms of the surface geometry error using Eq. 3.120:

$$\Sigma_1^N \frac{\Delta h_k/h_o}{j\pi h_o} = \Sigma_1^N N a_d \sin(j\pi/N).\sin(j\omega t + (2k-1)j\pi/N) \quad 1 \le k \le N \quad 1 \le j \le \infty \quad [3.125]$$

Using Eq. 3.11 to form the summation of $\sin(j\omega t + (2k-1)j\pi/N)$ results in:

$$\Sigma_1^N \Delta h_i/h_o = \frac{N a_d}{j\pi h_o} \sin(j\pi/N).\sin(j\omega t).\sin(j\pi)/\sin(j\pi/N) \qquad 1 \le j \le \omega$$
$$\Sigma_1^N \Delta h_i/h_o = \frac{N a_d}{j\pi h_o} \sin(j\omega t).\sin(j\pi) \qquad 1 \le j \le \infty \qquad [3.126]$$

Since $\sin(j\pi)$ is zero for all values of j, Eq. 3.126 states that the net effect of disc surface geometry errors on the bearing's axial force is theoretically averaged to zero. It should be noted that the summation of pocket pressures in Eq. 3.126 is only valid if the static pressures in all pockets are the same. Thus an annular thrust bearing that has a uniform pressure distribution is extremely effective in averaging out the net effect of surface geometry errors on axial force and hence axial motion error.

In the event that pocket pressures are not all the same, then Eq. 3.126 should be rewritten as:

$$F_a = -3\frac{A'}{P'_s}[\Sigma_1^N P'^2_{ok}\ \Delta h_k/h_o + \Sigma_1^N P'^2_{ok}(\Delta h_k/h_o)^2)] \qquad 1 \leq k \leq N \qquad [3.127]$$

where P'_{ok} is variable and part of the summation and $\Delta h_k/h_o$ is given by Eq. 3.125. Consider now the situation where surface geometry errors exist on both the stationary housing and the rotating disc. The effect of housing errors is to create an imbalance between pocket pressures and this in turn prevents errors on the disc from being averaged out so that they result in dynamic forces or motion error. For an annular thrust bearing with four pockets and surface geometry errors of 10% of the equilibrium bearing gap on both housing and disc surfaces, figure 3.36 shows how the vibration force depends on surface error wavelength. Where 1 to 12 lobes/circumference on the housing and 2, 3, 4 and 10 lobes/circumference on the disc are considered. It should be noted that harmonics of 4 lobes/circumference on either the housing or the disc are averaged to zero by a bearing with 4 pockets and that maximum dynamic force occurs for intermediate wavelengths corresponding to 2, 6 and 10 lobes/circumference. In general, the higher the number of lobes/circumference the lower the dynamic force produced as errors are averaged over a greater number of cycles. Force levels are significant, in this case being as high as 0.75% of the bearing's load capacity for surface error amplitudes of 10% of the bearing gap on both housing and disc surfaces. Frequencies of vibration relate to the wavelength of surface

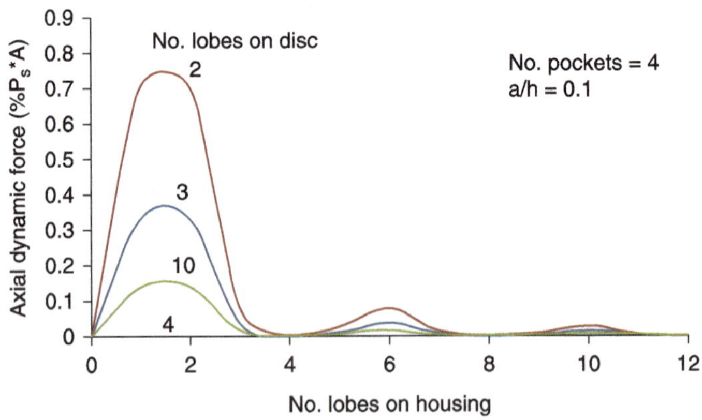

Figure 3.36 **Axial dynamic force due to housing and disc surface lobing errors**

errors on the disc, i.e. 2, 3 or 6 lobes/circumference produce frequencies of 2, 3 or 6 times the disc's rotational speed.

Figure 3.36 is an example of how surface errors on both the stationary and rotating surfaces of an annular thrust bearing interact to produce motion error. Of central importance to the mechanism of motion error generation is an imbalance in pocket pressures, preventing the effect of errors being averaged out over the full circumference of the bearing. It follows that other sources of pocket pressure imbalance may also enable surface errors on the disc to create motion error. Pressure differences due to small differences in the fluid resistance of compensation elements, moment or offset axial loads all increase the potential of the bearing to create motion error.

If the results shown in Figure 3.36 are applied to the annular thrust bearing of Section 3.10.5, supplied with oil at a pressure of 50 Bar and set up with a 50 μm designed clearance, then for a 5 μm peak amplitude of lobing present on both housing and disc surfaces the maximum motion error, corresponding to 2-point lobing on both surfaces, is 0.5 μm peak. This bearing is 150 mm outside diameter, 70 mm bore and has four pockets. Its axial stiffness is 940 N/μm and it has a bearing area of 0.0138 m^2. If the 0.5 μm motion error is attributed only to the 5 μm amplitude disc surface errors, this represents an attenuation of a factor of 10. For all other orders of lobing, attenuation factors are significantly higher.

3.12.3 Synchronous motion error of journal bearings

The same mechanism of motion error creation described above also applies to journal bearings. For a bearing with all pocket pressures equal, the effect of surface geometry errors on the shaft or housing lands are theoretically averaged to zero.

As for the annular thrust bearing, the journal bearing considered has a static equilibrium gap of h_o and shaft and housing are assumed to be fixed in position so that any surface geometry errors on the shaft and housing create variations in pocket pressures and hence dynamic force. For a journal bearing with simple harmonic surface geometry errors on both shaft and housing land surfaces (Figure 3.37), the average static gap variation due to errors on housing lands is given by Eq. 3.120:

$$\Delta h_k/h_o = \frac{-Na_h}{j\pi h_o}\sin(j\pi/N).\sin((2k-1)j\pi/N) \quad 1 \le k \le N \quad 1 \le j \le \infty \quad [3.128]$$

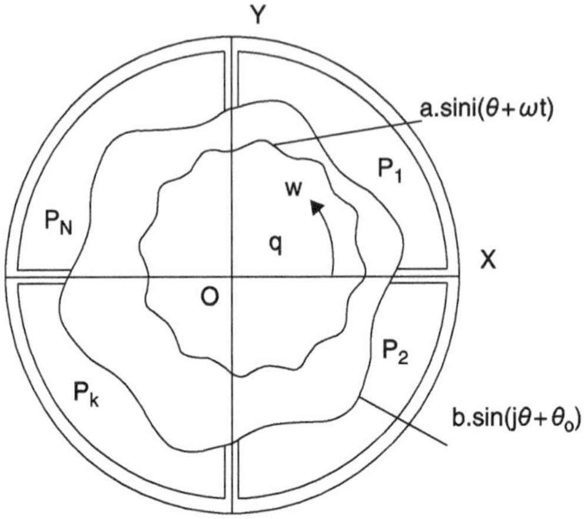

Figure 3.37 Hydrostatic journal bearing with surface errors on shaft and housing lands

and the effect of this is to create a static variation in pocket pressures given by Eq. 3.116:

$$\Delta P'_k = \frac{-3P'^2_o}{P'_s}[\Delta h_k/h_o + (\Delta h_k/h_o)^2)] \qquad [3.129]$$

The gap variation due to errors on the shaft is given by Eq. 3.125:

$$\Delta h_k/h_o = \frac{Na_d}{j\pi h_o}\sin(j\pi/N).\sin(j\omega t + (2k-1)j\pi/N) \quad 1 \le k \le N \quad 1 \le j \le \infty \qquad [3.130]$$

and this creates a radial dynamic force that has components:

$$F_X = -3\frac{A'}{P'_s}[\Sigma_1^N P'^2_{ok}\ \Delta h_k/h_o\ \cos((2k-1)\pi/N) + \Sigma_1^N P'^2_{ok}(\Delta h_k/h_o)^2)\cos((2k\text{-}1)\pi/N)] \quad 1 \le k \le N \qquad [3.131]$$

and:

$$F_Y = -3\frac{A'}{P'_s}[\Sigma_1^N P'^2_{ok}\ \Delta h_k/h_o \sin((2k-1)\pi/N) + \Sigma_1^N P'^2_{ok}\ (\Delta h_k/h_o)^2)\sin((2k-1)\pi/N)] \quad 1 \le k \le N \qquad [3.132]$$

where pocket number k has an angular position of $(2k-1)\pi/N$ with respect to the *x*-axis of the coordinate system used.

The journal bearing exhibits similar motion error characteristics to the annular thrust bearing, in as much as that surface geometry errors

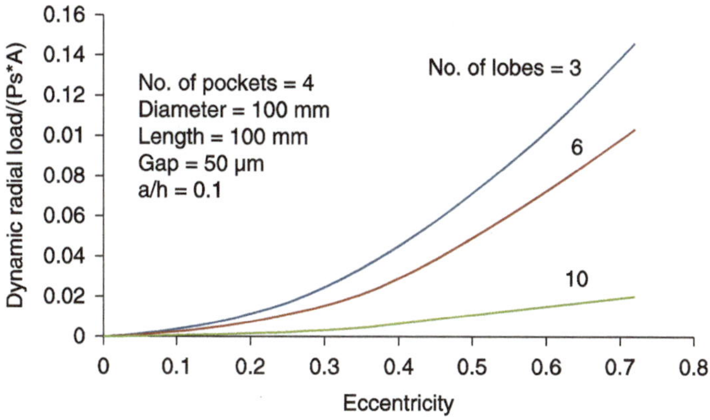

Figure 3.38 **Effect of bearing load on dynamic force for a journal bearing**

on the rotating component create vibration frequencies harmonic of rotational speed and that particular surface wavelengths are prone to creating relatively high motion errors, whereas others for which the number of lobes/circumference is an integer multiple of the number of bearing pockets generate low or zero vibration.

The application of a radial load to the journal bearing creates a static pressure imbalance between the pockets that can interact with surface geometry errors on the shaft enabling them to create a radial dynamic component of force. Figure 3.38 shows the dynamic force produced by lobing on the shaft of amplitude equal to 10% of the bearing gap. Under no load, pocket pressures are equal and the effect of the surface geometry error is averaged out. As load and hence eccentricity is increased, the pressure fluctuations produced in the loaded pockets become much larger than those in the unloaded pockets, reducing the ability of the bearing to average out the effect of the errors. Shorter wavelength features have less effect on dynamic force, as the average gap variation for each pocket is itself averaged over a higher number of lobes.

3.12.4 Asynchronous motion error

The conventional designs of hydrostatic bearing described in the preceding sections are stable and well damped and are not prone to generating asynchronous motion error. However, they do require a source of high pressure oil and asynchronous motion error may arise if the oil supply generates fluctuations in delivery pressure or flow rate. These may be expected to relate to features of the supply and hence bear no relationship to bearing speed of rotation.

In ultra precision hydrostatic bearing applications, the high pressure oil supply is an important part of the system and must be designed to deliver oil at the required pressure free from any pressure pulsations. Gear pumps are normally used to generate the high oil pressures required and they are prone to generating pulses at the gear meshing frequency and higher-order harmonics. Such pressure pulses are minimised by the use of precision gear pumps and high meshing frequencies for which damping is more effective. This is usually sufficient but where extremely low motion errors are essential, purpose built hydraulic pressure pulsation dampers can provide further attenuation levels of more than a factor of 10.

Pressure regulating valves are another possible source of disturbance. Passive devices for which the regulating pressure is set by a spring load are prone to instability and should be avoided. Regulating system pressure by using variable speed pumps driven by frequency inverters not only avoids the problem but provides a versatile oil supply.

3.13 Power consumption and heat generation

Hydrostatic bearings have large areas of oil subject to viscous shear and require oil to be delivered at high pressure. Both sliding friction and pumping losses end up as heat in the machine structure and the circulating oil. Any increase in oil temperature reduces viscosity, in turn reducing the bearing's frictional losses but at the expense of increased pumping power. In this section, simple calculations for the frictional and pumping losses are presented and used to determine an estimate of oil temperature rise sufficient for optimising oil viscosity selection and sizing the power supply.

3.13.1 Transition from laminar to turbulent flow

In this chapter, emphasis is given to low and moderate speed hydrostatic bearings, where friction is due to laminar flow of oil in the relatively deep pockets and the narrow lands. A feature of hydrostatic bearing operation is a transition from laminar to turbulent flow as speed is increased. Turbulent flow not only creates much higher friction losses, but it can also affect oil pressures in the pockets leading to a reduction in load capacity and stiffness.

It is therefore useful to have a criterion for the transition speed. The oil temperature rise was measured across conventional and 'Total Cross Flow' hydrostatic journal bearings as a function of speed (Figure 3.39) [12] and a distinct 'knee' observed in the temperature-speed relationship, where at speeds below the knee the slope of the curve was low and above which it was high. The knee was assumed to be indicative of the transition from laminar to turbulent flow and for the conventional bearing was found to correspond to a Reynolds number of 1071 in the pocket. Reynolds number is the ratio of inertial to viscous forces occurring in a fluid and is defined to be [9]:

$$Re = \text{inertia/viscous} = \rho Ud/\eta \qquad [3.133]$$

where ρ is oil density, U is relative velocity of the surfaces, d is a characteristic dimension of the surfaces, normally gap and η is absolute viscosity. Pocket depth was used as the characteristic dimension [12].

Another criterion for the transition from laminar to turbulent flow may be attributed to Taylor [13], who in 1923, studied the phenomenon of flow between two rotating cylinders and observed that just before the onset of turbulence, vortices were formed at a Reynolds number of:

$$Re > 41.3\ (D/2h_o)^{1/2}$$

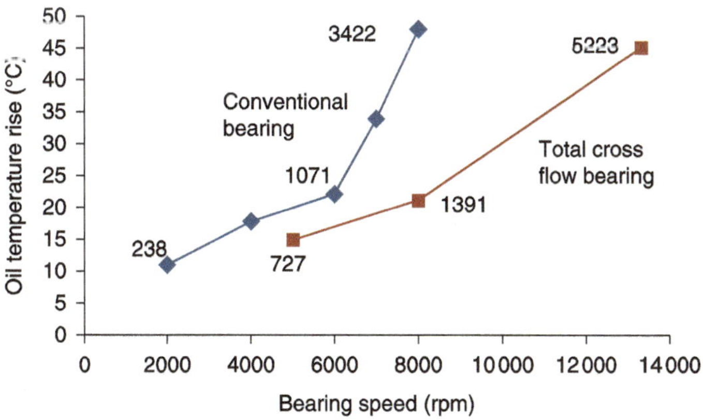

Figure 3.39 Oil temperature rise versus speed for hydrostatic bearing

where D is the cylinder diameter and h_o is the gap between them. Taylor developed a criterion for the onset of vortex flow known as the Taylor number, T_a, and expressed in terms of bearing related parameters is given by:

$$T_a = 2\rho U^2 h_o{}^3/(\eta^2 D) \geq 1700 \qquad [3.134]$$

This criterion, applied to the investigation of [12] is equivalent to a Reynolds number of 1131 and compares reasonably well to their observation for the onset of turbulence at a Reynolds number of 1071.

3.13.2 Bearing friction

Assuming laminar flow, then for two plates separated by a parallel gap of width h_o and area A, for which the opposing surfaces have a relative speed of U, shearing of the oil produces a force given by:

$$f = \eta A U/h_o \qquad [3.135]$$

Hydrostatic bearings have relatively deep pockets of large area surrounded by narrow lands of relatively small area, so the total friction force is:

$$f = \eta U[A_p/h_p + A_o/h_o] \qquad [3.136]$$

where the subscript p refers to the pocket and o the land. Pockets are made relatively deep to generate a uniform oil pressure and hence maximise load carrying capacity and stiffness, but this also has an advantage of reducing sliding friction and on higher speed bearings may be the more important criterion.

Friction in rectangular pad bearings

Applied to rectangular pad bearings and using the notation shown in Figure 3.15:

$$f = \eta U[(B-2W)(L-2W) + (4-\pi)(R-W)^2)]/h_p + \eta U[2LW + 2BW + \pi(R^2 - (R-W)^2]/h_o \qquad [3.137]$$

where h_p is pocket depth and h_o is the gap under the surrounding lands. The power loss is:

$$III = fU = \eta U^2[(B-2W)(L-2W) + (4-\pi)(R-W)^2)]/h_p + \eta U^2[2LW + 2BW + \pi(R^2 - (R-W)^2]/h_o \qquad [3.138]$$

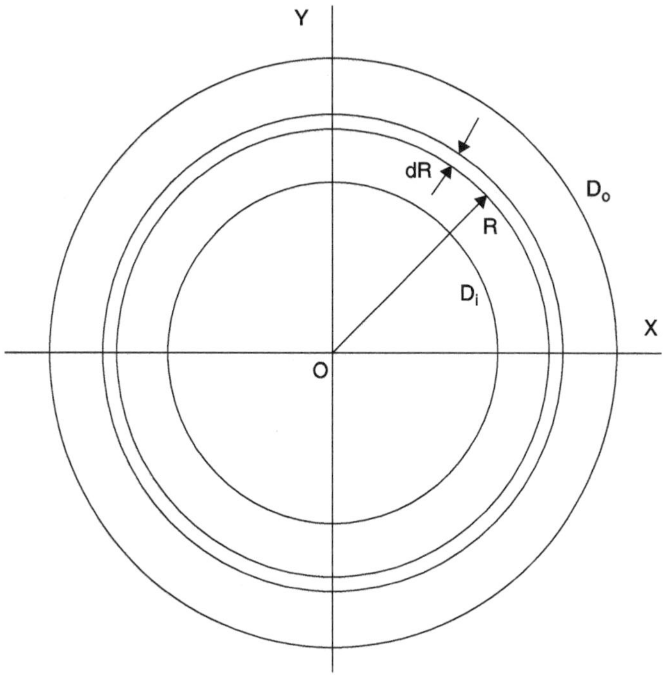

Figure 3.40 Elemental area on annular discs

Friction in annular thrust bearings

For annular thrust bearings, torque and power loss are of interest. First consider a pair of discs of bore diameter D_i and outside diameter D_o, for which one disc is rotating at a constant annular speed of ω (Figure 3.40). The torque generated by an annular element of width dR at radius R is:

$$dT = R.\ \tau.2\pi RdR = 2\pi R^2.dR.\eta.\ dU/dh \qquad [3.139]$$

where $dU/dh = \omega R/h_o$. Integrating over the bearing area:

$$T = 2\pi\int_{Ri}^{Ro} \omega R^3 \eta/h_o\ dR$$

$$T = \pi\eta\omega[D_o^4 - D_i^4]/32h_o \text{ where } D_o = 2R_o \text{ and } D_i = 2R_i \qquad [3.140]$$

The power dissipated by rotation is:

$$\text{Ш} = T\omega = \pi\eta\omega^2[D_o^4 - D_i^4]/32h_o \qquad [3.141]$$

Using the notation for the annular thrust bearing shown in Figure 3.20:

$$T_t = \pi\eta\omega\{[D_o^4 - (D_o - 2W_o)^4]/32h_o + [(D_i + 2W_i)^4 - D_i^4]/32h_o + [(D_o - 2W_o)^4 - (D_i + 2W_i)^4]\, NW_d/(\pi(D_o - 2W_o + D_i + 2W_i)16h_l) + [(D_o - 2W_o)^4 - (D_i + 2W_i)^4][1 - 2NW_d/\pi(D_o - 2W_o + D_i + 2W_i)\ 32h_p)]\} \quad [3.142]$$

which is essentially the sum of the torques due to outer and inner annular lands, the radial lands and the pockets. The angular width ratio of the radial lands is approximated to:

$$NW_d/\pi(D_o - 2W_o + D_i + 2W_i)$$

and hence the angular width ratio of the pockets is:

$$1 - NW_d/(\pi(D_o - 2W_o + D_i + 2W_i)$$

The power consumption is:

$$Ш = T_t\omega \quad [3.143]$$

Friction in journal bearing

Torque generated by the journal bearing of Figure 3.25 may also be determined by first considering an element of the surface. For a journal bearing of diameter D and length L, operating with constant gap, an annular surface element of width dZ has a length πD and for a shaft rotating at constant angular speed ω, then the torque generated by the element is:

$$dT = (D/2).(\eta dU/dh).\pi D.dZ \quad [3.144]$$

where $dU/dh = \omega D/(2h)$. Integrating from 0 to L gives the total friction torque of:

$$T = \pi\omega\eta LD^3/(4h_o) \quad [3.145]$$

Applying this to the end lands, axial lands and pockets of the bearing of Figure 3.25 gives:

$$T_t = \pi\omega\eta D^3\{[2W + (L - 2W)NW_d/(\pi D)]/(4h_o) + [(L - 2W)(1 - NW_d/(\pi D))]/(4h_p)\} \quad [3.146]$$

and a power consumption as given by Eq. 3.143.

3.13.3 Pumping power

Pumping power is simply the product of supply pressure and volume flow rate:

$$\text{Ш}_p = P_s.dV_i/dt \qquad [3.147]$$

where dV_i/dt is the volume flow rate associated with bearing number, i. The total power delivered by the supply is therefore:

$$\text{Ш}_p = P_s.\Sigma_1^N dV_i/dt \qquad [3.148]$$

where there are N bearings as part of the system.

3.13.4 Temperature effects

Both the pumping power and the friction power end up as heat in the bearing, the surrounding machine structure and the oil. An upper bound estimate of the oil temperature rise in passing through the bearing may be made by assuming all of the heat is dissipated in the oil:

$$\Delta T_i (\text{Ш}_p + \text{Ш})/(\rho s dV_i/dt) \qquad [3.149]$$

where ρ is the density of the oil, s is its specific heat, Ш is the friction loss, Ш_p the pumping loss and dV_i/dt the flow rate associated with bearing. The use of an upper bound estimate to oil temperature is supported in part by [14], in which bearing temperature effects were experimentally investigated for the case of a hydrostatic journal bearing. One of the findings was that the oil accounted for approximately twice the heat removal from the bearing than the surrounding structure, and that this did not vary much with bearing speed. The delivery temperature of the oil is controlled by the power supply and for ultra precision bearing applications will be set at a value required to maintain bearing running temperatures as close to ambient as is practical. In some cases, this may require an oil delivery temperature of less than 20 °C, in which case heat returned to the supply in the oil is removed by a cooler that is normally part of a refrigeration circuit.

Temperature rise due to pumping losses alone is relatively low, being approximately 6 °C per 100 Bar of supply pressure. The rise is proportional to delivery pressure but more or less independent of flow rate, as both pumping power and the heat capacity of the oil are proportional to flow rate and balance each other out. For moderate and high speed hydrostatic bearings, friction power can be high and produce a significant temperature rise in the oil. Although the temperature of the oil delivered to the bearing is controlled by the supply, the question is what is the effective temperature and hence oil viscosity to use for bearing performance calculations, as oil viscosity has a direct effect on both damping and power consumption. Using Eq. 3.149 to estimate oil temperature rise and assuming this defines the bearing operating condition, enables an approximate equilibrium temperature and viscosity to be found by iteration.

3.13.5 Bearing power consumption

The designed operating gap of a hydrostatic bearing affects many of the performance parameters important to ultra precision bearings – both static stiffness and damping increase with reducing gap, conversely motion error and bearing friction reduce with increasing gap. However, the motion error is not inherently dependent on gap, it also depends on manufacturing precision – how round or flat bearing surfaces are machined and how well pocket pressures are balanced. Thus the primary trade off is between a low gap giving high stiffness and damping and a high gap reducing bearing friction.

Figure 3.41 is an example of how total power loss of a given design of journal bearing design depends on gap. Total power loss is the sum of the bearing friction power and the pumping power. At low speeds, bearing friction is negligible and the power loss is predominantly pumping loss. In this case, reducing gap not only improves damping and stiffness but also minimises power loss by reducing oil flow rate. Power losses are also low, being tens of watts for gaps of less than 20 μm on the bearing considered. In the case of a journal bearing, power loss increases with speed squared, so operating with small gaps results in an impractically high temperature rise even at modest speeds. Figure 3.42 shows the

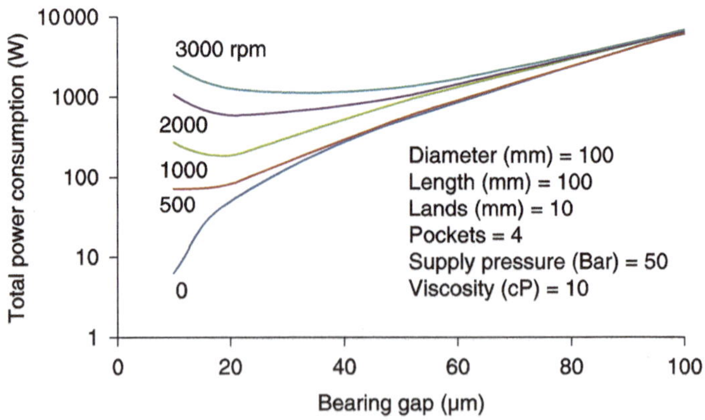

Figure 3.41 Example of the effect of bearing clearance on power consumption

temperature rise associated with the bearing of Figure 3.41, where a gap of 20 μm produces a temperature rise of 76.5 °C at a speed of 3000 rpm.

When bearing speeds are moderate, total power consumption can be minimised by choosing a gap to give the best trade off between pumping and bearing friction losses. The higher the bearing speed, the higher the optimum gap. For the bearing of Figure 3.41, a gap of 35 μm minimises the total power loss at 3000 rpm to a value of 1000 Watts, giving a temperature rise of just over 10 °C. At 2000 rpm, the optimum gap is 20 μm, giving a lower power consumption of 600 Watts but a higher temperature rise of 36 °C.

The example considered illustrates the significance of power consumption on hydrostatic bearings required to operate at speed, but does not present an optimum bearing design, this being strongly dependent on the specific requirements of the application. Where high or even moderate speeds are an important requirement, power consumption can be further reduced by attention to type of compensation, land and pocket design. The reader is referred to the work of Mohsin [6] and Mohsin and Ashley [15–17], aimed at increasing the maximum operating speed of hydrostatic bearings, for a more comprehensive explanation of this aspect of performance. Mohsin used diaphragm valves [6], to increase bearing stiffness over and above that obtainable with fixed designs of compensation element; serrated lands to reduce viscous drag and pocket inserts to reduce turbulence and cavitation in the bearing pockets. Using these developments, Abdolmaleki et al. [12] developed a machining spindle with a 75 mm diameter shaft capable of running at 13 300 rpm.

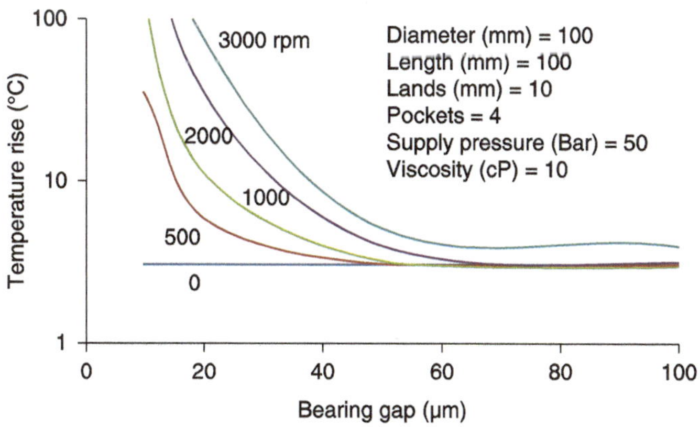

Figure 3.42 Example of the effect of bearing gap on temperature rise

3.13.6 *Thermal model*

For the purpose of estimating thermal distortion in machine structures, assuming a bearing temperature equal to the oil temperature given by Eq. 3.149 provides a reasonable starting point. When oil flow rates are high, the heat removal capacity of the oil will be high compared to that of the structure surrounding the bearing, so the effect of the latter may be neglected.

For low oil flow rates, a reasonable estimate of thermal distortion is found by iteration, first assuming the temperature given by Eq. 3.149 is the bearing temperature, determining the power dissipated by conduction through the structure produced by this temperature, subtracting the power lost by conduction from the total power loss and recalculating the bearing temperature.

3.14 Reliability

Hydrostatic bearings provide a robust solution to precision bearing applications. They have high load carrying capacity that provides a wide margin of safety against accidental loading, so much so that they are not usually the weakest element in the system. Oil supplies are also fitted with a number of safeguards to ensure reliable operation. Accumulators guard against oil pressure failure and filters prevent contamination. Accumulators allow time for the system to be shut down and bearing motion to cease before metal–metal contact occurs. Temperature and pressure sensors are also used to detect any drift in operating condition beyond permissible limits.

With oil pressure supplied there is no metal–metal contact and hence no bearing wear. The greatest risk to bearing failure in most cases is incidental damage when not in use. Without oil pressure supplied, metal–metal contact does occur and surface damage or wear can be produced by small relative movements. In this event bearing material selection minimises risk.

3.15 References

1. Stansfield, F.M. (1970), *Hydrostatic Bearings for Machine Tools*, Brighton, UK: The Machinery Publishing Company Co. Ltd.
2. Rowe, B.W. (2012), *Hydrostatic, Aerostatic and Hybrid Bearing Design*, UK: Butterworth-Heinemann.

3. Durazo-Cardenas, I.S., Corbett, J. and Stephenson, D.J. (2010), 'The performance of a porous hydrostatic journal bearing', *Proceedings of the Institute of Mechanical Engineers: Part J, Journal of Engineering Tribology*, **224(1)**: 81–9.
4. Kane, N.R., Sihler, A.H. and Slocum, A.H. (2003), 'A hydrostatic rotary bearing with angled surface compensation', *Precision Engineering*, **27**: 125–39.
5. Royle, J.K. (1962) 'Applications of automatic control to pressurised oil-film bearings', *Proceedings of the Institute of Mechanical Engineers*, **176(22)**: 532–41.
6. Mohsin, M.E. (1962) *The use of Controlled Restrictors for Compensating Hydrostatic Bearings, Advances in Machine Tool Design and Research*, Oxford: Pergamon Press.
7. *www.hyprostatik.de*
8. Reynolds, O. (1886) 'On the theory of lubrication and its application to Mr Beauchamp Tower's experiments, including an experimental determination of the viscosity of olive oil', *Philosophical Transactions of the Royal Society*, **177**: 157–234.
9. Hamrock, B.J. (1994), *Fundamentals of Fluid Film Lubrication*, McGraw Hill.
10. Roelands, C.J.A. (1966), 'Correlation aspects of the viscosity-temperature-pressure relationship of lubricating oils', PhD thesis, The Netherlands: Technisch Hogeschool.
11. Dwight, H.B. (1968), *Tables of Integrals and Other Mathematical Data*, 4th edition, New York: The MacMillan Company.
12. Abdolmaleki, M., Skorin, A., Wardle, F.P. and Wood, R.A.E. (1989), 'An experimental comparison between the performance of a "total cross flow" and equivalent conventional design hydrostatic journal bearing', *Paper XIV(v)*, Leeds, UK: Leeds-Lyon Tribology Symposium, 445–53.
13. Taylor, G.I. (1923), 'Stability of a viscous liquid contained between two rotating cylinders', *Philosophical Transactions of the Royal Society, London Series A*, **223**: 289–43.
14. Kher, A.K. and Cowley, A. (1970), 'An experimental investigation into the temperature effects in hydrostatic journal bearings', *Tribology*, **3(1)**: 165–228.
15. Mohsin, M. and Sharrat, A. (1981), 'The behaviour of hydrostatic pads with grooved lands', *Tribology*, **14(1)**: 33–45.
16. Mohsin, M. and Sharrat, A. (1980), 'The behaviour of a total cross flow hydrostatic thrust bearing', *Proceedings of the 21st International Machine Tool Design Conference*, Kampur.
17. Mohsin, M. and Sharrat, A. (1982), 'The behaviour of a total cross flow hydrostatic journal bearing', *Proceedings of the 23rd International Machine Tool Design Conference*. (Q. ref. 16 above)

3.16 Notation

A – Area
A_d – Area of radial land

A_p – Pocket area
a – Width
a_d – Amplitude of lobing or waviness on rotating surface
a_h – Amplitude of lobing or waviness on housing surface
B – Width of rectangular pocket
b_j – Coefficient
C – Damping coefficient
C_1,C_2 – Constants
D,d – Diameter
D_i – Bore diameter
D_o – Outside diameter
e – Eccentricity
F_a – Axial load
F_c – Load capacity
F_x,F_y – Load in direction x, y
G_o – Constant
h – Land gap
h_p – Pocket depth
h_o – Static equilibrium land gap
K – Static stiffness
L – Length of rectangular pocket or cylindrical journal bearing
L_T – Total length of rectangular lands around a pocket
M_d – Moment due to damping force
M_{xx}, M_{yy} – Moment about *x,y* axis
N – Number of pockets
P – Pressure
P_a – Ambient pressure
P_{av} – Average pressure
P_{avd} – Average pressure on axial or radial land
P_c – Pressure on circular land
P_p – Pocket pressure
P_s – Supply pressure
Q – Load
Q_a – Axial load
R – Radius
R_c – Capillary resistance
R_i – Bearing Bore Radius
R_o – Bearing's external radius
r_i – Radius of circular land's inner edge
r_o – Radius of circular land's outer edge
s – specific heat
T – Torque
t – time
t_m – Temperature
U,u – Relative surface speed
V – Volume
W – Width of land
W_d – Width of axial or radial land

W_e – Width of circular land on journal bearing
W_i – Width of inner land on annular thrust pad
W_o – Width of outer land on annular thrust pad
X,Y,Z – Cartesian coordinates

Symbols

Á – Fluid admittance associated with inner, outer or end lands
Ác – Fluid admittance associated with capillary restrictor
á – Fluid admittance associated with axial or radial land
Δh – Perturbation to bearing gap
ΔT – Temperature rise
ξ – Load Coefficient
η – Dynamic viscosity
f – Friction force
φ – Angular position
Ŕ – Fluid resistance associated with inner, outer or end lands
ŕ – Fluid resistance associated with axial or radial lands
θ – Angular tilt
ρ – Density
τ – Shear stress
Ш – Power

Subscripts

I, j, k, – integers
c – capillary

4

Aerostatic bearings

DOI: 10.1533/9780857092182.227

Abstract: Aerostatic bearings, often referred to as 'air bearings', not only require small clearances to function satisfactorily but the clearances also need to be accurately controlled to a small fraction of their value to achieve designed levels of performance. This imposes stringent demands on the manufacture of bearing components as well as the housings, shafts and other parts controlling their set-up. The term 'air bearing' is usually understood to mean 'air spindle', a complete system in which all of the critical components have been integrated by the bearing manufacturer. Despite the demands of manufacture, there is still a requirement for highly specialised air bearing designs in instrumentation, scientific equipment and bespoke manufacturing machinery. In recent years, air bearings have become available in the form of components or inserts simplifying ultra precision equipment design and assembly. This chapter briefly reviews the types of air bearing available and provides basic design calculations for those that are more widely used, namely the rectangular pad, annular thrust and cylindrical journal.

Key words: externally pressurised bearings, aerostatic bearings, air bearings, air spindles, fluid film bearings, high speed bearings, low motion error.

4.1 Introduction

Aerostatic bearings rely on an external source of high pressure gas, normally air, to generate load carrying capacity. In the vast majority of aerostatic bearing applications, the gas is compressed air sourced from a factory air line for which delivery pressures are typically 5 to 6 Bar. This is a relatively low pressure to supply to a bearing surface and as a consequence the load carrying capacity of aerostatic bearings is low compared to rolling elements or hydrostatic bearings of a similar diameter. Occasionally, in applications where load capacity or stiffness is critically important, aerostatic bearings may be used with dedicated compressors and operated at pressures of up to 10 Bar. Higher pressures can be generated by using bottled gas but in general this source of air is not practical or economic in industrial environments and is generally confined to the laboratory.

The main advantages in lubricating a bearing with compressed air are low friction, almost zero at low speed, and good averaging of the effect of bearing surface imperfections on the distribution of pressure within the bearing. As a result of the latter, aerostatic bearings produce the lowest motion errors of any bearing type. Bearings with a total motion error of less than 10 nm peak-peak are routinely produced by some manufacturers, thus enabling air bearings to compete very favourably with other bearing types in ultra precision machining applications, where high dimensional precision and optical quality surfaces are required. The low friction characteristics also give them an extremely high speed capability. Presently air bearing spindles used in the printed circuit board drilling industry are attaining rotational speeds in excess of 400 000 rpm.

As the viscosity of air is very low, the lubricating film thickness, or internal clearance within an aerostatic bearing, must be kept low in order that a reasonable working pressure can be maintained within the bearing, and the air consumption can be kept down to an acceptable level. Typically film thicknesses are in the 5 to 20 μm range, but occasionally may be as low as 2 μm or as high as 50 μm. With such small clearances, aerostatic bearing parts must be accurately made, bearing surfaces well finished and assembly expertly performed. Thus in the aerostatic bearing industry, there is a tendency for manufacturers to produce bearing assemblies such as spindles, rotary tables and linear slides designed to meet specific market requirements and to avoid the need for customers to deal with some highly specialised aspects of ultra precision engineering. Such assemblies not only include aerostatic bearings but may also have drive motors, encoders, sensors and work or tool holding devices integrated into an overall package. Whilst this has been a development trend within the industry, there is still a requirement for highly specialised aerostatic bearings for use in bespoke machinery, instrumentation and scientific equipment. Here manufacturers can choose to design and build their own bearings or to use bearing inserts available from some of the air bearing manufacturers.

With very low motion errors and low friction resulting in low thermal distortion, aerostatic bearings are used in extreme precision applications such as metrology equipment (roundness measuring and coordinate measuring machines), diamond machining of optical components, ultra precision grinding, dicing of brittle materials such as silicon wafers and testing of hard disc drives. They are also used at extremely high speeds in bore grinding and printed circuit board drilling machines. Each of these applications imposes a different set of requirements upon the bearing design and the spindles optimised for each application are quite distinctive.

In this chapter, the basic types of aerostatic bearing in common use are described and their operating principles explained. A method for calculating their load capacity and static stiffness is given and the mechanisms that generate motion error and damping also described.

4.2 Bearing geometries and arrangements

Air bearings are manufactured in a variety of geometries having cylindrical, conical spherical, flat annular, circular or rectangular forms. High manufacturing precision of the air bearing components is needed in order to achieve the essential small air gap separating moving and stationary surfaces. If the gap is typically 10 μm, then bearing dimensions must be controlled to better than 1 μm to maintain the gap at a near constant value necessary to consistently achieve the designed performance. In practice, this level of precision restricts bearing geometries to the simple forms for which high precision machining techniques are well developed.

Figure 4.1 gives examples of air bearing arrangements in current use. By far the most common arrangement is the combination of cylindrical journal and annular thrust bearings (Figures 4.1(a and b)). Figure 4.1(a) shows the bearing arrangement traditionally used in air bearing spindles for high precision machining. It comprises two cylindrical journals that support radial and moment loads and a double-sided thrust bearing that supports axial load in two directions, as well as providing axial location of the rotor. The thrust bearing is usually placed close to the working end of the spindle to minimise the length of shaft for which heating and

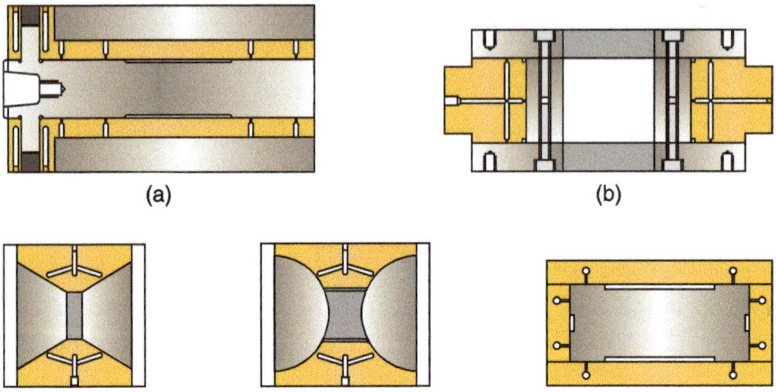

Figure 4.1 **Examples of air bearing geometries and arrangements: (a) spindle with cylindrical journals and annular thrust bearings; (b) rotary table with cylindrical journal and annular thrust bearings; (c) spindle with conical bearings; (d) spindle with spherical bearings; and (e) slide with rectangular flat pads**

thermal growth affects machining accuracy. For ease of manufacture, the two journal bearings are often made the same in diameter and machined into a single sleeve, thus minimising misalignment between their axes. For this reason, ultra precision air spindles with integral drive motors usually have them placed outboard of the journal bearings at the spindle's non-working end.

There are many variations of air spindles with double thrust and journal bearings. Figure 4.1(b) shows an arrangement often used in rotary tables designed to support predominantly axial loads. In this case a single journal supports any incidental radial loads and provides radial location of the rotor. As the journal is short in length, a thrust bearing can be placed at either end to take the axial load and provide axial location. As for the spindle in Figure 4.1(a), this bearing arrangement is chosen for its ease of manufacture. In this case the large diameter thrust bearings means that it is critically important to control the squareness of locating faces with respect to the axis of the journal bearing. Grinding machines and manufacturing techniques for achieving micron level tolerances on squareness are widely available for this task.

Air bearings having tapered (Figure 4.1(c)) or spherical (Figure 4.1(d)) bearing geometries tend to be confined to highly specialised applications such as disc verification spindles used in the hard disc drive industry. These bearings have 'cup' and 'cone' forms that can be lapped against each other to achieve an extremely parallel gap. They are normally used in a back–back arrangement (Figure 4.1), where the magnitude of the gap can be accurately controlled simply by grinding a flat spacer positioned centrally between the cones. The two bearings are normally positioned close together, so that the effect of thermal expansion on bearing clearance is kept to a minimum. The spherical bearing form (Figure 4.1(d)) has an added advantage in that it is completely unaffected by misalignment of any of the bearing surfaces, but this benefit is offset by limited load capacity and stiffness resulting from a length to diameter ratio, which must be less than one. For both conical and spherical bearing types, the back–back arrangement is sensitive to axial thermal expansion, even if the two bearings are placed close together. In practice, this confines their use to relatively low speed applications where heat generation is negligible.

Figure 4.1(e) shows a 'box' section bearing geometry frequently used on air bearing slides. The bearing arrangement comprises four pairs of flat rectangular pads positioned adjacent to the corner edges of the rectangular beam. Pads on opposite sides of the beam are identical in order to centre the beam. To maintain a constant bearing gap, components must be manufactured with good flatness, parallelism and squareness of the critical surfaces. Furthermore, the gap for the bearing pads on upper and lower surfaces is set by matching the width of the beam to the height of the side plates and this requires high precision machining.

Whilst air bearings are made in a variety of geometries, their basic principle of operation remains the same and the factors that influence their design and performance are similar. In the following sections, the focus is on cylindrical journal and flat annular bearing forms.

4.3 Principle of operation

Most aerostatic bearings are supplied with pressurised air through feed holes (orifices) arranged in a single row located at the bearing centre or two rows, each offset from the centre. The outer edges of the bearing are open to the atmosphere and the pressurised air is intentionally allowed to leak out. Because the air is a viscous fluid and the bearing gap is small, a significant pressure is maintained in the central region of the bearing that enables it to support load. Figure 4.2(a) shows a typical pressure profile in an unloaded cylindrical journal bearing fed with air

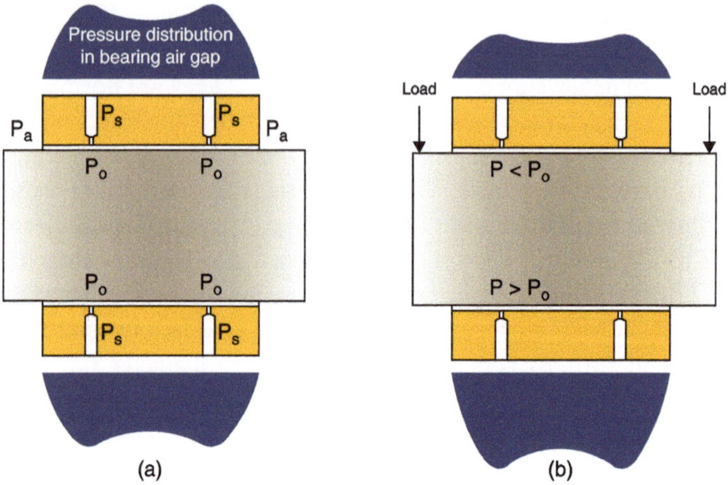

Figure 4.2 Pressure distribution in a journal bearing

through two rows of orifices. As the shaft is located centrally in the journal, the air gap is constant around and along the bearing's surface, so the pressurised air flows from the orifices substantially in an axial direction towards the ends of the bearing, setting up a relatively steep pressure gradient in this region. In the central region of the bearing, between the two rows of orifices, the air pressure is almost constant at an average pressure slightly lower than the inlet pressure. The reason is due to a locally divergent air flow pattern and hence a local pressure drop around each orifice.

To enable the bearing to support the load and to generate static stiffness, the resistance to air flow of the two rows of orifices is matched to the resistance to air flow of the bearing gap. This is done by designing the unloaded bearing to have a known pressure P_o in the bearing gap immediately downstream of the orifices. Common practice is to make P_o equal to or slightly greater than half of the air supply pressure P_s.

The application of a radial load to the bearing displaces the shaft in a radial direction, (Figure 4.2(b)), producing an air gap that varies with angular position around the journal. Where the air gap is closed down on the side of the bearing beneath the load vector, its resistance to air flow is locally increased, so most of the pressure drop occurs in the bearing gap, increasing P_o. Conversely, on the opposite side of the bearing where the gap is increased, the local resistance to air flow is reduced and the local value of the pressure P_o is decreased. The magnitude of P_o changes with the radial displacement of the shaft and provided the applied load is within the design limits of the bearing, equilibrium is reached where the pressure difference, integrated over the bearing area, is equal to the load. In a well-designed bearing, the pressure P_o approaches the air supply pressure P_s as the bearing gap locally reduces to zero. Conversely, P_o approaches atmospheric pressure P_a as the bearing gap locally increases to twice the design gap.

The mechanism that creates a pressure difference across the bearing as the shaft is displaced is often termed 'compensation'. It is an important aspect of aerostatic bearing design, determining the static load capacity of the bearing, its static stiffness and influencing its motion error.

4.4 Types of compensation

There is no one ideal compensation system and many different designs are in current use, each having its own particular set of advantages and disadvantages. That described in the preceding section based on one or two rows of orifices is used on a large majority of aerostatic bearings, as it is relatively straightforward to design and manufacture. The main objective in designing a compensation system is to maximise the static load carrying capacity or the static stiffness of the bearing without compromising its stability or incurring an unacceptably high air consumption. These requirements are approached when the pressure distribution in an unloaded bearing is uniform over as much of its area as possible and when the change in average bearing pressure with applied load is a maximum. Ideally, the average pressure in the bearing, as the gap is closed to zero, should approach the air supply pressure. Conversely, as the gap is increased to twice the designed or 'no load' gap, the average pressure should fall to zero. If these conditions were to be achieved, then the load capacity of the bearing would simply be given by the product of air supply pressure and bearing area. However, practical designs of compensation systems fall some way short of the ideal.

4.4.1 Inlet restriction

Inlet restriction is the term used to describe a compensation system in which a resistance to air flow is placed upstream of the bearing (Figure 4.3). The inlet restriction R_r is designed to reduce the air supply

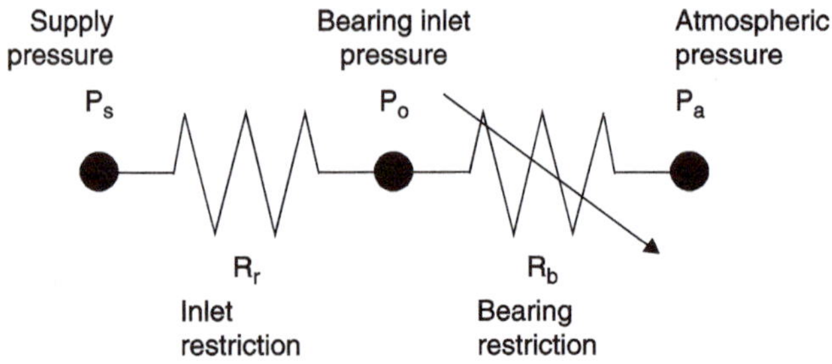

Figure 4.3 Inlet restriction compensation

pressure P_s to P_o at inlet to the bearing gap for some specified bearing load condition, usually no load. As the bearing restriction R_b is load dependent, then the inlet pressure P_o increases or decreases with load, providing the required compensation.

Figure 4.4 shows three designs of compensation systems based on inlet restriction that have proved successful in practice. The diagrams each show a 90-degree arc of a journal bearing unwrapped and the axial and angular pressure profiles generated by the compensation arrangement. The axial pressure profiles are for loaded bearings and show the pressure distribution on opposite sides of the bearing in line with the load vector.

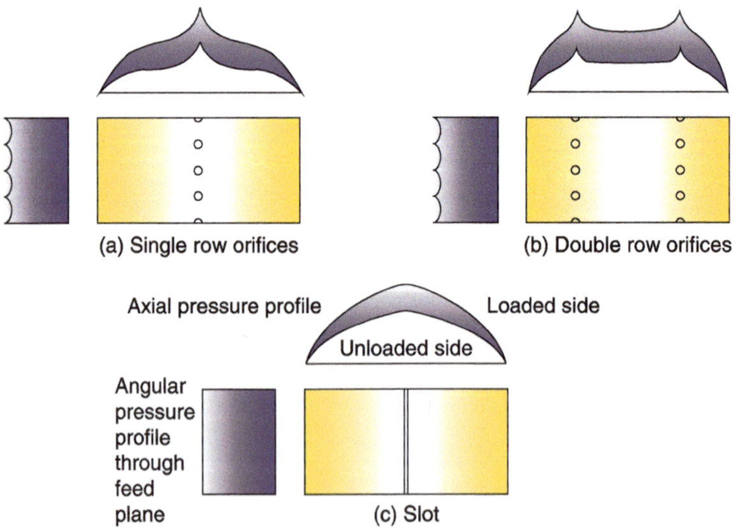

Figure 4.4 Practical designs of inlet restriction compensation

The shaded area is therefore proportional to the applied load. The angular pressure profiles are for an unloaded bearing and show how the compensation arrangement affects pressure distribution in the direction of motion.

Perhaps the simplest design and least expensive to manufacture is a single row of orifices positioned in a central radial plane (Figure 4.4(a)). However, in this arrangement the pressure in the bearing gap decays steadily towards the ends of the bearing, as shown by the axial pressure profiles. Local to the orifices, the rate of decay is particularly rapid as the air flow is divergent and this limits the average pressure in the bearing and hence its load capacity as represented by the shaded area between the two axial pressure profiles. Around the bearing, divergent air flow local to the orifices creates a series of pressure spikes, with the position of maximum pressure coincident with the orifices. This has some affect on bearing load capacity, but of more significance can be the effect on motion error.

Compared to the single row, the double row of orifices (Figure 4.4(b)) adds a small amount to bearing manufacturing costs but considerably improves both the bearing load capacity and static stiffness. The trade off is well worthwhile and this is by far the most widely used compensation arrangement in current use. Row position is optimised but is typically within 25% of the bearing length away from the end of the bearing. Whilst pressure decays rapidly between the row and bearing end, it remains nearly constant between the rows. The area between the two axial pressure profiles shown in Figure 4.4(b) and hence the bearing's load capacity, is significantly improved compared to that of Figure 4.4(a). However, increasing the number of rows of orifices has little effect on the pressure variation around the bearing where pressure spikes occur in the plane of the orifices.

One compensation system that provides a constant pressure in the feed plane of an unloaded bearing is the slot fed arrangement shown in Figure 4.4(c). However, slot fed bearings are significantly more expensive to manufacture than the orifice fed bearings shown in Figures 4.4(a) and (b). The reason is that to match the air flow resistance of the bearing gap, the slot must be extremely narrow, of the order of 10 μm, and several mm in depth. To achieve such a narrow gap requires the bearing journal to be manufactured in two pieces and butted together with the slot machined into one of the abutment faces. Slot fed bearings are not considered further in this chapter and the reader is referred to [1] for a more comprehensive description of this type of bearing and how it should be designed.

4.4.2 Porous materials

Porous media bearings are another form of inlet restrictor bearing that overcome the pressure profile limitations of orifice fed bearings. By manufacturing the complete bearing journal from a porous material (Figure 4.5(a)) and supplying pressurised air to its external surface, enables very uniform pressure profiles to be generated in both axial and angular directions. Porous air bearings are known for their high load capacity, static stiffness and low motion errors [2]. Their main disadvantage is that porous materials, normally graphite or ceramic, are not ideal bearing materials. They do not tolerate overload and touchdown as well as conventional materials and the fine pores required to match the bearing restriction demand high standards of air cleanliness.

The touchdown limitation is largely overcome by using porous inserts (Figure 4.5(b)). The load bearing part of the surface is manufactured from a conventional air bearing material such as lead bronze and the porous material is moulded into two narrow slots machined into the bearing journal. The arrangement is equivalent to the double-row orifice bearing of Figure 4.4(b) but now pressurised air is introduced into the bearing gap uniformly around its circumference, eliminating the pressure spikes produced by the divergent air flow associated with orifices. The axial and angular pressure profiles are similar to the full porous media bearing of Figure 4.5(a) and therefore it also has the potential for high load capacity, static stiffness and low motion error.

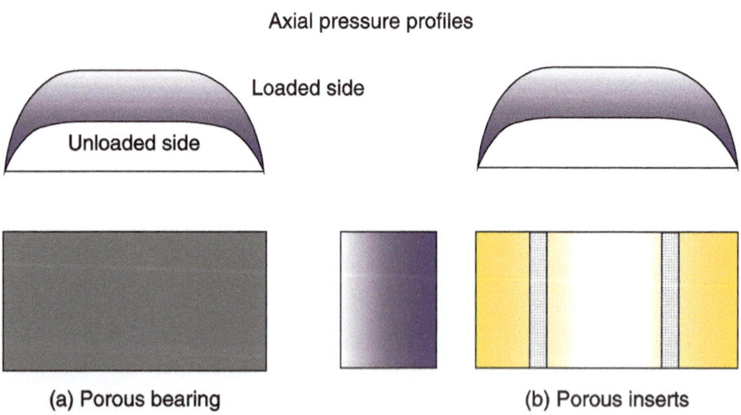

Figure 4.5 Compensation by use of porous materials

4.4.3 Outlet restriction

Compensation can also be achieved by placing a restriction to air flow downstream of the bearing, in which case the full air supply pressure P_s is introduced to the bearing inlet. The downstream restriction may take the form of a narrow land around the end of the bearing, where the average gap is designed to be less than that within the central region of the bearing (Figure 4.6). Any change in gap produced by the application of a load changes the resistance to air flow of the central region of the bearing as well as that of the land. However, since the gap over the land is less than that within the bearing, then any change in gap has a proportionately larger effect on the land's resistance to air flow than on that within the bearing. As the gap is reduced, the land's resistance to air flow increases, resulting in the pressure at inlet to the land, P_l and hence the average pressure within the loaded side of the bearing to increase. Conversely, as the land gap increases, the resistance to air flow reduces, allowing pressure to fall within the unloaded side of the bearing.

In effect, for outlet restrictor bearings, the restriction is built into a bearing surface and this requires a surface that is not perfectly cylindrical

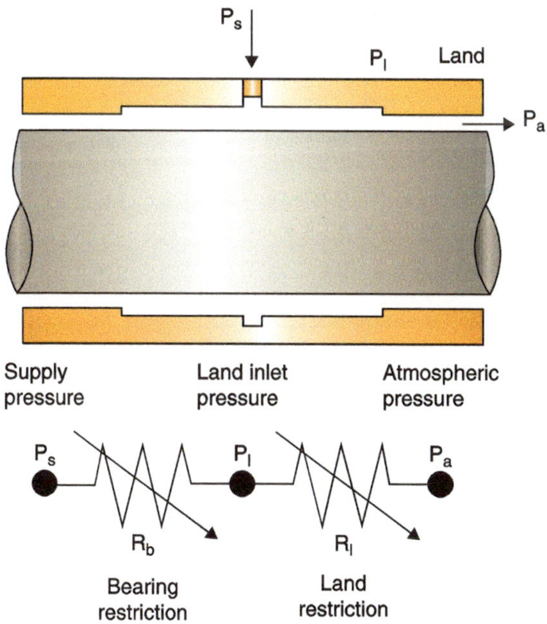

Figure 4.6 Compensation by outlet restriction

or flat in the case of annular thrust or slide bearings. Instead the surface contains carefully designed features, such as steps or grooves that are shallow in depth and require specialised manufacturing techniques. There are a variety of techniques in use, individual bearing manufacturers having refined the technique best suited to their requirements.

Three types of compensation arrangement are shown in Figure 4.7. The bearing diagrams each show a 90-degree arc of a journal bearing unwrapped. Above the diagrams are the axial pressure profiles on the loaded and unloaded side of the bearing and to the left-hand side of the diagrams is the no load angular pressure profile.

Figure 4.7(a) shows a stepped bearing that has a narrow annular slot at its geometric centre into which air at the supply pressure P_s is introduced. The bearing is designed to have a particular operating gap over its central region, and this gap is 'stepped' down or reduced in level towards the ends of the bearing. Air pressure within the central feed slot remains at the supply pressure P_s for all load conditions. However, when load is applied to the bearing, the land gap reduces on the loaded side of the bearing locally, increasing its resistance to air flow and building up pressure within the adjacent central region of the bearing. The axial pressure profile is almost constant at a level approaching the supply pressure P_s. On the unloaded side of the bearing the land gap is increased and the local

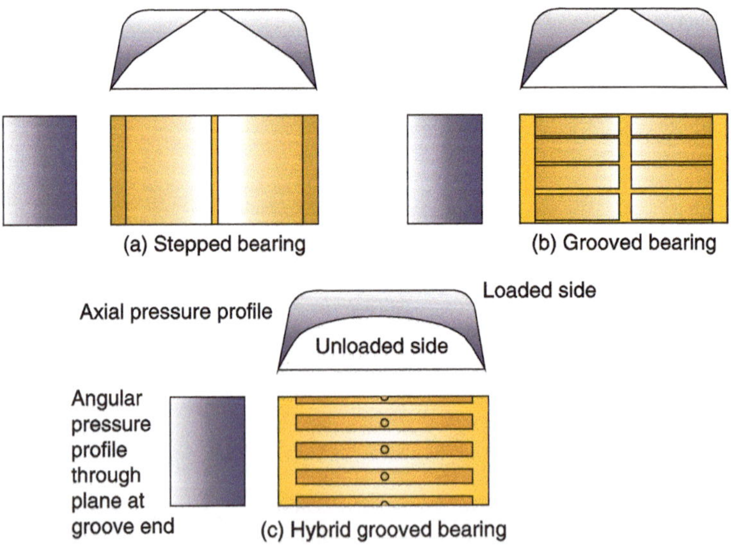

Figure 4.7 Compensation employing outlet restriction

resistance to air flow is reduced. The pressure at entry to the land is also locally reduced, creating a pressure profile on the unloaded side of the bearing that reduces steadily from the supply pressure P_s at the centre of the bearing to atmospheric pressure P_a at the end of the bearing. The shaded area between the two axial pressure profiles in Figure 4.7(a) is proportional to the bearing's load capacity. The no load angular pressure profile in any radial plane is constant in level.

In having a relatively large gap over its central region compared to that over the lands at its ends, the stepped outlet restrictor bearing of Figure 4.7(a) does not attain a high load capacity. Under load, the high pressure air on the loaded side of the bearing is relatively free to flow around the bearing's circumference to the low pressure region on its unloaded side, thus reducing the average pressure difference across the bearing. This phenomenon is termed 'cross flow' and is a generally unwanted feature of most air bearings.

The grooved outlet restrictor bearing of Figure 4.7(b) considerably reduces cross flow and can support much higher loads than the stepped bearing of Figure 4.7(a). The bearing was invented by Arneson in 1966 and was the subject of a patent [3]. This bearing also has a central annular groove supplied with pressurised air and lands at each of its ends providing outlet restriction. However, the main difference is that it has a series of fine axial grooves extending from the central annular groove to the inboard edges of the lands. The restriction to air flow over the central region of the bearing is largely determined by the depth and width of the axial grooves and is matched to the land's restriction to air flow determined by the land gap and width. Under applied load, the air pressure in the grooves develops similar axial pressure profiles to that of the stepped bearing. Only now the ridges between the grooves restrict cross flow of air, enabling a much higher average pressure difference to occur. The no load angular pressure profile of the grooved outlet restrictor bearing is not perfectly constant in level, but exhibits a small ripple where, over most of the bearing's length, the pressure over the centre of the ridge is slightly less than the pressure in the grooves.

In both of the bearing designs of Figures 4.7(a) and (b), the pressure at the centre of the bearing is constrained to be the supply pressure P_s and this leads to an axial pressure profile on the unloaded side of the bearing that reduces almost linearly over the half length of the bearing. If, on this type of bearing, the supply pressure was allowed to reduce on its unloaded side, then the bearing could support a significantly higher load.

The bearing of Figure 4.7(c) combines inlet and outlet restriction. It has axial grooves and outlet restrictor lands as for the bearing of

Figure 4.7(b), but instead of the air being fed into a central annular groove at the supply pressure P_s, it is fed through a central row of orifices. A shallow axial groove extends in both directions from each orifice to lands at the end of the bearing. The orifices provide inlet restriction so that under load, the pressure at inlet to the bearing gap may change in level. On the loaded side of the bearing, the inlet pressure approaches the supply pressure P_s and the axial pressure profile is similar to that of the bearing of Figure 4.7(b). However, on the unloaded side of the bearing, the inlet pressure falls and the axial pressure profile has a lower average value than for the bearing of Figure 4.7(b). The larger pressure difference across the bearing thus generates a significantly higher load carrying capacity. But the no load angular pressure profile is similar, exhibiting a low amplitude ripple superimposed on a large static value.

4.5 Approximate load capacity and static stiffness

It is often useful to have an estimate of the static stiffness and load capacity of an air bearing prior to embarking upon its detailed design. This allows the machine designer to evaluate the bearing–shaft system's performance and find a near optimum bearing size and arrangement before committing to lengthy bearing design calculations. Load capacity can be simply expressed in terms of air supply pressure, P_s, bearing area and an empirical constant, ξ. For an annular thrust bearing, the load capacity is given by:

$$F_c = \xi.P_s.\pi(D_o^2 - D_i^2)/4 \qquad [4.1]$$

where D_o and D_i are the outer and inner diameters of the bearing respectively. ξ is often referred to as a 'Load Coefficient' but may be regarded as a measure of the effectiveness of the compensation system. In general, $0.3 < \xi < 0.4$, where 0.4 applies to a well-designed bearing for which the compensation system has been optimised. The bearing's static stiffness requires the 'no load' equilibrium gap, h_o to be known and is given by:

$$K = \xi.P_s.\pi(D_o^2 - D_i^2)/(4h_o) \qquad [4.2]$$

For a journal bearing:

$$F_c\ \xi.P_s.\pi.L.D \qquad [4.3]$$

$$K = \xi.P_s.\pi.L.D/h_o \qquad [4.4]$$

where the projected bearing area is the product of diameter, D and length, L and $0.3 < \xi < 0.4$ as for the thrust bearing. For a slide bearing with a rectangular pad of length L and width W:

$$F_c\ \xi.P_s.\pi.L.W \qquad [4.5]$$

$$K = \xi.P_s.\pi.L.W/h \qquad [4.6]$$

again $0.3 < \xi < 0.4$.

It should be noted that the dominant variables that determine the static load capacity of an aerostatic bearing are its area and the air supply pressure. Similarly for static stiffness, it is bearing area, air supply pressure and no load equilibrium gap that are dominant. It follows that bearings designed for high load capacity are large in area and bearings designed for high static stiffness may be large in area but must also have a small gap. Whilst ξ has an effect, the difference between a bearing with a carefully optimised compensation system and one for which the compensation system has not been optimised at all is relatively modest.

4.6 Load–deflection analysis

A more accurate calculation of bearing load capacity and static stiffness and optimisation of the compensation system requires the pressure distribution within the bearing gap to be determined at several different gaps produced by the action of an applied load. This is done by consideration of viscous fluid flow in a narrow gap. In 1886, Reynolds [4] derived the differential equations governing the pressure distribution of fluid flowing through a narrow gap and, in fact, the equations are still known as Reynolds equations today. Although in the original work Reynolds confined his attention to incompressible fluids, in 1913 Harrison [5] added in the effects of compressibility to make the equations more generally applicable. A full derivation of Reynolds equation and its application to fluid film bearing lubrication is given in [6].

Early in the development of air bearing technology, the air flow in the bearing gap was assumed to be one-dimensional (1-D), thus enabling the differential equations derived by Reynolds to be reduced to simple engineering formulae expressing pressure explicitly in terms of the physical properties of air, the bearing geometry and the boundary conditions of supply and ambient air pressure. This is described in the work by Powell [7] and used to determine the internal design of annular thrust and cylindrical bearings. However, in most practical air bearing designs, the air flow is not strictly 1-D and the simple formulae provide only an approximation to bearing performance. Attempts at applying empirical factors based on the work of Dudgeon [8] to compensate for two-dimensional (2-D) flow met with some success, but the rapid developments in computer technology and numerical analysis occurring at the same time that air bearings were being developed made it unnecessary to pursue this approach. Instead Reynolds equation is now solved accurately in two dimensions using standard numerical techniques. Stout et al. [9–11] were among the first investigators to publish this type of analysis.

4.6.1 Reduced forms of Reynolds equation

Reynolds considered fluid flow through a narrow gap between two parallel surfaces and developed general equations describing the relationship between fluid properties, pressure gradients, gap size and surface velocities. The equations are based on the assumptions:

- Inertia forces in the fluid, due to acceleration, can be neglected compared to forces produced by viscous shearing.
- Laminar flow conditions exist at all points in the fluid film.
- Pressure is constant over any section normal to the direction of fluid flow.
- The boundaries are solid and impervious.
- There is no slip in the boundaries between the fluid and the surfaces.

Reynolds equation is derived and explained in [6]. Here a standard, reduced form of Reynolds equation is taken as a starting point for determining the pressure distribution between surfaces separated by a small distance h:

$$\frac{\partial}{\partial x}\left[\frac{\rho h^3}{\eta}.\frac{\partial p}{\partial x}\right]+\frac{\partial}{\partial y}\left[\frac{\rho h^3}{\eta}.\frac{\partial p}{\partial y}\right]=12\left[u\frac{\partial}{\partial x}(\rho h)+v\frac{\partial}{\partial y}(\rho h)\right]+\frac{\partial(\rho h)}{\partial t} \qquad [4.7]$$

Equation 4.7 is the most general form of Reynolds equation used in this chapter. It expresses the pressure p in a narrow gap of width h in terms of Cartesian coordinates x and y, taking into account the effect of the surface velocity $u = (u_a + u_b)/2$, $v = (v_a + v_b)/2$ between the surfaces and squeeze film effects, $\partial h/\partial t$. In Eq. 4.7, ρ is density and η is viscosity. Written in terms of Cartesian coordinates, Eq. 4.7 is relevant to flat, rectangular bearing surfaces such as those used in linear slides. It can be simplified by using the standard equation for isentropic compression of a gas:

$$\rho = p/RT \qquad [4.8]$$

and by normalising pressure with respect to ambient pressure:

$$P' = p/p_a$$

where R is the gas constant and T is absolute temperature, both of which can be assumed to be constant for normal air bearing applications:

$$\frac{\partial}{\partial x}\left[P'h^3\frac{\partial P'}{\partial x}\right]+\frac{\partial}{\partial y}\left[P'h^3\frac{\partial P}{\partial y}\right]=12\eta\left[u\frac{\partial(P'h)}{\partial x}+v\frac{\partial(P'h)}{\partial y}\right]+\frac{\partial(P'h)}{\partial t} \qquad [4.9]$$

Further simplification of Eq. 4.9 can be made for the purpose of determining the static load–deflection characteristics of a linear slide subject to a centrally applied normal load. For this situation, the gap h is constant and relative surface velocities are low so that the terms on the right-hand side of the equation may be neglected:

$$P'.\frac{\partial^2 P'}{\partial x^2}+\left[\frac{\partial P'}{\partial x}\right]^2+P'\frac{\partial^2 P'}{\partial y^2}+\left[\frac{\partial P'}{\partial y}\right]^2=0 \qquad [4.10]$$

By putting $P=(P')^2$, Eq. 4.10 becomes:

$$\frac{\partial^2 P}{\partial x^2}+\frac{\partial^2 P}{\partial y^2}=0 \qquad [4.11]$$

Equation 4.7 may also be applied to annular thrust bearings; however, for the purpose of deriving a numerical solution, it is convenient to work in terms of radial polar coordinates:

$$r = x\cos\theta + y\sin\theta \qquad \text{and} \qquad \theta = \mathrm{Tan}(y/x)$$

that transform Eq. 4.7 to:

$$\frac{\partial}{\partial r}\left[rh^3 p\frac{\partial p}{\partial r}\right]+\frac{1}{r}\frac{\partial}{\partial\theta}\left[h^3 p\frac{\partial p}{\partial\theta}\right]=12\eta\left[v_r\frac{\partial(prh)}{\partial r}+v_\theta\frac{\partial(ph)}{\partial\theta}\right]+\frac{\partial(ph)}{\partial t} \qquad [4.12]$$

where $v_r=(v_{ra}+v_{rb})/2$ and $v_\theta=(v_{\theta a}+v_{\theta b})/2$ are the surface velocities in radial and circumferential directions. For the purpose of determining the axial static stiffness of circular and annular bearing geometries rotating about a central axis, h is constant, p is independent of time, $v_r = 0$ and v_θ is assumed to be small. Thus the terms on the right-hand side of Eq. 4.12 may be neglected. Putting $P = p^2$:

$$\frac{\partial P}{\partial r} + r\frac{\partial^2 P}{\partial r^2} + \frac{1}{r}\frac{\partial^2 P}{\partial \theta^2} = 0 \qquad [4.13]$$

For cylindrical journal bearings subject to a central radial load, the gap is not constant but varies simple harmonically with angular position. Normalising with respect to the designed bearing clearance h_o, the gap may be expressed as:

$$H = h/h_o = 1 + \varepsilon.\cos\theta \qquad [4.14]$$

where $0 < \varepsilon < 1$ is the eccentricity of the bearing. By transforming the Cartesian coordinates to cylindrical polar coordinates:

$$\theta = x/r \quad Y = y/r$$

Replacing Y by Z, ρ by p/RT and substituting in Eq. 4.7 gives

$$\frac{\partial}{\partial Z}\left[h^3 p\frac{\partial p}{\partial Z}\right] + \frac{1}{r^2}\frac{\partial}{\partial \theta}\left[h^3 p\frac{\partial p}{\partial \theta}\right] = 12\eta\left[\frac{r\omega}{2}\frac{\partial}{\partial \theta}(ph) + \frac{\partial}{\partial t}(ph)\right] \qquad [4.15]$$

where ω is the angular speed and $r\omega$ is the relative surface speed of the bearing. For low speeds and for h independent of time, the terms on the right-hand side of Eq. 4.15 may be neglected. Furthermore, for small eccentricities, $h^3 = h_o^3(1 + 3\varepsilon\cos\theta)$. Thus for the purpose of determining small amplitude static load–deflection characteristics, Eq. 4.15 can be simplified to:

$$\frac{\partial^2 P}{\partial \theta^2} + r^2\frac{\partial^2 P}{\partial Z^2} - \frac{\partial P}{\partial \theta}\frac{3\varepsilon.\sin\theta}{(1+3\varepsilon.\cos\theta)} = 0 \qquad [4.16]$$

Equations 4.10, 4.13 and 4.16 can be solved numerically to determine the pressure distribution in rectangular pad, annular thrust and cylindrical journal bearings respectively. The method of solution is the same for all three types of bearing and in the following two sections the pressure distribution in annular thrust and cylindrical journal bearings is derived as part of an overall solution scheme for determining their load capacity and static stiffness.

4.6.2 Axial load capacity and static stiffness of annular thrust bearings

Equation 4.13 describes the fluid pressure distribution between two flat surfaces separated by a small, constant gap, h. It can be used as part of a numerical solution scheme to determine the axial load capacity and static stiffness of circular and annular thrust bearings. The overall solution scheme entails:

a. guessing the pressure in the bearing film just downstream of the compensation element(s) (i.e. orifice);
b. thus enabling the pressures at inlet to, and outlet from, the bearing film to be defined;
c. solving Eq. 4.13 numerically to determine the complete pressure distribution in the film;
d. from the pressure distribution, determining the mass flow rate of fluid through the bearing film;
e. defining the pressure drop across the compensation element(s);
f. determining the mass flow rate of fluid through the compensation element(s);
g. comparing the mass flow rates through the bearing film and compensation element(s) and adjusting the guessed value of the downstream pressure if they are not equal;
h. repeating steps b to g, until the two mass flow rates are equal;
i. repeating steps a to h, for a pre-defined range of bearing gaps.

A straightforward numerical solution to Eq. 4.13 is obtained using a standard relaxation technique. This entails superimposing a fine grid over the bearing surface and determining pressure at each of the grid's node points. Not all of the bearing's surface needs to be considered, as for a concentrically loaded bearing there are axes of symmetry. For example, Figure 4.8 shows a mesh suitable for finding the pressure distribution in an annular thrust bearing supplied with air at pressure P_f through an array of equally spaced orifices located on a pitch circle diameter.

Through symmetry, it is only necessary to consider a segment of the bearing equal to 180 degrees/(2 × no. of orifices) where the orifice is positioned on one of the radial boundaries. For this situation the pressure distribution on adjacent segments is the mirror image of that for the segment shown. Equation 4.13 expresses pressure as a continuous

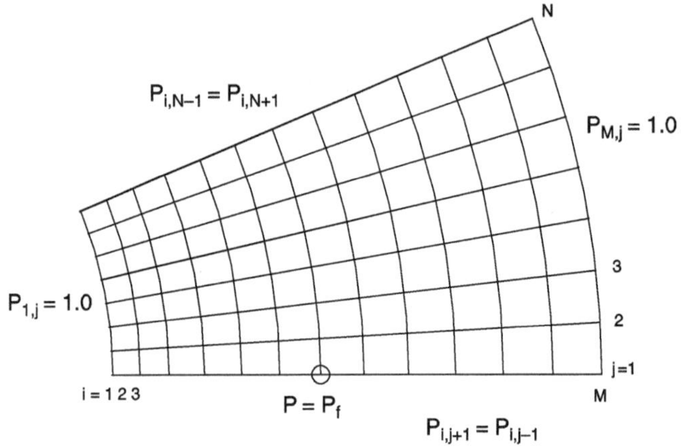

Figure 4.8 **Meshing strategy for annular thrust bearing**

function of position and needs to be put into a finite difference form to enable pressure at discrete points, namely the mesh nodes, to be calculated. If Δr and $r\Delta\theta$ are the radial and angular mesh increments, then:

$$\frac{\partial P}{\partial r}=\frac{\Delta P}{\Delta r}=(P_{i+1,j}-P_{i-1,j})/2\Delta r \qquad [4.17]$$

$$\frac{\partial P}{\partial \theta}=\frac{\Delta P}{\Delta \theta}=(P_{i,j+1}-P_{i,j-1})/(2\Delta\theta) \qquad [4.18]$$

$$\frac{\partial^2 P}{\partial r^2}=[(P_{i+1,j}-P_{ij})/\Delta r-(P_{i,j}-P_{1-1j})/\Delta r]/\Delta r=[P_{i+1,j}+P_{i-1j}-2P_{ij}]/(\Delta r)^2 \qquad [4.19]$$

and:

$$\frac{\partial^2 P}{\partial \theta^2}=[P_{i,j+1}+P_{i,j-1}-2P_{ij}]/(\Delta\theta)^2 \qquad [4.20]$$

By substituting the above into Eq. 4.13, and choosing a mesh such that $\Delta r = r\Delta\theta$, the required finite difference equation is obtained:

$$P_{ij}=[P_{i+1,j}+P_{i-1,j}+P_{i,j+1}+P_{i,j-1}]/4-\Delta\theta[P_{i+1,j}-P_{i-1,j}]/8 \qquad [4.21]$$

which gives the pressure at any node, i,j in the mesh in terms of the pressures at the four neighbouring nodes. The method of solution is to first define the boundary conditions at the bore and outside diameter of the bearing that constitute the outlet boundaries to the fluid film:

$P_{1,j} = 1.0$ and $P_{M,j} = 1.0$ (pressure = 1 Bar at bore and outside diameter)

and to guess the pressure at the orifice-bearing film junction that is the inlet boundary to the fluid film:

$$P_{a,b} = P_o$$

Sensible initial values of pressure are assigned to all other nodes, i.e. $1 < P_{ij} < P_o$ and Eq. 4.21 is used in an iteration loop to recalculate pressure at all non-boundary nodes in terms of the pressures at neighbouring nodes. The iterations are stopped when new values of pressure do not differ significantly from preceding values. At the radial boundaries, it is necessary to know $P_{i,-1}$ and $P_{i,N+1}$, in order to iterate pressure values on the radial boundaries. From symmetry:

$$P_{i,-1} = P_{i,+1} \text{ and } P_{i,N+1} = P_{i,N-1} \qquad [4.22]$$

For the guessed value of inlet pressure to the bearing film, P_o, the mass flow rate of air through the film is determined by applying the standard formula for 1-D gas flow through a narrow gap [7], to the bore and outside diameter edge elements of the mesh:

$$dm/dt = \Sigma^{N}_{j=1,}\ [(P_{2,j} - P_{1,j}) + (P_{M-1,j} - P_{M,j})]h^3 r\Delta\theta/(24\eta RT\Delta r) \qquad [4.23]$$

The assumption of 1-D flow at the bearing boundaries is reasonable for practical air bearing designs and a small grid size.

Mass flow rate determined from Eq. 4.23 is compared to that determined for the orifice using Eqs 4.33 or 4.55 in Section 4.7, depending on whether the pressure drop across the orifice exceeds a critical value at which it becomes choked. These equations are rewritten here but the reader is referred to Section 4.7 for a full definition of the symbols used.

For air, the critical pressure ratio is:

$$P^*/P_s = [2/(\gamma + 1)]^{\gamma/(\gamma+1)} = 0.528 \qquad [4.24]$$

For $P^*/P_s = P_o/P_s > 0.528$

$$dm/dt = \frac{C_{orf} C_d \pi d_o^2}{4(1+\delta^2)^{1/2}} P_s[2\gamma/(\gamma - 1)RT\{(P_o/P_s)^{2/\gamma} - (P_o/P_s)^{(\gamma+1)/\gamma}\}]^{1/2} \qquad [4.25]$$

For $P^*/P_s = P_o/P_s \leq 0.528$

$$dm/dt = 0.25 C_d^* \pi d_o^2 P_s [2\gamma/(\gamma - 1)RT\{(P^*/P_s)^{2/\gamma} - (P^*/P_s)^{(\gamma+1)/\gamma}\}]^{1/2} \qquad [4.26]$$

If the mass flow rates from Eqs 4.25 or 4.26 are not equal to within a specified tolerance, the pressure P_o is adjusted and the mass flow rates recalculated. Once the mass flow rates are balanced, the load supported by the bearing is determined from:

$$F_k = r\Delta r\Delta\theta\, \Sigma_1^{\ M}\Sigma_1^{\ N}\, P_{ij}^{\ 0.5} \qquad [4.27]$$

where F_k is the load corresponding to a bearing film thickness of h_k and $1 < k < k_{max}$.

The static stiffness of the bearing is defined for incremental values of the bearing film thickness:

$$K_s = [F_{k+1} - F_k]/[h_{k+1} - h_k] \qquad 1 < k < k_{max} \qquad [4.28]$$

4.6.3 Example of load–deflection characteristics of annular thrust bearings

Annular thrust bearings may be used singularly or in pairs, depending on the application. For single bearings, the load–deflection characteristics and hence the static stiffness is highly non-linear. Figure 4.9 is an example of the load–gap relationship for an orifice compensated bearing. For moderate and high gaps, the load supported by the bearing increases exponentially as the bearing gap reduces. It reaches a maximum at some defined value of gap and at lower gaps steadily reduces as the gap closes down to zero.

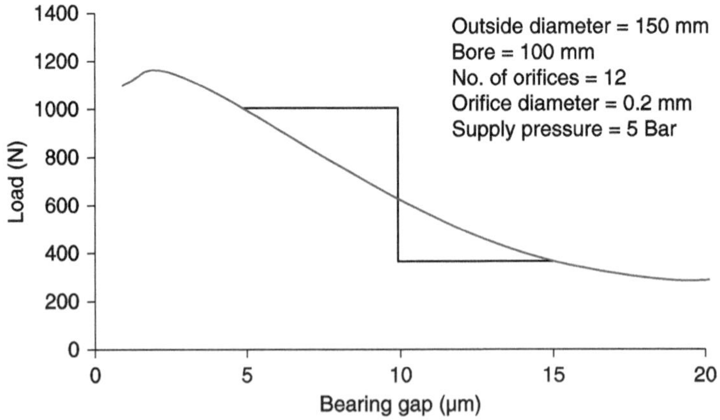

Figure 4.9 Typical load–deflection relationship for single-sided annular thrust bearing

As the bearing gap is reduced from a high value, the resistance to air flow of the bearing film increases relative to that of the orifice, so that an increasing proportion of the supply pressure occurs in the film enabling it to support higher loads. But at low gaps the resistance to air flow also becomes influenced by the orifice or pocket edge effect and this restricts the air flow into the bearing film. The latter becomes the dominant effect at very low gaps, producing a loss in load capacity and a negative static stiffness.

The load–deflection characteristics of a double-sided thrust bearing can be deduced from the load–gap relationship of a single-sided bearing. A double-sided bearing is set up to have a defined clearance so that under no load the operating gaps for each side of the bearing are the same and equal to half the total clearance. Under load, the gap on one side of the bearing is reduced, whilst that on the other side of the bearing is increased, the changes in gap being equal in magnitude. By choosing a no load equilibrium gap for a single bearing and considering changes in gap and the corresponding changes in load, the double-sided bearing characteristics are obtained. For example, in Figure 4.9, an equilibrium gap of 10 μm is chosen for which the bearing load is 610 N. For a deflection of 5 μm, one side of a double-sided bearing experiences an increase in load to 1000 N, whilst the other experiences a decrease in load to 380 N, the difference in load 620 N being the net external load supported by a double-sided bearing. For a well-designed double-sided annular thrust bearing, the load–deflection relationship is almost linear over most part of its range. Figure 4.10 shows the load–deflection characteristic of the double-sided equivalent to the bearing of Figure 4.9.

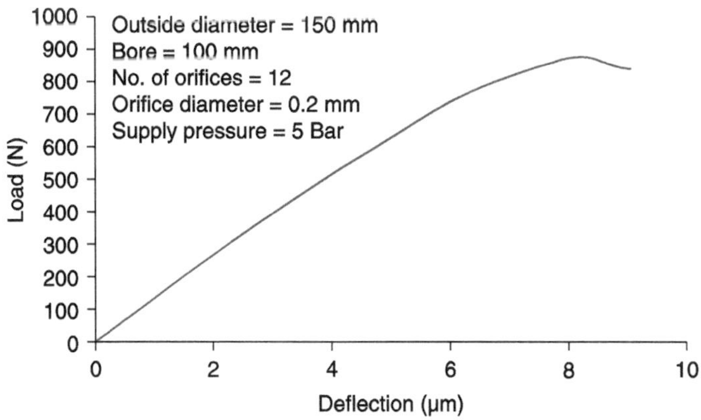

Figure 4.10 **Load–deflection characteristic of a double-sided thrust bearing**

Axial load capacity of a thrust bearing may be determined directly from the load deflection relationships (Figures 4.9 and 4.10). Usual practice is to define load capacity at a minimum gap that allows for manufacturing tolerances. The allowance varies from one manufacturer to another and indeed from one bearing type to another. In the example of Figure 4.10, defining load capacity at a gap of 5 μm, half of the no load gap gives a conservative estimate of load capacity.

4.6.4 Approximate tilt stiffness of annular thrust bearings

In a majority of air bearing arrangements, annular thrust bearings are used to support axial loads and two journal bearings are used to support radial and moment loads, the journal bearings being suitably spaced for the level of moment load expected. But there are bearing arrangements for which axial length is limited, such as the rotary table shown in Figure 4.1(b) and only a single journal bearing is used to take radial load. In such an arrangement, the annular thrust bearings are required to take both axial and moment loading. An accurate calculation of tilt stiffness and load capacity of annular thrust and circular pad bearings requires Eq. 4.13 to be solved having first expressed the gap, h, as a function of radial and angular position. Under moment loading:

$$h = h_o + r\psi\sin\theta \qquad [4.29]$$

where ψ is the level of tilt and θ is angular position. The numerical solution requires at least half of the bearing to be meshed, as now there is only one axis of symmetry. If on a double-sided thrust bearing there is also the need to take into account an axial load, the solution entails considerably more calculation than to determine axial stiffness alone. In this chapter, an approximate calculation of tilt stiffness of annular thrust bearings is made, deduced from the axial stiffness determined by solution of Eq. 4.13.

In the approximate calculation, it is assumed that the axial stiffness of the bearing is the sum of incremental stiffnesses uniformly distributed around the bearing's pitch circle diameter (Figure 4.11), such that:

$$dK = K_a.d\theta/2\pi \qquad [4.30]$$

where $d\theta$ is the angular increment. If the bearing is subject to a tilt, ψ, the moment generated over $d\theta$ is:

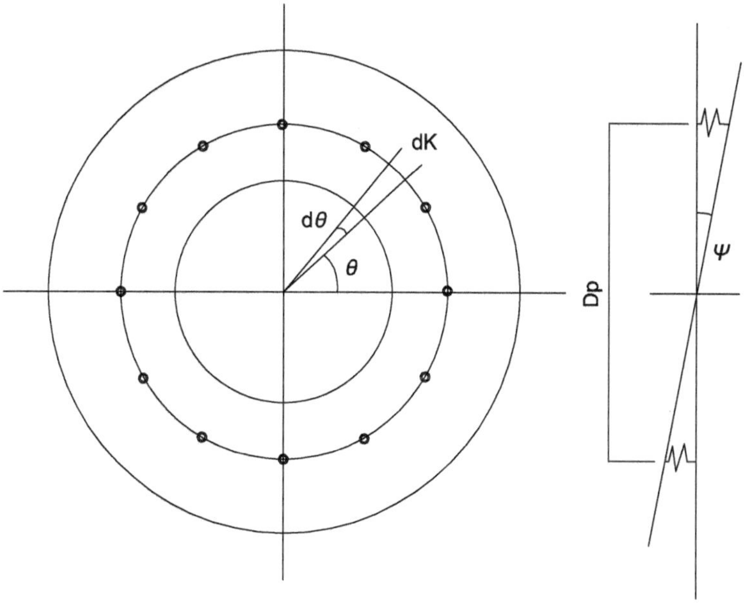

Figure 4.11 Tilt of an annular thrust bearing

$$dM = K_a (D_p/2)^2 \psi \sin^2\theta d\theta/(2\pi) \quad [4.31]$$

Integrating with respect to θ over an interval 0 to π, i.e. half the bearing:

$$M/2 = K_a D_p^2 \psi/(8\pi) \ [\textstyle\int_0^\pi \sin^2\theta . d\theta] \quad [4.32]$$

For the complete bearing:

$$M = K_a D_p^2 \psi/2 \quad [4.33]$$

which gives a tilt stiffness of:

$$K_\psi = M/\psi = K_a D_p^2/2 \quad [4.34]$$

4.6.5 Load capacity and static stiffness of cylindrical journal bearings

Equation 4.16 describes the pressure distribution in a cylindrical journal bearing rotating at low or moderate speeds. As for the annular thrust bearing, it forms an essential part of an overall solution scheme for determining the static load–deflection characteristics of the bearing. Steps a to i listed at the beginning of Section 4.6.2 also apply to the journal bearing. However, an essential difference between a thrust bearing subject to a concentric axial load and a journal bearing subject to a centrally applied radial load is that for the former, the gap remains constant in magnitude whereas for the latter, it varies simple harmonically with angular position. The difference is significant, so the solution scheme considered in Section 4.6.2 is applied to a journal bearing using Eq. 4.16 to determine pressure distribution.

As for the thrust bearing, a standard relaxation technique is used to determine the pressure at the nodes of a fine grid superimposed over the bearing's surface. Figure 4.12 shows the mesh used for a cylindrical journal bearing compensated by two rows of orifices and subject to a centrally applied radial load.

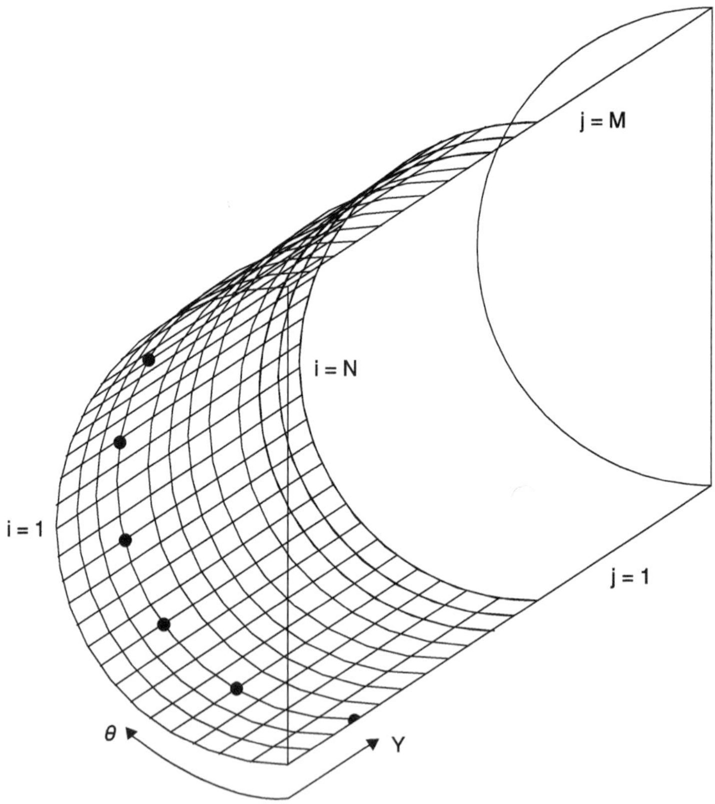

Figure 4.12 Meshing strategy for cylindrical journal bearing

Through symmetry, it is only necessary to consider a quarter of the bearing as pressure distribution is mirrored about central axial and radial planes. In Figure 4.12, radial load is applied through the geometric centre of the bearing in a vertical direction and orifices are located at node points on the mesh. Pressure at the end of the bearing is assumed to be ambient pressure = 1 Bar, whilst pressure at the orifice nodes is determined from equilibrium of air mass flow rate through the orifices and the bearing.

Equation 4.16 expresses pressure as a continuous function of position and needs to be put into a finite difference form to enable pressure at discrete points, namely the mesh nodes, to be calculated. If ΔZ and $\Delta\theta$ are the axial and angular mesh increments, and i and j are node coordinates, then:

$$\frac{\partial P}{\partial Z}=\frac{\Delta P}{\Delta Z}=(P_{i+1,j}-P_{i-1j})/2\Delta Z \quad [4.35]$$

$$\frac{\partial P}{\partial \theta}=\frac{\Delta P}{\Delta \theta}=(P_{ij+1}-P_{ij-1})/(2\Delta\theta) \quad [4.36]$$

$$\frac{\partial^2 P}{\partial Z^2}=[(P_{i+1,j}-P_{ij})/\Delta Z-(P_{i,j}-P_{I-1j})/\Delta Z]/\Delta Z=[P_{i+1,j}+P_{i-1j}-2P_{ij}]/(\Delta Z)^2 \quad [4.37]$$

and:

$$\frac{\partial^2 P}{\partial \theta^2}=[P_{i,j+1}+P_{i,j-1}-2P_{ij}]/(\Delta\theta)^2 \quad [4.38]$$

By substituting the above into Eq. 4.16, the required finite difference equation is obtained:

$$2P_{ij}[\Delta Z^2+\Delta\theta^2]=r^2[P_{i+1,j}+P_{i-1,j}]\Delta\theta^2+[P_{i,j+1}+P_{i,j-1}]\Delta Z^2\frac{-3\varepsilon.\sin\theta}{1+\varepsilon.\cos\theta} \quad [4.39]$$

$$[P_{i,j+1}-P_{i,j-1}]\Delta\theta\Delta Z^2$$

which gives the pressure at any node, i,j in the mesh in terms of the pressures at the four neighbouring nodes. The method of solution is to first impose boundary conditions at the end of the bearing, at the orifice locations and along the boundaries of symmetry:

- $P_{1,j}=1.0$ (pressure = 1 Bar at bearing end)
- $P_{a,b}=P_{oj}$ (pressure = orifice downstream pressure)
- $P_{N-1,j}=P_{N+1,j}$ (symmetry along the radial boundary)
- $P_{i,M-1}=P_{i,M+1}$ and $P_{i,1}=P_{i,-1}$ (symmetry along the axial boundaries)

then to guess initial values of pressure at all other nodes and finally to use Eq. 4.39 in an iteration loop recalculating pressure at node i,j in terms of pressures at neighbouring nodes. The iterations are stopped when new values of pressure do not differ significantly from preceding values.

For a bearing with N_o orifices per row, the number of orifices over half the circumference is the integer value of $(N_o+1)/2$. This is the number of orifices included in the solution scheme and for which the downstream pressures are required as boundary conditions. In order to calculate the bearing film pressure, the downstream pressures under each of the orifices are initially guessed and then modified along with the bearing film node pressures in an iterative loop. At each stage of the iteration, more accurate values of $P_{oj}(j \leq N_o+1)$ are determined by balancing the mass flow rate of air through the orifice with that exhausting locally from the bearing end plus that crossing boundaries midway between the orifices. The latter is determined by applying Eq. 4.23, used for the annular thrust bearing, to the appropriate boundary elements of the cylindrical journal mesh:

$$\begin{aligned} dm/dt = {} & \Sigma_{j=U}^{L} (P_{2,j} - P_{1,j})\, h^3.r.\Delta\theta/(24.\eta.RT.\Delta Z) + \\ & \Sigma_{i=1}^{N} (P_{i,U} - P_{i,U-1})\, h^3.\Delta Z/(24.\eta.RT.r\Delta\theta) + \\ & \Sigma_{i=1}^{N} (P_{iL} - P_{i,L-1})\, h^3.\Delta Z/(24.\eta.RT.r\Delta\theta) \end{aligned} \qquad [4.40]$$

where U and L are the coordinates of the axial boundaries either side of an orifice.

4.6.6 Example of load–deflection characteristics of cylindrical journal bearings

A typical load–deflection characteristic of a journal bearing calculated using the above solution scheme is shown in Figure 4.13. For small deflections, the curve generally exhibits good linearity and high stiffness but for larger deflections, the curve becomes non-linear, generally exhibiting a lower bearing stiffness. The extent of the linear range and the degree of non-linearity vary with bearing design.

Static stiffness and load capacity are determined from the load–deflection characteristic. Stiffness is usually the slope of the load–deflection curve local to zero deflection. As for the thrust bearing, load capacity is defined at some minimum value of gap that is dependent on manufacturing tolerances. In the case of the bearing shown in Figure 4.13, defining load capacity at 5 μm deflection (minimum gap = 5 μm) gives a conservative estimate of load capacity of 200 N.

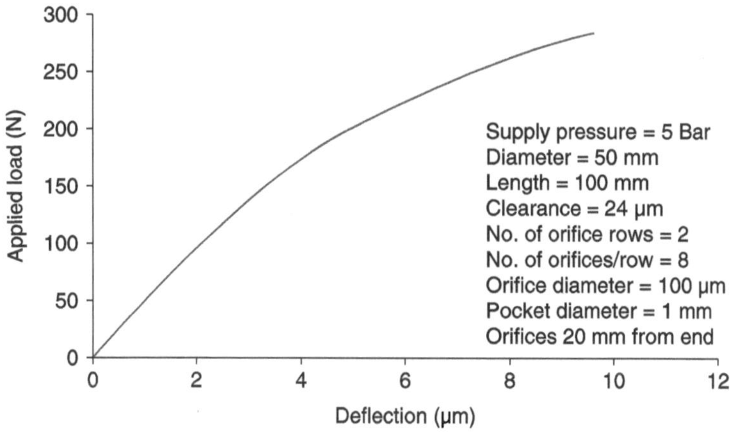

Figure 4.13 Typical radial load–deflection characteristic of an aerostatic journal bearing

4.6.7 Approximate tilt stiffness of cylindrical journal bearings

In applications where significant moment loads occur, it is normal practice to use two journal bearings and ensure they are spaced sufficiently far apart to provide adequate tilt stiffness and moment load capacity for the arrangement as a whole. A single journal bearing has a relatively low tilt stiffness and moment load capacity and is rarely considered. As for the thrust bearing, an accurate calculation of tilt characteristics adds significantly to the mesh size needed to obtain a solution and only an approximate calculation is included here.

Tilt stiffness is derived from the radial stiffness calculated earlier in this section, by assuming the radial stiffness is the sum of elemental stiffness uniformly distributed over the length of the bearing.

In Figure 4.14, the incremental radial stiffness $dK = K_r.dx/L$, where L is the length of the bearing, K_r is the bearing's total radial stiffness and dx is a length increment. If the bearing is subject to a small tilt ψ, then the moment generated by this element is:

$$dM = \psi.(K_r/L).x^2dx \quad [4.41]$$

Integrating from $-L/2$ to $L/2$ gives the tilt stiffness:

$$M/\psi = K_r.L^2/12 \quad [4.42]$$

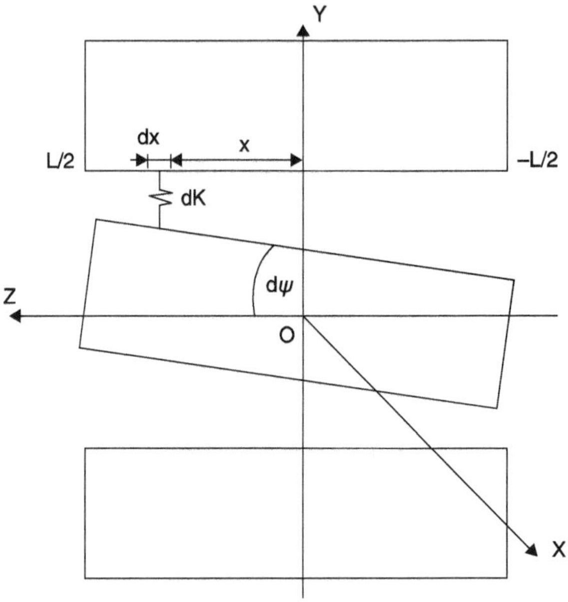

Figure 4.14 Tilt of a journal bearing

The moment capacity is defined at a minimum gap occurring at the end of the bearing. Equation 4.42 applies to a journal bearing under no load, in which case the tilt stiffness is the same about the *x*- and *y*-axes (Figure 4.14) or to a loaded bearing in which case the tilt stiffness is that about the *x*-axis.

4.7 Mass flow rate of air through an orifice

Figure 4.15 shows a practical design of orifice used in an air bearing. The diagram is more or less to scale, showing a typical orifice of diameter, d_o, 100 μm used with a typical bearing film thickness, h, 10 μm. In this case, it is evident that the main restriction to air flow is not the orifice of area, $\pi d_o^2/4 = 7.85 \times 10^3\ \mu m^2$, but the perimeter area at the junction of the orifice and bearing film, $\pi d_o h = 3.14 \times 10^3\ \mu m^2$. Furthermore, as the latter is dependent on film thickness, it is not constant in value but changes with bearing load. This characteristic often, but not always, limits bearing

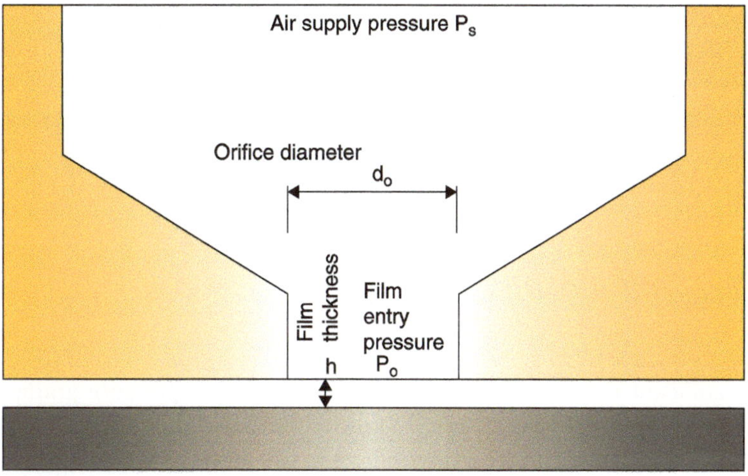

Figure 4.15 Orifice at entry to air bearing film

load carrying capacity, so to reduce its effect orifices are sometimes used with pockets.

Figure 4.16 shows an orifice of diameter d_o used with a pocket of diameter d_r at entry to the bearing film. Pocket diameters of around 1 mm are practical, therefore increasing the perimeter area, by as much as a factor of ten compared to that for a typical plain orifice. Pocket diameters

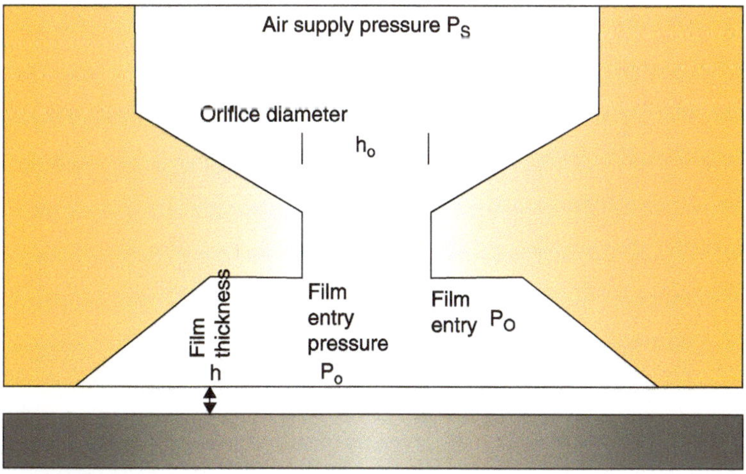

Figure 4.16 Pocketed orifice at entry to bearing film

are not unlimited. Pockets increase the internal volume of the bearing and because air is compressible, a bearing with a large internal volume needs a finite time to adjust to a change in bearing load. The air pressure in the bearing has to build or reduce to a new equilibrium value to compensate for the change in load. This time delay can, in the extreme, lead to a self excited dynamic instability known as 'pneumatic hammer'.

4.7.1 Plain orifice

An orifice considered in isolation has a negligible length, so that there are no significant frictional or viscous effects that affect pressure gradients. Pressure drop across an orifice is due entirely to a dynamic effect, in which the air is accelerated to a high speed where the associated kinetic energy is derived from a loss in static pressure across the orifice. The relationship between mass flow rate of air and pressure drop across an orifice is well established in the field of fluid mechanics, and it is known that the velocity of the air in a converging orifice cannot exceed Mach 1. The effect of this is to impose a maximum limit on the mass flow rate of air that can pass through the orifice, a phenomenon known as choking. For air velocities of less than Mach 1 at the throat of the orifice, the mass flow rate is [9]:

$$dm/dt = 0.25\ C_d\ \pi\ d_o^2\ P_s\ [2\gamma/(\gamma-1)RT\{(P_o/P_s)^{2/\gamma} - (P_o/P_s)^{(\gamma+1)/\gamma}\}]^{1/2} \qquad [4.43]$$

where γ is the ratio of specific heats for air and is equal to 1.4. C_d, the coefficient of discharge is an empirical factor relating actual mass flow rate to theoretical mass flow rate for an orifice. P_s and P_o are the pressures at orifice inlet and outlet respectively and d_o is the orifice diameter.

Experimental data shows C_d to vary with orifice diameter and pressure drop across the orifice [9]. In both cases, there is considerable experimental scatter, even for orifices of the same nominal design. This experimental data has been curve fitted to determine empirical equations suitable for use in an air bearing design program. To account for orifice diameter effects:

$$C_d = 0.52 + 2.1 d_o - 3.0\ d_o^2 \qquad [4.44]$$

where the units of d_o are mm. To account for pressure drop effects:

$$C_d = C_d/[1.174 - 0.327(P_o/P_s)] \qquad [4.45]$$

where:

$$P_o/P_s > [2/(\gamma+1)]^{\gamma/(\gamma+1)} = 0.528 \text{ for air} \qquad [4.46]$$

It is at this pressure ratio that air velocities reach Mach 1 in the throat of the orifice and choking occurs. Higher air supply pressures can increase mass flow rate through the orifice by increasing air density but reducing the downstream pressure below this critical ratio has no effect on air mass flow rate.

4.7.2 Orifice with pocket

Whether a plain orifice, such as that shown in Figure 4.15, or a pocketed orifice, as in Figure 4.16, is used, there is a need to account for the restriction to air flow occurring at the junction of the bearing film and the orifice or pocket. From an analytical viewpoint, it is convenient to consider the plain orifice as a special case of a pocketed orifice, where the pocket diameter is equal to the orifice diameter.

The restriction to air flow from the orifice and the pocket edge act in series. Pressure drop across the orifice is given by Eq. 4.16 above. It is assumed that the pressure drop across the pocket edge takes the same form [3]:

$$dm/dt = \pi\, d_r\, h\, P_p\, [2\gamma/(\gamma-1)RT\{(P_o/P_p)^{2/\gamma} - (P_o/P_p)^{(\gamma+1)/\gamma}\}]^{1/2} \qquad [4.47]$$

with the exception that the orifice area is replaced by the curtain area of the pocket, $\pi\, d_r\, h$, where d_r is the pocket diameter. P_p is the pressure in the pocket and P_o the pressure in the bearing film. In [3], a coefficient of discharge is not used to describe pocket edge losses. Instead, they are accounted for separately by considering dynamic pressure losses in the bearing film local to the pocket edge. It is convenient to eliminate pocket pressure P_p from Eqs 4.6 and 4.20. The following approximation was derived for the mass flow rate of air through an orifice and pocket in series [3]:

$$dm/dt = \frac{C_{orf} C_d\, \pi d_o^2}{4(1+\delta^2)^{1/2}} P_s[2\gamma/(\gamma-1)RT\{(P_o/P_s)^{2/\gamma} - (P_o/P_s)^{(\gamma+1)/\gamma}\}]^{1/2} \qquad [4.48]$$

where:

$$C_{orf} = [(1+\delta^2)/(1+K_d\delta^2 C_d^2)]^{1/2} \qquad [4.49]$$

and:

$$\delta = \pi d_o^2/4\pi d_r h = d_o^2/4 d_r h \text{ (ratio of orifice to pocket edge areas)} \qquad [4.50]$$

K_d is an entrance loss coefficient determined from [10.]

4.7.3 Pocket edge loss effects

For typical supply pressures and bearing geometries, the air flow rate across a pocket edge leads to high and sometimes even supersonic air velocities. As a result, a significant proportion of the static pressure is converted into kinetic energy. Associated with this is a dynamic pressure P_{dyn} that subtracts from the static pressure in the bearing film local to the pocket. Vohr's [10] entrance loss coefficient relates the static pressure in the pocket, P_p, the static pressure local to the pocket edge, P_c, and the dynamic pressure:

$$P_p - P_c = K_d \, P_{dyn} \qquad [4.51]$$

From experimental work, Vohr showed that the entrance loss coefficient K_d was dependent on Reynolds number:

$$K_d = 0.15 + 0.000225\, R_e \qquad \text{for} \qquad 0 < R_e < 2000 \qquad [4.52]$$

or:

$$K_d = 0.6 \text{ for } 2000 < R_e$$

The Reynolds number R_e is given by:

$$R_e = dm/dt/(\pi \, \eta \, d_r/2) \qquad [4.53]$$

McCabe et al, [11] also investigated pressure loss effects around the edges of pockets and presented a comprehensive set of experimental measurements of pressure profiles in the pocket and the bearing film for a centrally-fed thrust disc. Figure 4.17 is one example of McCabe's results that clearly shows the dynamic pressure loss local to the pocket edge. However, at a small, further distance into the film, the static pressure recovers before decaying in a viscous manner, as described by Reynolds equation. The effect was shown to be strongly dependent on bearing clearance for McCabe's experimental arrangement. It was significant at 10 μm clearance and became dominant for clearances greater than 20 μm. At 5 μm clearance, the pressure profile was close to the theoretical viscous profile. The results also showed the dynamic pressure loss to increase with air supply pressure.

Dynamic pressure loss local to the edge of a pocket or indeed a plain orifice can lead to significant errors in calculating air bearing performance, as it affects the pocket edge pressure that defines both the pressure drop across the bearing film and the orifice–pocket combination. The dynamic

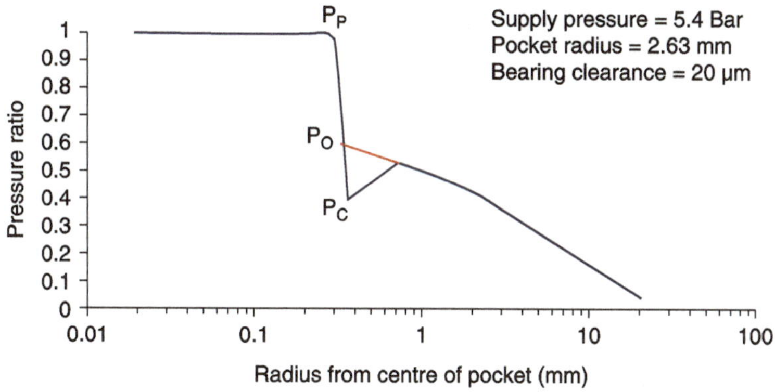

Figure 4.17 Experimental measurement of pressure profile across pocket edge

effects were accounted for by using Eq. 4.25 above, putting $P_{dyn} = P_p - P_c$ (Figure 4.17) [9]:

$$P_p - P_o = K_d(P_p - P_c) \qquad [4.54]$$

where P_c is the minimum static pressure in the bearing film at some point just outside of the pocket edge. P_o is the theoretical film pressure at the pocket edge and is the pressure required to balance calculated air mass flow rates through the bearing and orifice–pocket combination.

4.7.4 Choked flow

Choked flow in the orifice occurs at the limiting pressure ratio given by Eq. 4.46 above. This pressure ratio is known as the critical pressure ratio, for which lower values are physically impossible. When choking occurs:

$$dm/dt = 0.25\, C_d^* \,\pi\, d_o^2\, P_s\, [2\gamma/(\gamma-1)RT\{(P^*/P_s)^{2/\gamma} - (P^*/P_s)^{(\gamma+1)/\gamma}\}]^{1/2} \qquad [4.55]$$

where:

$$P^*/P_s = [2/(\gamma+1)]^{\gamma/(\gamma+1)} = 0.528 \text{ for air} \qquad [4.56]$$

In the event of choked flow, P^* is the pressure at inlet to the bearing film.

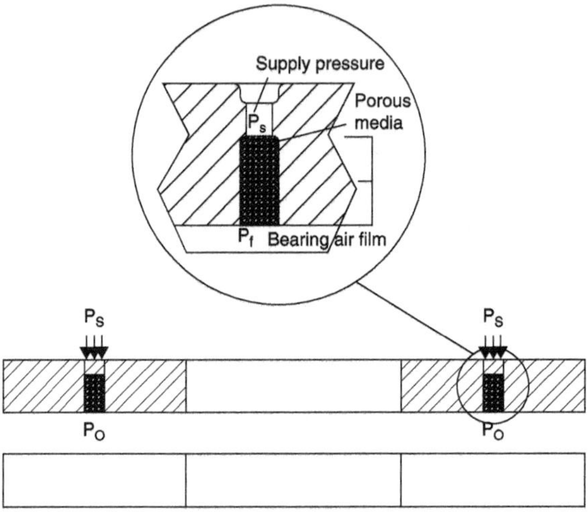

Figure 4.18 Annular thrust bearing with porous insert

4.8 Mass flow rate of air through a porous insert

Porous inserts, for example, as shown in Figure 4.18, allow pressurised air to be introduced into the bearing gap along a continuous boundary, thereby overcoming the inherent limitations associated with orifices and locally divergent flow. In order to determine the load capacity and stiffness of bearings with porous inserts, it is necessary to be able to calculate the mass flow rate of air through the insert. For this, Darcy's Law can be used as a starting point [12]. For 1.D flow through a porous media of cross-sectional area A and length L, Darcy's law relates the volume flow rate Q of a fluid to the pressure gradient $P_s - P_f$ across the media, the permeability of the media k_p and the fluid's viscosity η:

$$Q = -\ k_p A\ (P_s - P_o)/\eta L \qquad [4.57]$$

Originally developed for the flow of liquids through soil, the equation is applicable to the flow of gases through media with fine pores. However, since air is compressible, variation in air density through the media must be accounted for. This is done by considering flow through a thin layer of length dL, for which:

$$Q = -k_p A\ dP/\eta dL$$

The mass flow rate of air is given by:

$$\frac{dm}{dt} = \rho Q = -\frac{\rho k_p A}{RT\eta}\frac{dP}{dL} \qquad [4.58]$$

where ρ is the density of the air, which for isentropic flow can be expressed in terms of the gas constant R, absolute temperature T and pressure P:

$$\rho = P/RT$$

Integrating Eq. 4.58 with respect to L gives:

$$\frac{dm}{dt} = \frac{k_p A}{RT\eta L}(P_s^2 - P_o^2) \qquad [4.59]$$

where P_s is the air supply pressure and P_o is the bearing film pressure (Figure 4.18).

The insert is assumed to be narrow compared to the width or length of the bearing, so that for the case of annular thrust bearings subject to a centric axial load, the pressure at the end of the insert in contact with the bearing film can be considered constant and equal to P_f. If the insert has a mean diameter of D_p and a width w, then replacing A in Eq. 4.32 gives the mass flow rate of air into the bearing:

$$\frac{dm}{dt} = -\frac{\pi w D_p k_p}{RT\eta L}(P_s^2 - P_o^2) \qquad [4.60]$$

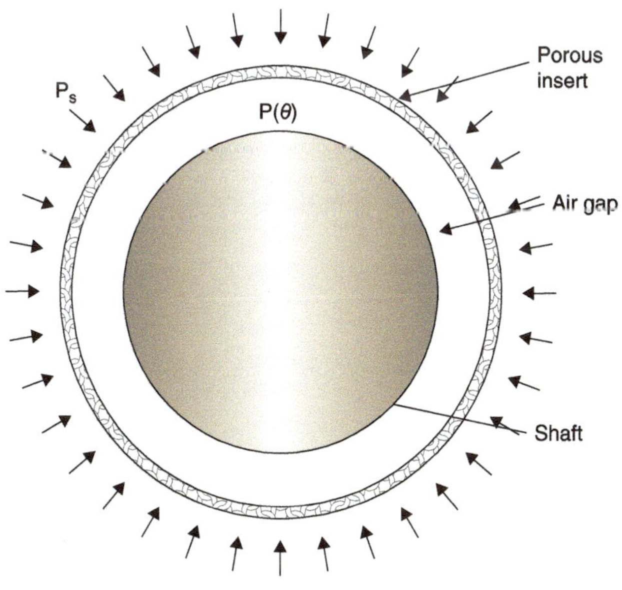

Figure 4.19 Porous insert used on journal bearing

For journal bearings subject to a radial load, the pressure at the end of the insert in contact with the bearing film varies with angular position (Figure 4.19).

Assuming the difference between outside and bore diameters of the insert is small compared to its mean diameter, the air flow through the insert can still be considered as 1-D. However, since the bearing film pressure is not constant with angular position, the pressure at the inlet boundary to the film is derived from a consideration of air mass flow rate through elements of the insert and the inlet boundary.

If the porous insert is of width, w, and the difference between its outer and bore diameters is 2L, then considering an element of the insert of length $rd\theta$, the mass flow rate of air through it is:

$$\frac{dm}{dt} = -\frac{wrd\theta k_p}{RT\eta L}(P_s^2 - P_{ij}^2) \qquad 1 < j < M \tag{4.61}$$

where P_s is the air supply pressure and P_{ij} is the local pressure at inlet to the bearing film. The air mass flow rate through elements of the bearing film, either side of and adjacent to the insert is (Figure 4.20):

$$\frac{dm}{dt} = \frac{-h^3 rd\theta}{12\eta RT.dz}[(P_{ij}^2 - P_{i+1j}^2) + (P_{ij}^2 - P_{i-1j}^2)]\ 1 < j < M \tag{4.62}$$

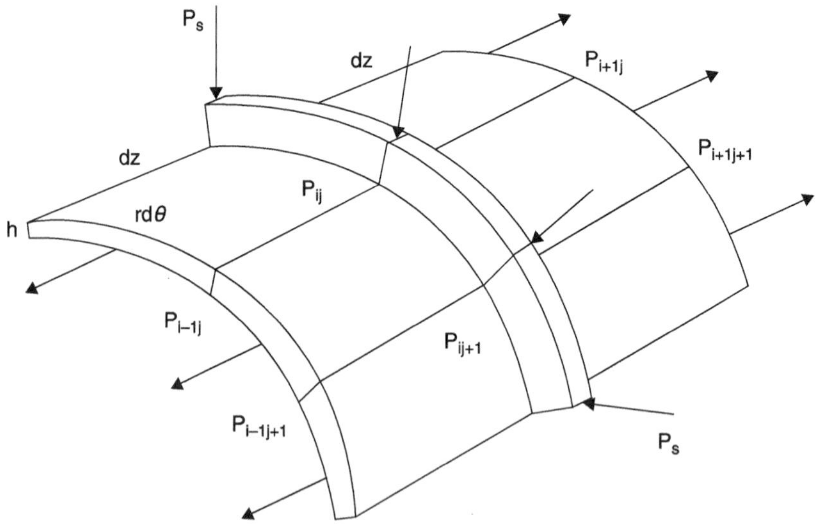

Figure 4.20 Porous insert and inlet boundary elements

where dz is the axial length of the mesh element used for the bearing film. Equating the two mass flow rates and simplifying gives the required expression for the air film pressures at the inlet boundary:

$$P_{ij}^2 = [wk_p dz/Lh^3.P_s^2 + P_{i+1j}^2 + P_{i-1j}^2]/[2 + wk_p dz/Lh^3] \qquad 1 < j < M \qquad [4.63]$$

The pressures P_{ij} are found by integrating the above expression into the numerical solution scheme used to determine film pressure at mesh nodes (Section 4.6.5).

4.9 Grooved and recessed bearings

Grooves and recesses with a variety of geometries may be used to enhance the pressure distribution in an air bearing. In Figure 4.21, the boundary (A) represents the edges of a recess placed around an orifice. The recess depth is constant and in this case its effect will be to generate a more uniform radial pressure distribution in the plane of the orifice, generally

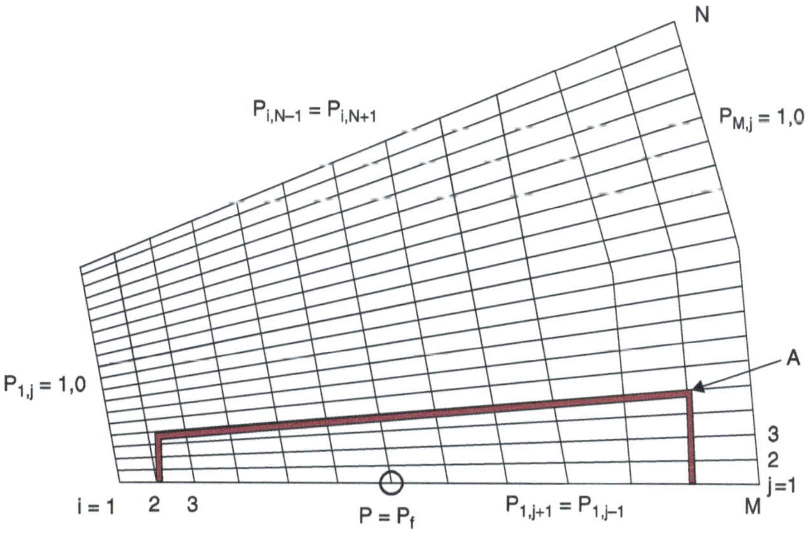

Figure 4.21 Meshing strategy for annular thrust bearing with recess

improving its load capacity and static stiffness. However, for benefits in air bearing performance, it is important to optimise the recess width, length and depth for the particular bearing geometry being considered. The effect of recesses and grooves on pressure distribution may be determined analytically by the numerical solution schemes described in the previous sections. A convenient solution is obtained by using a fine mesh to calculate the bearing pressure distribution and by defining groove or recess edges that are coincident with mesh boundaries. Across the boundaries, there must be continuity of air mass flow. From [7], the 1-D mass flow rate of air in a narrow gap between two flat plates is given by:

$$\frac{dm}{dt} = \frac{wh^3 p}{12\eta RT}\frac{dp}{dx} \qquad [4.64]$$

where h is the width of the gap, w is the plate width and x is a coordinate measured in the direction of air flow. Applying this to small mesh elements either side of a circumferential recess boundary (Figure 4.21) gives:

$$h_r^3 \frac{dP}{dr_r} = h_l^3 \frac{dP}{dr_l} \qquad [4.65]$$

and to a radial boundary:

$$h_r^3 \frac{dP}{rd\theta_r} = h_l^3 \frac{dP}{rd\theta_l} \qquad [4.66]$$

where h_r and h_l are the gaps over the recess and land respectively and $P = p^2$. The corresponding finite difference formulae for the pressures at nodes along the recess edges are:

$$P_{bj} = [h_r^3 P_{b-1j} + h_l^3 P_{b+1j}]/[h_r^3 + h_l^3] \qquad [4.67]$$

or:

$$P_{bj} = [h_r^3 P_{b+1j} + h_l^3 P_{b-1j}]/[h_r^3 + h_l^3]$$

depending on an inner or outer recess edge, and:

$$P_{ib} = [h_r^3 P_{ib-1} + h_l^3 P_{ib+1}]/[h_r^3 + h_l^3] \qquad [4.68]$$

for circumferential and radial boundaries respectively. The subscript 'b' represents the boundary node integer. The same solution scheme may be applied to journal bearings and the same Eqs 4.67 and 4.68 can be used to obtain the pressures at the recess or groove boundaries.

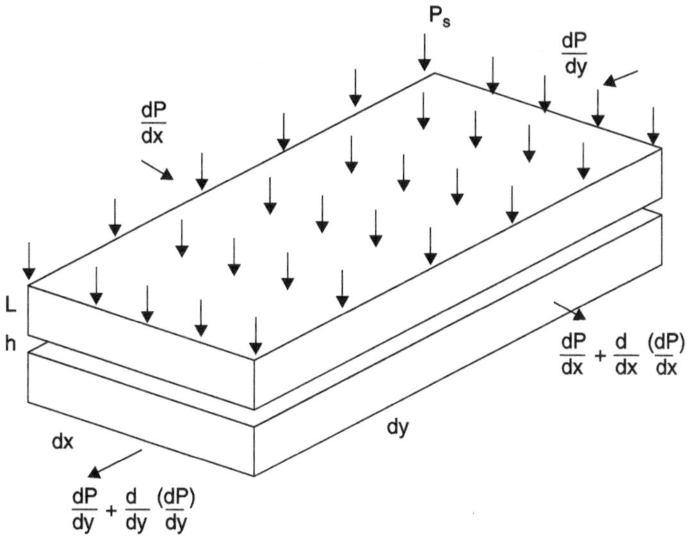

Figure 4.22 Element of porous air bearing

4.10 Porous air bearings

One side of a porous air bearing contains a continuous layer of porous material, the external surface of which is supplied with pressurised air. If the thickness of the layer is small compared to its length and width, then air flow through the layer can be considered as 1-D and Darcy's law can be used to determine the partial differential equation describing the pressure distribution within the bearing film. This equation is derived by considering equilibrium of air mass flow rate through an element of the bearing (Figure 4.22). In this case, a rectangular element of length dx and width dy is considered. Air at pressure P_s is supplied to the external surface of the porous layer on one side of the bearing, so that from Darcy's law the air mass flow rate entering the air film of the element from the air supply is:

$$dm_s/dt = k_p \, dx.dy \, (P_s^2 - P^2)/(\eta LRT) \qquad [4.69]$$

where L is the thickness of the porous layer and P is the average pressure in the air film of the element. This adds to the air mass flow rate entering the element driven by pressure gradients in the air film given by:

$$\frac{dm_i}{dt} = -\frac{dy\,h^3p}{12\eta RT}\frac{\partial p}{\partial x} - \frac{dx\,h^3p}{12\eta RT}\frac{\partial p}{\partial y} \qquad [4.70]$$

The net mass flow rate of air into the element's film must equal the net mass flow rate out as given by:

$$\frac{dm_o}{dt} = -\frac{dy\,h^3}{12\eta RT}\left[p + \frac{\partial p}{\partial x}dx\right]\left[\frac{\partial p}{\partial x} + \frac{\partial}{\partial x}\left(\frac{\partial p}{\partial x}\right)dx\right]$$

$$-\frac{dx\,h^3}{12\eta RT}\left[p + \frac{\partial p}{\partial y}dy\right]\left[\frac{\partial p}{\partial y} + \frac{\partial}{\partial y}\left(\frac{\partial p}{\partial y}dy\right)\right] \qquad [4.71]$$

Equating the net inflow of air given by Eqs 4.69 and 4.70 to the net out flow of air given by Eq. 4.71, putting $P = p^2$ and simplifying results in:

$$\frac{6k_p}{Lh^3}(P_s^2 - P^2) = \frac{\partial^2 P}{\partial x^2} + \frac{\partial^2 P}{\partial y^2} \qquad [4.72]$$

Equation 4.72 describes the pressure distribution in the bearing's air film and can be solved using the relaxation technique described in Sections 4.6.2 and 4.6.3. Devitt [2] develops the analytical equations describing pressure distribution in flat annular and cylindrical journal bearings, whilst [12,13] present the results of analytical and experimental investigations of cylindrical journal bearings. The reader is referred to these for a more detailed treatment of porous bearings. In the remainder of this chapter, attention is focused on orifice fed annular thrust and cylindrical journal bearings.

4.11 Air bearing static stiffness characteristics

The analyses presented in Sections 4.6.2 and 4.6.5 can be used to optimise the internal design of air bearings having orifice, groove/recess or porous material compensation. Depending on the application, optimisation usually involves determining the bearing design that gives maximum static stiffness, maximum load capacity, minimum air consumption or minimum power consumption at some specified operating condition(s). The analyses allow this to be achieved by enabling individual design parameters to be systematically varied and their effect on bearing performance to be quantified. In ultra precision applications it is usually maximum static stiffness that is required and optimisation then involves

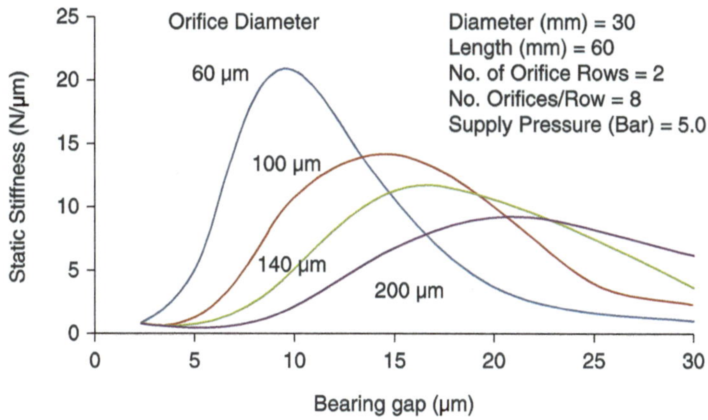

Figure 4.23 Example of static stiffness characteristics of orifice compensated journal bearing

determining the maximum allowable size of bearing, the minimum practical operating clearance and the best design of compensation system.

Figure 4.23 is an example of the static stiffness characteristics of an orifice compensated journal bearing. In this case, the bearing is 30 mm in diameter, has a length to diameter ratio of two and 2 rows of 8 pocketed orifices situated 15mm from the ends of bearing. The effect of orifice diameters of 60, 100, 140 and 200 µm upon static stiffness are considered over a range of bearing clearances, varying from 2.5 to 30 µm.

The curves for each orifice diameter show that there is a maximum static stiffness at a particular bearing clearance and the larger the orifice diameter, the larger the bearing clearance at which maximum stiffness occurs. However, the larger the bearing clearance, the lower the value of the maximum stiffness. The fact that each curve exhibits a maximum verifies that to achieve high stiffness, the design of the compensation system must be matched to the bearing design and more specifically, the resistance to air flow of the compensation system must relate to that of the bearing's air film. Highest bearing stiffness is achieved at small bearing clearances and requires the use of small diameter orifices. For the bearing of Figure 4.23, 60 µm diameter orifices produce a maximum stiffness of 21 N/µm at a gap of 18 µm. Higher stiffness may be achieved by the use of even smaller diameter orifices, but there is a practical lower limit to orifice diameter beyond which the risk of blockage by contamination and the difficulty in cleaning become unacceptable. This limit is presently around 50 μm, but the vast majority of bearings use orifices of 100 μm diameter or larger.

Figure 4.23 illustrates the importance of matching the compensation system to the bearing design. It does not present fully optimised bearing designs. Further improvements in static stiffness may be achieved by adjusting the number of orifices in each row (usually increasing them) or by moving the position of the rows. However, the practical difficulties with small diameter orifices remains a major obstacle towards achieving bearings with high static stiffness. Orifice compensated bearings tend to be the preferred choice in medium to high speed applications, where larger bearing clearances are required in any case to reduce power loss, allow for thermal or centrifugal growth of the shaft and where aerodynamic effects enhance stiffness.

For air bearings, high static stiffness is more easily achieved with compensation systems that feed air into the bearing along a line or over an area. Figure 4.24 is an example of the static stiffness achieved on a journal bearing with groove compensation. In this case, there are 12 grooves and the groove/(groove + land) width ratio is 40%. The four curves are for different groove depths and as for the orifice compensated bearing, the compensation system must be matched to the bearing in order to achieve maximum static stiffness. In this case, the lower the bearing clearance, the shallower the groove required for maximum stiffness. Shallow grooves, of the order of 5 μm in depth, are readily machined into the bearing surface, so the principle benefit of the grooved bearing is that it enables low bearing clearances to be used and therefore

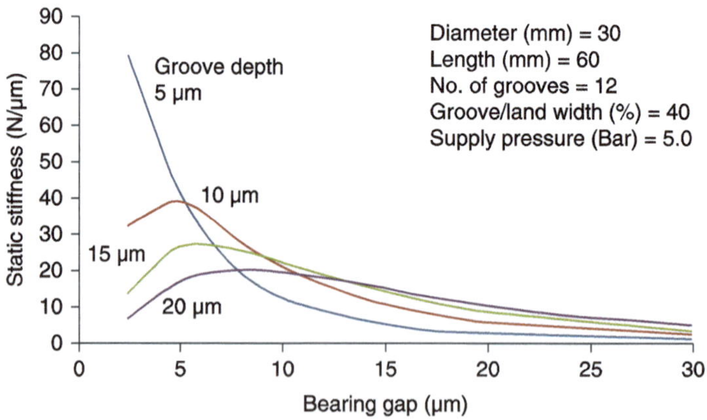

Figure 4.24 **Example of static stiffness characteristics of groove compensated journal bearing**

much higher static stiffness to be achieved than for the orifice compensated bearing. Similar benefits in stiffness are achieved with bearing designs employing porous media.

4.12 Damping and dynamic stability

Damping exhibited by aerostatic bearings is almost always low in magnitude, in terms of damping ratios, usually less than 0.15. Their damping characteristics are also complex and unlike any other bearing type, aerostatic bearings may even exhibit negative damping. That is to say they have the ability to feed energy into a vibration and in so doing create a self-excited instability known as 'pneumatic hammer'. Aerostatic bearing damping factors are therefore important performance parameters, not only quantifying the bearing's ability to limit vibration amplitudes but also its degree of stability.

The reason that aerostatic bearing damping characteristics are complex is that they depend on two interacting mechanisms. Squeeze film damping occurs in all types of fluid film bearing and provides a valuable source of positive energy dissipation. If two parallel surfaces separated by a narrow gap approach each other, they displace the fluid in the gap, causing viscous friction and a build up of pressure that resists the relative motion of the surfaces. When air is the fluid in the gap, the resisting pressure P is proportional to the velocity of approach of the surfaces V, the density of the air in the gap ρ, the area of the surfaces A, but inversely proportional to gap h squared:

$$P \propto V.A.\rho/h^2$$

If this were the only mechanism present, then a well damped air bearing might be expected to have a small gap, large surface area and high pressure (high air density) within the gap. However, in an air bearing, squeeze film damping interacts with the effects of compensation delay. Air bearings are designed to have a precise relationship between the magnitude of their gap and the pressure distribution within it. So that when an increase in applied load reduces the bearing gap, the internal pressure increases to compensate for the change in load. The compensation is usually achieved by restricting the air flow rate into the bearing by first passing it through fine orifices. With the bearing having a finite volume defined by its area and gap and with air being compressible and its flow rate restricted, there is a small time delay between a rapid change in gap and the internal pressure reaching a new equilibrium value. The net effect of this delay is to create negative damping. For example, if the bearing gap is rapidly increased and there is a delay in pressure reducing to a new equilibrium value, there is an over pressure within the bearing that actually assists the relative motion of its surfaces. As for squeeze film damping, the effect is proportional to the velocity of approach (or separation) of the surfaces. Compensation delay reduces the apparent damping level of an air bearing and when its effect exceeds the effect of squeeze film damping, the bearing becomes dynamically unstable.

Relevant past work has been concerned more with the phenomenon of stability rather than damping and in general a distinction has been made between flat pad bearings, rectangular and annular types and cylindrical journal bearings. The former are prone to a mode of instability known as 'pneumatic hammer', a self-excited instability creating large amplitude oscillations between the bearing surfaces. whereas the latter are more prone to 'whirl'. where the shaft or rotor takes up an orbital motion in the journal.

For flat pad bearings, several approaches to the analysis of stability have been published. Blondeel et al. [16] and Roblee et al. [17] favoured an approach based on control system technology, modelling the bearing as a passive static compliance having an active dynamic feedback loop consisting of restrictor and gap transfer functions. Mishra [18] used a perturbation technique for investigating the stability of a rectangular bearing pad subject to offset load, whilst Chen et al. [19] used time step integration to determine the dynamic response of a rectangular bearing pad.

For cylindrical journal bearings, the analysis is more complex and most investigators have favoured a perturbation approach [20–23]. [20] considered the case of orifice fed bearings, [21,23] porous bearings and [22] slot fed bearings. In all cases, the Reynolds equation was used to describe the static pressure distribution in the bearing gap and the dynamic pressure determined by considering a small time-dependent disturbance to the static equilibrium gap and pressure field.

4.12.1 Damping of annular thrust bearings

The analytical approach favoured here is similar to that of [19]. All published stability analyses have required a numerical solution and advantages of the time step integration approach are that it is a straightforward extension of the finite difference analysis used for

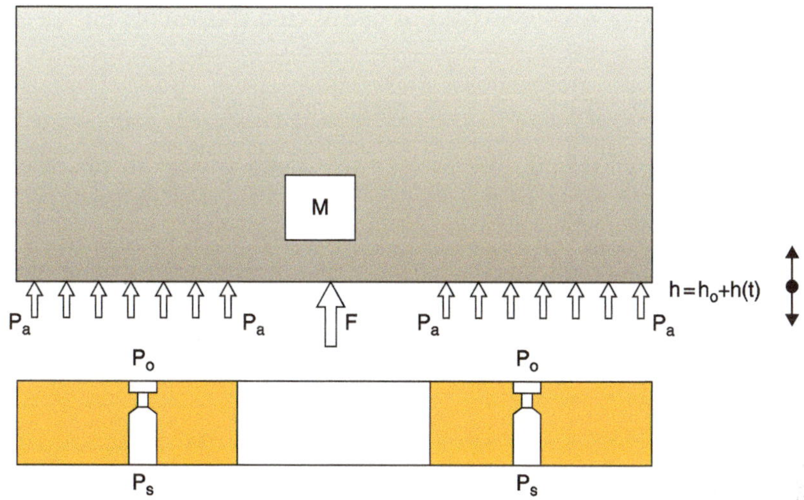

Figure 4.25 Annular thrust bearing as part of a simple dynamic system

determining the static bearing characteristics (Section 4.6.2), and it can account for large amplitude non-linearity observed to be a feature of 'pneumatic hammer'.

The bearing is considered to be part of a simple dynamic system in which it supports a mass M similar to the actual mass it will support in the application of interest (Figure 4.25). The bearing provides stiffness and damping, so under the application of an impulse, the mass will exhibit an oscillatory transient response. If the bearing is well damped, the transient will decay to zero, whereas if it is unstable, the oscillation will persist and even increase in amplitude. Damping coefficients can be determined from the rate of decay of the transient response knowing the system parameters, mass M, natural frequency of oscillation and/or the bearing stiffness.

The thrust bearing considered is supplied with pressurised air via a number of equally spaced orifices located on a pitch circle diameter, and the orifices may be plain or pocketed. Bearing gap may be constant or it may vary with radius and angular position if grooves or recesses are present. At static equilibrium, the bearing gap is denoted by $h_o = h_o(r,\theta)$ and the pressure by $P_o = P_o(r,\theta)$. P_o is determined from the static analysis in Section 4.6.2 by solving:

$$\frac{\partial P}{\partial r} + r\frac{\partial^2 P}{\partial r^2} + \frac{1}{r}\cdot\frac{\partial^2 P}{\partial \theta^2} = 0$$

A finite difference scheme is used to find P_o and this entails the use of Eq. 4.21:

$$P_{ij} = [P_{i+1,j} + P_{i-1,j} + P_{i,j+1} + P_{i,j-1}]/4 - \Delta\theta[P_{i+1,j} - P_{i-1,j}]/8$$

to obtain pressures at the nodes of a fine mesh subject to boundary conditions at the edges of the bearing and the recesses or grooves, if present. Note that in Eq. 4.13, $P = P_o^{\ 2}$ and in Eq. 4.21, $P_{ij} = P^2_{oij}$.

For dynamic equilibrium, the effect of the bearing gap varying with time must be taken into account. Thus, Eq. 4.12, the reduced form of Reynolds equation with the squeeze film term is used, ignoring the influence of surface speed, v_θ, i.e. for low speed bearings, Eq. 4.12 is reduced to:

$$\frac{\partial}{\partial r}\left[rh^3 p\frac{\partial p}{\partial r}\right] + \frac{-1}{r}\cdot\frac{\partial}{\partial\theta}\left[h^3 p\frac{\partial p}{\partial\theta}\right] = 12\eta\frac{\partial(ph)}{\partial t} \qquad [4.73]$$

Eq. 4.73 applies to both annular thrust bearings with flat surfaces or surfaces containing flat bottomed grooves or recesses. Fo the latter it is applied seperately to the regions inboard and outboard of the groove or recess boundaries subject to continuity of mass flow at the boundary. Also, putting $P = p^2$, Eq. 4.73 may be rewritten as:

$$\frac{\partial P}{\partial r} + r\frac{\partial^2 P}{\partial r^2} + \frac{1}{r}\frac{\partial^2 P}{\partial\theta^2} = \frac{24\eta}{h^3}\left[h\frac{\partial p}{\partial t} + p\frac{\partial h}{\partial t}\right] \qquad [4.74]$$

Using Eqs 4.17 to 4.20 and putting $\partial p/\partial t = (p_{ij,k+1} - p_{ij,k})/\Delta t$, where the integer subscript k denotes time $k\Delta t$, the finite difference form of Eq. 4.74 is:

$$p_{ij,k+1} = p_{ijk} + \left\{\left[\frac{2(P_{i+1,jk} + P_{i-1,jk} + P_{i,j+1k} + P_{i,j-1k} - 4P_{ijk}) + \Delta\theta(P_{i,j+1k} - P_{i,j-1k})}{48\eta r_i(\Delta\theta)^2}\right]h^3_{ij} - p_{ijk}\frac{\partial h}{\partial t}\right\}\Delta t \qquad [4.75]$$

Before solving Eq. 4.75, the static pressure distribution P_o in the bearing is determined for a given static gap h_o, using the procedure described in Section 4.6.2 and Eq. 4.21. For the static analysis, any variation in gap due to recesses or grooves is taken into account by considering continuity of mass flow at the recess or groove boundaries. However, if recesses or grooves are present, the gap h_{ij} must be defined at all mesh nodes. Furthermore, if a pocketed orifice is used, the pocket volume must also be accounted for. As pockets are small in diameter, this may be done simply by approximating the gap at the orifice node to:

$$h_{ab} = h_{oab} + V_p/(r_a^{\ 2}\Delta\theta) \qquad [4.76]$$

where a,b are the orifice node coordinates, V_p is the pocket volume and $r_a^2\Delta\theta$ is the area of the mesh grid local to the orifice.

As for the static analysis, the solution of Eq. 4.75 requires boundary conditions to be defined; however, now initial conditions must also be defined. The boundary conditions are $p_{1,j}=P_{1,j}=1.0$ and $p_{Mj}=P_{M,j}=1.0$ (dynamic and static pressure = 1 Bar at bore and outside diameter) and at the radial boundaries $p_{i,-1}=p_{i,+1}$, $p_{i,N+1}=p_{i,N-1}$ and $p_{i,-1}=p_{i,+1}$, $p_{i,N+1}=p_{i,N-1}$.

Remembering that $P=p^2$, the initial conditions are $p_{ij}=p_{oij}$ and $h_{ij}=h_{oij}$ at all nodes other than boundary or orifice nodes where subscript o refers to time zero. The term $\partial h/\partial t$ is the velocity of approach of the bearing surfaces and is the same at all nodes. For a transient, time dependent analysis, it is considered to be an impulse and assigned an initial value. This value is not critical and for a stable bearing is chosen to generate an initial amplitude of motion that is between 10 and 50% of the static equilibrium gap.

At the orifice, the initial condition is $p_{ab}=p_{oab}$, as determined from the static analysis. For subsequent instances in time, equilibrium of air mass flow rate through the orifice and through the bearing gap local to the orifice must be maintained. For successive increments in time, the changes in mass flow rate through the orifice and the gap must be equal:

$$\Delta(dm_o/dt)=\Delta(dm_g/dt) \qquad \text{i.e. } (d^2m_o/dt^2)\Delta t=(d^2m_g/dt^2)\Delta t$$

For the orifice:

$$\frac{d^2m_o}{dt^2}=\frac{d}{dp}\left(\frac{dm_o}{dt}\right)\frac{dp}{dt} \qquad [4.77]$$

For a pocketed orifice, differentiating Eq. 4.48 with respect to p and substituting into Eq. 4.77 gives:

$$\frac{d^2m_o}{dt^2}=\frac{C_{orf}C_d\,\pi d_o^2}{8(1+\delta^2)^{1/2}}P_s[2\gamma/(\gamma-1)RT]^{1/2}$$

$$\frac{[(2/\gamma)(p/P_s)^{(2-\gamma)/\gamma}-((\gamma+1)/\gamma)(p/P_s)^{1/\gamma}]}{[(p/P_s)^{2/\gamma}-(p/P_s)^{(\gamma+1)/\gamma}]^{1/2}}\frac{dp}{dt} \qquad [4.78]$$

which may be put in the form:

$$(d^2m_o/dt^2)\Delta t=C_o(p)\;dp \qquad [4.79]$$

where $C_o(p)$ represents the right-hand side of Eq. 4.78 and dp is the pressure change in the bearing gap under the orifice. For successive increments in time:

$$(d^2m_o/dt^2)\Delta t = C_o(p)\ (p_{ab1} - p_{abo}) \qquad [4.80]$$

In the bearing gap local to the orifice, the mass flow rate of air is given to a first approximation by:

$$\frac{dm_g}{dt} = \left[\frac{\partial P}{\partial r}\frac{r_a d\theta}{dr_a} + \frac{\partial P}{\partial \theta}\frac{dr_a}{r_a d\theta}\right]\frac{h_{ab}^3}{12\eta RT} \qquad [4.81]$$

where $P = p^2$ and the mesh grid is chosen such that $dr_a = r_a d\theta$ and a, b are the node coordinates of the orifice. Differentiating Eq. 4.81 with respect to time and using the latter relationships gives:

$$\frac{d^2m_g}{dt^2} = \left[\frac{\partial P}{\partial r} + \frac{\partial P}{\partial \theta}\right]\frac{h_{ab}^2}{4\eta RT}\frac{dh}{dt} + \left[\frac{\partial p}{\partial r} + \frac{\partial p}{\partial \theta}\right]\frac{h_{ab}^3}{6\eta RT}\frac{\partial p}{dt} \qquad [4.82]$$

In terms of discrete pressures at nodes local to the orifice Eq. 4.82 may be rewritten as:

$$\Delta t\frac{d^2m_g}{dt^2} = [P_{ab+1} - P_{ab-1} + P_{a+1b} - P_{a-1b}]\frac{h_{ab}^2}{4\eta RT}dh$$

$$+ [p_{ab+1} - p_{ab-1} + p_{a+1b} - p_{a-1b}][p_{ab,1} - p_{ab,0}]\frac{h^3}{3\eta RT} \qquad [4.83]$$

For equilibrium of air mass flow rate at the node corresponding to the orifice:

$$\frac{d^2m_g}{dt^2}\Delta t + \frac{d^2m_o}{dt^2}\Delta t = 0 \qquad [4.84]$$

Thus combining Eq. 4.80 and Eq. 4.83 gives:

$$p_{ab,1} - p_{ab,o} = -\{[P_{ab+1,o} - P_{ab-1,o} + P_{a+1b,o} - P_{a-1b,o}]h_{ab}^2\ dh/[4\eta RT]\}/ \qquad [4.85]$$

$$\{C_o + [p_{ab+1,o} - p_{ab-1,o} + P_{a+1b,o} - p_{a-1b,o}]h_{ab}^3/[3\eta RT]\}$$

where the third subscript refers to time and $t_1 = t_o + \Delta t$.

The discrete pressures p_{ij} are determined at small increments in time, using the Runga–Kutta numerical integration scheme. The net force generated by the bearing is the integral of pressure over the bearing area. At time $t = k\Delta t$:

$$F_k = 2\,N\Sigma_1^{\,M}\Sigma_1^{\,N}\ p_{ijk}\ r_i\Delta\theta\Delta r_i \qquad [4.86]$$

where N is the number of orifices. For dynamic equilibrium of the mass M:

$$F_k - F_o = M\ d^2h/dt^2 \qquad [4.87]$$

where F_o is the static equilibrium force generated by the bearing at time $t = 0$. Equation 4.87 is also solved to find h(t) using the Runga–Kutta technique. Expressed in discrete form:

$$dh/dt = (h_{ij,k+1} - h_{ij,k})/\Delta t$$

$$d^2h/dt^2 = (dh/dt)_{k+1} - (dh/dt)_k = (h_{ij,k+2} - 2h_{ij,k+1} + h_{ij,k})/\Delta t^2$$

Thus:

$$h_{k+2} = 2\,h_{k+1} - h_k + [(F_{k+2} - F_o)/M]\Delta t^2 \qquad [4.88]$$

The solution of Eq. 4.88 is a transient time series (Figure 4.26).

For stable bearing designs, the series typically takes the form of a damped oscillation, whereas for unstable bearings, the oscillations continue to grow in amplitude with time until they equal or theoretically exceed the bearing clearance. Damping levels may be determined from the rate of change of successive maxima or minima [24]. The logarithmic decrement is defined to be:

$$\delta_a = \ln\,(x_{mi}/x_{mi+1})$$

or averaged over j cycles:

$$\delta_a = [\ln\,(x_{mi}/x_{mi+j})]/j \qquad [4.89]$$

[24] shows δ to be related to the damping ratio by:

$$\gamma_d = \delta_a/(2\pi) \qquad [4.90]$$

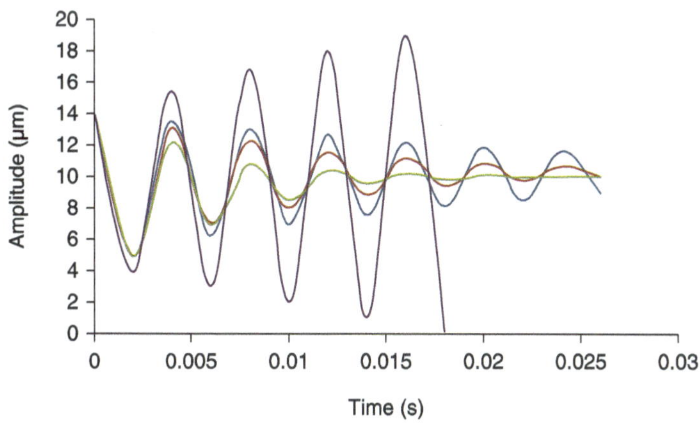

Figure 4.26 **Typical transient responses of annular thrust bearing**

4.12.2 Example of the effect of bearing volume on damping

Adding pockets or machining grooves or recesses to an air bearing increases its internal volume, but these features also affect the pressure (air density) distribution and the air flow rate through the bearing. In so doing, they can be designed to maintain or even increase damping levels, in addition to improving bearing load capacity and static stiffness. For example, Figure 4.27 shows the effect of recess volume on damping for four different air bearing designs – A, B, C and D. Bearing A represents a conventional bearing without any recess or pocket. B shows the effect of adding recesses to bearing A. Damping levels are reduced and in this example, when the dimensionless recess volume reaches 0.55, i.e. the recess volume is 0.55 times the volume of the bearing without recesses, damping becomes negative and the bearing becomes unstable. C shows the effect of changing the inlet restriction of bearing B, orifice diameters are increased to increase the air flow rate through the bearing, reducing compensation delay. Damping levels and the threshold of instability are both increased. Finally D shows the effect of changing the recess design used on B and C to reduce the restriction to air flow at outlet from the bearing, resulting in a further increase in damping capacity.

The optimum bearing design is usually a compromise. Adding features to bearing surfaces improves the air pressure distribution within the bearing, resulting in higher static stiffness and load capacity. However, for a given level of compensation they reduce damping levels and may

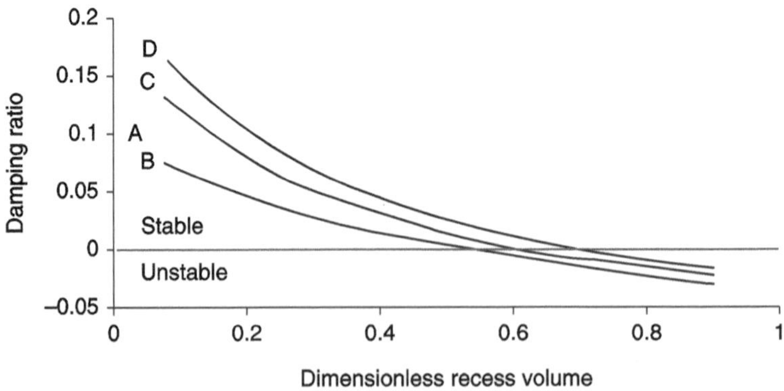

Figure 4.27 Example of the effect of bearing volume on damping ratio

even lead to instability. Damping levels may be recovered by adjusting the level of compensation but at the expense of air consumption. Thus the trade off is static stiffness v, damping v, air consumption.

Consider by way of example a rotary table with a double-sided thrust bearing of 200 mm outside diameter and 120 mm bore, required to have a high axial static stiffness. If orifices are used for compensation, the main challenge for the designer is to optimise the operating gap of the bearing, and the number and diameter of orifices. Figure 4.28 shows that static stiffness is inversely proportional to operating gap, so it is important to select a practical minimum value. The number and diameter of orifices has to be matched to the gap to achieve maximum bearing stiffness. For example, in Figure 4.28, a gap of 10 μm requires 24 orifices of 160 μm diameter, whereas a gap of 5 μm requires 16 orifices of 50 μm diameter to give maximum stiffness. Figure 4.28 also shows the static stiffness of the equivalent recessed bearing. For maximum static stiffness, the recess design must be optimised but it is not sensitive to bearing gap. In Figure 4.28, recess geometry and oriface diameter are kept constant. In this case, 12 recesses are used with 250 μm diameter orifices to give about a 50% improvement in stiffness over the conventional bearing.

The damping characteristics of both bearings are shown in Figure 4.29. For the orifice fed bearing damping increases with gap, whereas for the recessed bearing damping increases with reducing gap. At low gaps the air flow rate through an orifice fed bearing is restricted, so compensation delay is increased and damping reduced. In the recessed

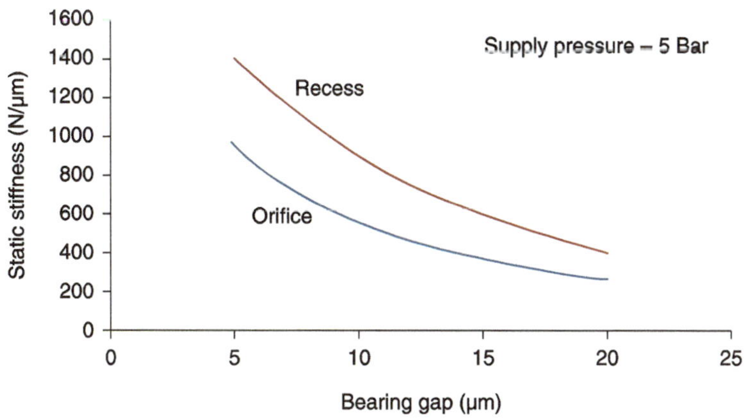

Figure 4.28 Effect of recesses on static stiffness

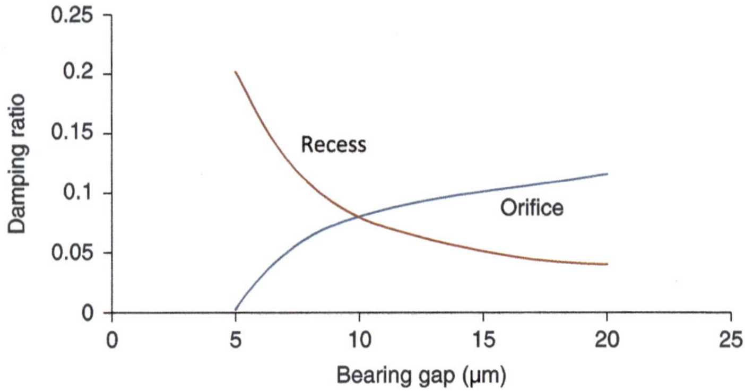

Figure 4.29 Effect of recesses on damping

bearing, compensation delay is controlled by the recess design so that at low bearing gaps, squeeze film damping has a far greater effect.

4.12.3 Damping of cylindrical journal bearings

As for annular thrust bearings, past investigations of the dynamic characteristics of aerostatic journal bearings have been primarily concerned with instability. Journal bearings are not only prone to the 'pneumatic hammer' mode of instability explained in the previous section, but may also be susceptible to a whirl mode of instability. The latter generally occurs at high speeds, where any eccentricity of the rotor within the bearing journal creates high aerodynamic forces. These forces tend to induce an orbital motion of the rotor and if speeds are high enough, a feedback mechanism is generated that causes both the amplitude of motion and the magnitude of aerodynamic forces to continue to increase until seizure occurs.

Larson and Richardson [25] observed whirl instability in rotors supported by aerostatic journal bearings and developed an empirical stability criterion as early as 1962. Later in 1967, Lund [20] presented a dynamic analysis of a rotor supported by aerostatic journal bearings, taking into account both whirl and pneumatic hammer modes of instability occurring on orifice fed bearings. This analysis was based on perturbation theory and valid for small eccentricity motions of the rotor. A similar approach was adopted in Rao and Majumdar in 1979 [21], for porous journal bearings soon after their introduction into the market and again in 2006 by Miyatake [23], in relation to ultra precision porous bearings.

In this section, the ability of aerostatic journal bearings to support externally applied dynamic forces occurring in high precision machinery applications is of primary interest. This ability is quantified in terms of a damping coefficient, where the higher the damping coefficient, the lower the bearing's motion amplitude in response to a given magnitude of dynamic force. Pneumatic instability is considered to be a special case of negative damping. However, the analysis is confined to low and moderate speed bearings, where aerodynamic forces may be neglected, thus excluding the whirl mode of instability.

The method of solution preferred follows that of [19] and the preceding section. The bearing is considered to be part of a simple dynamic system in which it supports a mass M similar to the actual mass it will support in the application of interest (Figure 4.25). The bearing provides stiffness and damping, so under the application of an impulse, the mass will exhibit an oscillatory transient response. If the bearing is well damped the transient will decay to zero, whereas if it is unstable the oscillation will persist and even increase in amplitude. Damping coefficients can be determined from the rate of decay of the transient response knowing the system parameters, mass M, natural frequency of oscillation and/or the bearing stiffness.

The starting point is the reduced form of Reynolds Eq 4.15; without the speed dependent term, this gives the pressure distribution in the bearing as a function of position and time when it is subject to a disturbance in the gap, h:

$$\frac{\partial}{\partial Z}\left[h^3 p\frac{\partial p}{\partial Z}\right]+\frac{1}{r^2}\frac{\partial}{\partial \theta}\left[h^3 p\frac{\partial p}{\partial \theta}\right]=12\eta\frac{\partial}{\partial t}(ph) \qquad [4.91]$$

In this analysis attention is confined to the case of a journal bearing fed by plain or pocketed orifices. The bearing may contain recesses, in which case the gap, h, must be defined as a function of position. Furthermore, if a pocketed orifice is used, the pocket volume must also be accounted for by locally adjusting the gap, h. For the dynamic analysis, only the limiting case of a bearing oscillating about zero eccentricity is with small amplitude is considered for which $h(t)=h_o+\Delta h(t)\cos\theta$.

Putting $P=p^2$ and assuming the bearing is subject to a disturbance $\partial h_o/\partial t$ in direction θ, i.e. $\partial h/\partial t=(\partial h_o/\partial t)\cos\theta$, Eq. 4.91 may be simplified to:

$$\frac{\partial^2 P}{\partial Z^2}+\frac{1}{r^2}\frac{\partial^2 P}{\partial \theta^2}=\frac{12\eta}{h_o^3}\left[h_o\frac{\partial p}{\partial t}+\frac{p\Delta h_o}{\partial t}\cos\theta\right] \qquad [4.92]$$

As in previous analyses, a finite difference scheme is used to obtain a numerical solution to Eq. 4.92. Putting $\partial p/\partial t = (p_{ij,k+1} - p_{ij,k})/\Delta t$, where the integer subscript k denotes time $k\Delta t$, the finite difference form of Eq. 4.92 is:

$$p_{ij,k+1} = p_{ijk,o} + \left\{ \left[\left(\frac{P_{i+1,jk} + P_{i-1,jk} - 2P_{ij}}{\Delta Z^2} \right) + \left(\frac{P_{i,j+1k} + P_{i,j-1k} - 2P_{ijk}}{(r\Delta\theta)^2} \right) \right] h_{ij}^3 - p_{ij} \frac{\Delta h}{\partial t} \cos\theta_j \right\} \Delta t \quad [4.93]$$

The solution scheme follows that of the annular thrust bearing described in the previous section. First, the static pressure distribution $P_o(Z,\theta)$ is determined by the method described in Section 4.6.5, using the mesh shown in Figure 4.12. The dynamic pressure distribution $p(Z,\theta,t)$ is then found by time step integration subject to boundary and initial conditions. As pockets are small in diameter, this may be done simply by approximating the gap at the orifice node to:

$$h_{ab} = h_{oab} + V_p/(r\Delta\theta\Delta Z) \quad [4.94]$$

where a,b are the orifice node coordinates, V_p is the pocket volume and $r\Delta\theta\Delta Z$ is the area of the mesh grid local to the orifice. The boundary conditions are:

- $p_{1,j} = 1.0$ (pressure = 1 Bar at bearing end)
- $p_{N-1,j} = p_{N+1,j}$ (symmetry along the central radial boundary)
- $p_{i,M-1} = p_{i,M+1}$ and $p_{i,1} = p_{i,-1}$ (symmetry along the axial boundaries)

and for equilibrium of mass flow rate of air at each orifice, Eq. 4.85 is modified to suit the mesh used for the cylindrical journal bearing:

$$\begin{aligned} p_{ab,1} - p_{ab,o} = -\{[(P_{ab+1,o} - P_{ab-1,o})\,(r\Delta\theta)^2 \\ + (P_{a+1b,o} - P_{a-1b,o})\,\Delta Z^2]h_{ab}^2\,dh/[4\eta RT]\}/ \\ \{C_o(r\Delta\theta)^2 + \Delta Z^2) + [(p_{ab+1,o} - p_{ab-1,o})\,(r\Delta\theta)^2 \\ + (p_{a+1b,o} - p_{a-1b,o})\,\Delta Z^2]h_{ab}^3/dh/[3\eta RT]\} \end{aligned} \quad [4.95]$$

For initial conditions $\partial h_o/\partial t$ is assigned a value sufficient to generate a maximum disturbance to the static equilibrium gap of approximately 50% and p_{ij} are equal to the static equilibrium values. Integration of Eq. 4.95 gives the pressure distribution within the bearing as a function of position. The net force acting on the bearing is:

$$F_k = 4\Sigma_1^{\,M}\Sigma_1^{\,N}\, p_{ijk}\, r\cos\theta_j\Delta\theta\Delta Z \quad [4.96]$$

and similar to the annular thrust bearing the dynamic equilibrium of the mass M:

$$F_k - F_o = M\cos\theta_j \, d^2h/dt^2 \qquad [4.97]$$

where F_o is the static equilibrium force generated by the bearing at time $t = 0$. Equation 4.89 is also solved to find h(t), using the Runga–Kutta technique. Expressed in discrete form:

$$dh/dt = (h_{ij,k+1} - h_{ij,k})/\Delta t$$

$$d^2h/dt^2 = (dh/dt)_{k+1} - (dh/dt)_k = (h_{ij,k+2} - 2h_{ij,k+1} + h_{ij,k})/\ \Delta t^2$$

Thus:

$$h_{k+2} = 2h_{k+1} - h_k + [(F_{k+2} - F_o)/M\cos\theta_j]\Delta t^2 \qquad [4.98]$$

Damping coefficients are determined from the transient response of h, using Eq. 4.92.

4.13 Motion error

Unlike ball bearings, published research on the motion error produced by air bearings is very limited. But the extremely low levels of motion error being achieved by some manufacturers in, for example, diamond turning and disc drive verification spindles, when compared to other types of air bearing spindle, is testament to considerable practical development within the industry. There are several well-known mechanisms by which air bearings can produce motion error and in this book they are broadly categorised as manufacturing errors, pneumatic hammer and half speed whirl.

Pneumatic hammer is a self-excited instability and occurs mainly on annular thrust and slide way bearings. It creates both a high amplitude vibration and loud audible noise at a characteristic frequency. The phenomenon can even occur on static bearings, so the frequency does not relate to rotational or linear speed, in other words the motion error is asynchronous. Bearings exhibiting pneumatic hammer are generally unusable, so if it occurs the bearing is considered to be incorrectly designed or is being used outside the range of operating conditions for which it was designed. Because it is a self-excited instability, pneumatic hammer is considered here to be negative damping and is covered more fully in Section 4.12.

Half speed whirl occurs on very high speed spindles and is normally associated with journal bearings. At high rotational speeds, aerostatic bearings develop aerodynamic lift and in fact if the speed is high enough, the aerodynamic lift becomes predominant, completely outweighing aerostatic effects. This adds considerably to the load capacity and static stiffness of the spindle; however, the spindle becomes prone to the shaft taking up an orbital motion within the bearing journal. Aerodynamic lift relies on the shaft being eccentric within the bearing journal, so that the lubricating fluid–air is dragged into a converging wedge where it creates an increase in pressure. The level of pressure increase and hence lift is strongly dependent on the average entrainment speed of the lubricant, which is equal to half the surface speed of the shaft. If now the shaft takes up an orbit and the orbital frequency approaches half the shaft's rotational speed, the entrainment speed approaches zero and the aerodynamic lift is lost. In practice, half speed whirl leads to a catastrophic bearing failure, i.e. seizure when the shaft's rotational speed reaches twice the first critical speed of the bearing–shaft system. Such high speed bearings fall outside the scope of this text and the reader is referred to [20] for a comprehensive explanation of half speed whirl.

For air bearings not prone to pneumatic instability or half speed whirl, motion errors are extremely low and practical measurements verify that the errors are predominantly synchronous with bearing speed. Motion errors arise from manufacturing errors, either associated with imperfections on the bearing surfaces or bearing alignment. In either case, motion errors are low due mainly to the effectiveness of a compressible air film in averaging them out. To put this into perspective, consider the present status of the industry. The very best air bearing products may have synchronous motion errors of 10 nm peak–peak or less and many manufacturers routinely achieve better than 50 nm peak–peak. Air bearing gaps are typically in the range 5 to 20 μm and it is reasonable to expect that manufacturing imperfections on the bearing surfaces must be kept low in comparison with gap size if the bearing is to exhibit low motion errors. Roundness (journal bearings) and flatness (annular thrust bearings) of air bearing surfaces are typically in the range 0.1 to 1 μm peak to valley, with the very best bearings being nearer 0.1 μm. Thus from a broad overview of present air bearing performance, the air film may be expected to attenuate the effect of surface imperfections on motion error by at the very least a factor of 10.

In this section, annular thrust and journal bearings with different types of manufacturing error are considered from the point of view of explaining the mechanisms by which motion error is generated and identifying the important variables that determine motion amplitudes. The lack of published work on this aspect of air bearing performance means that the explanation is largely theoretical and based on logical deduction.

Asynchronous motion errors produced by aerostatic bearings are low in comparison with synchronous errors, and there is even less in the way of published work. Discussion is limited to work performed by the author.

4.13.1 Air bearing model

As air bearings operate with small gaps, typically 5 to 20 μm, and as the gap must be accurately controlled if the bearing's designed level of load capacity and static stiffness is to be achieved, then not only are the dimensions of bearing components critical to function but so is the geometry and finish of the working surfaces. Whilst air bearings containing geometrical or roughness surface imperfections of several microns in height may well function with limited load capacity and stiffness, they will most certainly exhibit motion errors that may be unacceptable in ultra precision applications. Bearing manufacturers therefore endeavour to keep the maximum height of surface imperfections to a small fraction of the bearing gap, and for air bearings used in ultra precision applications this demands stringent specifications for surface finish, cylindricity, roundness or flatness. Surface finishes are generally lower than 0.05 μm R_a and roundness or flatness lower than 0.2 μm peak–valley. But even such well finished surfaces can create time-dependent fluctuations in air film pressure that in turn create motion errors. For the purpose of explaining how bearing surface imperfections create motion error, the simple structural model of Figure 4.30 is used.

As for the ball bearing, the structural model of the air bearing (Figure 4.30(a)) comprises a linear spring of stiffness K in series with a source of excitation $\Delta.\sin(\omega t)$, where Δ is the amplitude of the bearing's motion error arising from imperfections on its surfaces. Depending on the bearing arrangement and how the bearing is mounted, the excitation may generate a relative displacement between the shaft and housing and/or transmit a disturbing force. In a limiting case, the shaft and housing can be considered completely unrestrained, in which case the relative movement between the two will be $\Delta.\sin(\omega t)$. In another limiting case, both housing and shaft can be considered rigidly fixed in position and then the disturbance will generate a force between the two given by:

$$F = K.\Delta.\sin(\omega t) \qquad [4.99]$$

In determining the motion error of an air bearing, it is convenient to consider the latter. The shaft and housing are rigidly fixed in position so that any imperfections on the bearing surfaces only create a fluctuation in

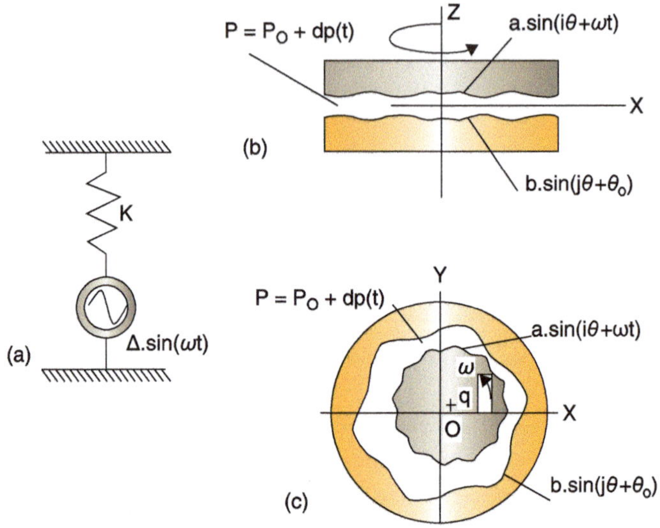

Figure 4.30 **(a) Structural model; (b) thrust bearing; and (c) journal bearing with surface imperfections**

the pressure within the air film. Integrating pressure over the bearing area gives the disturbing force, F and the amplitude of the motion error, and Δ is then simply given by F/K in Eq. 4.101.

Figures 4.30(b) and (c) show models of thrust and journal bearings to be considered. In each case, one of the surfaces is stationary and the other is rotating at a constant angular speed given by ω. For the thrust bearing, both surfaces are concentric (Figure 4.30(b)). For the journal bearing, the stationary outer surface is fixed with its centre at O, but the shaft and hence the inner surface of the bearing rotates at constant angular speed ω about its centre at q.

Both surfaces of each bearing contain small amplitude geometrical errors. In general, these will be a function of angular position θ and radius r in the case of a thrust bearing, or θ and axial position z in the case of a journal bearing. It is convenient to consider these geometrical errors to be simple harmonic in form and only vary in height in the direction of motion, i.e. for both thrust and journal bearings:

For the stationary surface:

$$b = b.\sin(j\theta + \theta_o) \qquad 2 < j < N \qquad [4.100]$$

where b is the amplitude of the error, j is the number of waves per circumference and θ_o is a phase angle defining angular position with respect to the frame of reference. θ_o has no physical significance in this analysis and may be put equal to zero. Similarly for the rotating surface:

$$a = a.\sin(i\theta' + \theta'_o) \qquad 2 < i < N \qquad [4.101]$$

where a is amplitude, θ' is defined with respect to a frame rotating at angular speed ω, θ'_o is a phase angle defining the position of the rotating frame with respect to a fixed frame at any point in time and I is the number of waves/circumference. Θ'_o may be expressed as a function of time, $\theta'_o = \omega t$. Thus the above equations may be expressed as:

$$b = b.\sin(j\theta) \qquad 2 < j < N$$

$$a = a.\sin(i\omega t + \theta') \qquad 2 < i < N \qquad [4.102]$$

For both annular and cylindrical surfaces, Fourier series can be used to describe the more complex geometries produced by practical manufacturing processes. Any circular or cyclic geometry can be considered to be the sum of a number of simple harmonic terms. For example, for the stationary bearing surface:

$$b(\theta) = b_o + \Sigma_1^N b_m \sin(m\theta) + \Sigma_1^N b_m \cos(m\theta) \qquad [4.103]$$

Furthermore, if it is necessary to describe a 2-D surface geometry, Eq. 4.103 can be extended to two dimensions. However, for the purpose of explaining how bearing surface imperfections create motion error, a 1-D form is sufficient. In this analysis, only small amplitude surface geometry errors and small amplitude motion errors are considered, where it is reasonable to assume that the relationship between the two is linear. Thus the effect of individual terms in Eq. 4.103 upon pressure in the bearing's air film can be considered separately (Eq. 4.102) and added linearly to determine the net effect produced by a complex surface geometry.

Since the analysis is confined to small amplitude motion errors and surface imperfections, then the relationship between the two can be determined by considering a perturbation to the bearing's static equilibrium pressure distribution. In practice, the latter is defined by the bearing geometry, applied load and the conditions at air inlet and outlet boundaries. In the case of a concentrically loaded thrust bearing, the static equilibrium pressure distribution can be determined theoretically from Eq. 4.13 and for a centrally loaded journal bearing from Eq. 4.16, as described in Sections 4.6.2 and 4.6.5 respectively. Whilst these equations are non-linear and may in general describe a complex pressure distribution, a perturbation analysis only considers first-order departures from equilibrium and in so doing effectively linearises the relationship between surface geometry and motion errors.

This amounts to assuming that perturbations in air film pressure are simply a function of the air film geometry and viscous shearing of the air, so that as the bearing rotates it creates time-dependent fluctuations in film geometry and hence air pressure. Dynamic effects dependent on air inertia and bearing volume are neglected, in other words the change in pressure produced by a change in film geometry under rotating conditions is assumed to be instantaneous.

Bearing surface errors are considered as small departures from an otherwise geometrically perfect surface that result in a small change to the bearing gap. On a bearing rotating at constant speed, the gap varies with position and time and is given by:

$$h = h_o(\theta) + \Delta h(\theta, t) \qquad [4.104]$$

where h_o is the static equilibrium gap and Δh is the perturbed value of the gap that varies with position and time. It is assumed that h, as defined by Eq. 4.104, produces a pressure distribution in the bearing gap given by:

$$p = p_o(\theta, Z) + \Delta p(\theta, Z, t) \qquad [4.105]$$

where p_o is the static equilibrium pressure field and Δp is the time dependent perturbation in pressure produced by Δh.

4.13.2 *Motion error produced by annular thrust bearings*

Equation 4.13 describes the static pressure distribution in an annular thrust bearing of constent gap. Taking into account variation in the magnitude of the gap and putting $P = p^2$, Eq. 4.13 becomes:

$$\frac{\partial}{\partial r}\left[\frac{rh^3 \partial P}{\partial r}\right] + \frac{1}{r}\frac{\partial}{\partial \theta}\left[\frac{h^3 \partial P}{\partial \theta}\right] = 0 \qquad [4.106]$$

In general, the gap h may vary with radius r and angular position θ, but for the purpose of explaining the effect of surface geometry on motion error, only variations with θ are considered. For $h = h(\theta)$ [4.106] may be rearranged to give:

$$\frac{\partial P}{\partial r} + \frac{r\partial^2 P}{\partial r^2} + \frac{3}{rh}\frac{\partial h}{\partial \theta}\frac{\partial P}{\partial \theta} + \frac{1}{r}\frac{\partial^2 P}{\partial \theta^2} = 0 \qquad [4.107]$$

If the static pressure distributed on a bearing with geometrically perfect surfaces is given by $p_o(\theta,r)$ and if imperfections that are only a function of θ are superimposed upon the surfaces, then the bearing gap is:

$$h = h_o + \Delta h(\theta)$$

and this produces a perturbation to the pressure distribution:

$$P = P_o(\theta,r) + \Delta P(\theta,r)$$

Substituting into Eq. 4.107 and separating out the first-order quantities gives:

$$\frac{\partial \Delta P}{\partial r} + \frac{r\,\partial^2 \Delta P}{\partial r^2} + \frac{1\partial^2 \Delta P}{r\partial \theta^2} + \frac{3}{hr}\frac{\partial \Delta h}{\partial \theta}\frac{\partial \Delta P}{\partial \theta} = 0 \qquad [4.108]$$

Equation 4.108 gives the pressure perturbation ΔP produced by a small variation $\Delta h(\theta)$ in the bearing gap. For simple harmonic surface imperfections on both inner and outer bearing surfaces (Figure 4.30(b)), the bearing gap at any angular position θ and at any instant in time may be expressed as:

$$h = h_o + \Delta h = h_o + a.\sin i(\omega t + \theta) + b.\sin(j\theta) \quad 2 < i < N, 2 < j < N \qquad [4.109]$$

for which:

$$\partial \Delta h/\partial \theta = ia.\cos i(\omega t + \theta) + jb.\cos(j\theta) \quad 2 < i < N, 2 < j < N \qquad [4.110]$$

Substituting Eqs 4.109 and 4.110 into Eq. 4.108 gives:

$$\frac{\partial \Delta P}{\partial r} + \frac{r\,\partial^2 \Delta P}{\partial r^2} + \frac{1\partial^2 \Delta P}{r\partial \theta^2} + \frac{3[ia.\cos i(\omega t+\theta) + jb.\cos(j\theta)]\partial \Delta P}{r[h_o + a.\sin i(\omega t+\theta) + b.\sin(j\theta)]\partial \theta} = 0 \qquad [4.111]$$

Thus the effect of gap shape upon air flow and pressure distribution is complex. Figure 4.31 illustrates the air flow in an orifice-fed annular thrust bearing, when one of its surfaces contains a simple harmonic gap variation, which is only a function of θ. When the gap is a minimum under the orifice, air flow local to the orifice is predominantly in a circumferential direction, whereas when the gap is a maximum under the orifice, flow is predominantly in a radial direction. Equation 4.111 can be used to determine these variations as a function of time, but this requires a combination of finite difference and time step integration numerical solution schemes similar to that used in Section 4.12.1 for the determination of damping coefficients. Only now a lack of symmetry requires the whole of the bearing to be meshed, leading to a time-consuming calculation.

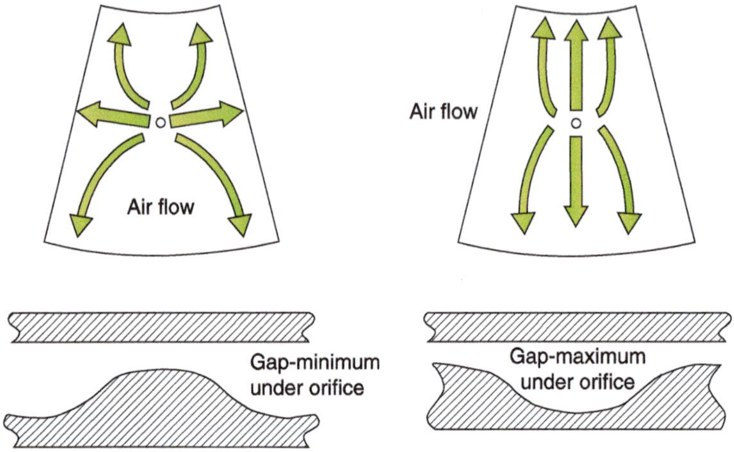

Figure 4.31 Example of the effect of gap shape on air flow

For the purpose of explaining the effect of surface geometry errors on motion error and to obtain a first estimate of the averaging effect of the air film, an approximate solution is developed here. For an orifice-fed annular thrust bearing subject to small amplitude surface errors, the dominant mechanism by which the surface errors create pressure variations in the bearing gap and hence motion error, is assumed to occur local to the orifices. Fluctuations in gap beneath the orifices regulate the mass flow rate of air into the gap and hence the pressure distribution within the complete film. Mass flow rate of air through a pocketed orifice, taking into account the curtain effect of the air film, is given by Eqs 4.48 to 4.50. Combining these, and eliminating C_{orf} and δ, gives the mass flow rate in terms of the gap h, the orifice and pocket geometry and the entry and exit pressures:

$$dm/dt = \frac{d_r h C_d \pi d_o^2 P_s}{(16 d_r^2 h^2 + K_d C_d^2 d_o^4)^{1/2}}[2\gamma/(\gamma-1)RT\{(P_o/P_s)^{2/\gamma}-(P_o/P_s)^{(\gamma+1)/\gamma}\}]^{1/2} \quad [4.112]$$

For simplicity consider a bearing for which the pressure ratio $P_o/P_s < 0.528$, the orifice is choked so the effect of a perturbation in the bearing gap on the mass flow rate of air entering the film may be written as:

$$\frac{dm/dt}{dm_o/dt} = \frac{(h_o+\Delta h)[16 d_r^2 h_o^2 + K_d C_d^2 d_o^4]^{1/2}}{h_o[16 d_r^2 (h_o+\Delta h)^2 + K_d C_d^2 d_o^4]^{1/2}} \quad [4.113]$$

where dm_o/dt is the mass flow rate of air for the static equilibrium gap h_o. Simplifying and rearranging gives:

$$\frac{dm/dt}{dm_o/dt} = 1 + \frac{\Delta h}{h_o[1+16d_r^2h_o^2/K_dC_d^2d_o^4]} \quad [4.114]$$

To a first approximation pressure at any point in the air film local to an orifice is assumed to obey the 1-D relationship:

$$\frac{P^2 - P_a^2}{P_o^2 - p_a^2} = \frac{dm/dt}{dm_o/dt} \quad [4.115]$$

If the perturbed pressure is $P = P_o + \Delta P$, then combining Eqs 4.114 and 4.115 gives:

$$\frac{\Delta P}{P_o} = \frac{\Delta h(1-P_a^2/P_o^2)}{2h_o[1+16d_r^2 h_o^2/K_dC_d^2d_o^4]} \quad [4.116]$$

Equation 4.116 gives the approximate fluctuation in pressure ΔP produced by a fluctuation in gap Δh local to a single orifice. The ability of the air film to average out pressure fluctuations is dependent on orifice and pocket characteristics. If it is assumed that entrance loss effects are negligible, i.e. $K_d = 1$ and the coefficient of discharge $C_d = 1$, then the averaging effect is primarily dependent on the compensation factor, d_rh_o/d_o^2, where d_r is pocket diameter, d_o is orifice diameter and h_o is bearing gap.

Figure 4.31 shows the effect of gap variation and compensation factor on the pressure variation local to an orifice. A compensation factor of 0.1 is typical of a plain orifice for which in % terms the pressure fluctuation is about 38% of the % gap variation. A compensation factor of 1 is typical of a pocketed orifice, for which the pressure fluctuation is only about 2.6% of the gap variation.

Equation 4.116 and Figure 4.32 apply locally to a single orifice, and to determine the effect this has on the bearing's motion error, the net effect of all orifices must be considered.

The load supported by a single-sided annular thrust bearing is proportional to the pressure at any point in its air film, thus local to orifice number k, the fluctuation in load $\Delta F_k/F_o$ is proportional to $\Delta P_k/P_o$. The total fluctuation in the bearing's axial load is therefore:

$$\Delta F/F_o = \Sigma_1^N \Delta F_k/F_o = \Sigma_1^N \Delta P_k/P_o = \alpha\, \Sigma_1^N \Delta h_k/h_o \quad [4.117]$$

where N is the number of orifices, h_k (Δ) is the local gap variation and α is an averaging factor determined from Eq. 4.116. For simple harmonic surface imperfections on both rotating and stationary bearing surfaces

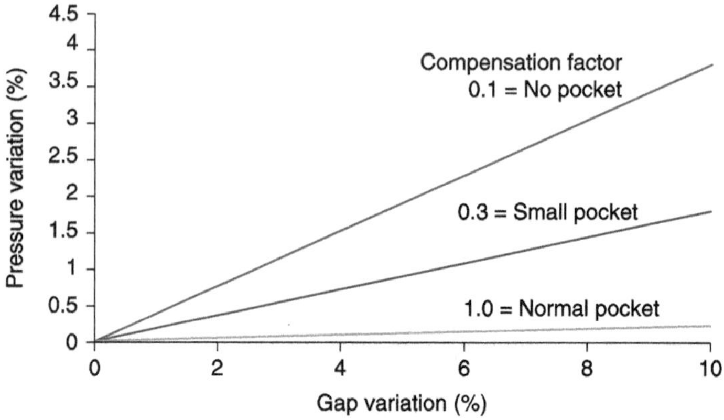

Figure 4.32 Effect of gap variations on pressure local to an orifice

(Figure 4.30(b)), the variation in bearing gap at any angular position θ and at any instant in time may be expressed as the ratio:

$$\frac{\Delta h_k}{h_o} = \frac{a}{h_o}.\sin i(\omega t + \theta_k) + \frac{b}{h_o}.\sin(j\theta_k) \qquad 2 < i < \infty,\ 2<j<\infty \qquad [4.118]$$

where θ_k is the angular position of orifice k and $\theta_k = 2\pi k/N$. Thus the fluctuation in bearing load is:

$$\frac{\Delta F}{F_o} = \alpha \Sigma_1^N \left[\frac{a}{h_o}.\sin i(\omega t + 2\pi k/N) + \frac{b}{h_o}.\sin(2\pi jk/N)\right] \quad 2<i<\infty,\ 2<j<\infty \quad [4.119]$$

Using the formulae in [26]:

$$\Sigma_1^{\ N} \sin(2\pi jk/N) = 0$$

$$\Sigma_1^{\ N} \sin i(\omega t + 2\pi k/N) = \sin(i\ \omega t + (N-1)i\pi/N).[\sin(i\pi)/\sin(i\pi/N)] = N \sin(i\ \omega t + (N-1)i\pi/N)$$

where the factor $\sin(i\pi)/\sin(i\pi/N)$ takes the value N for i equal to an integer multiple of N or zero for all other values. Simplifying Eq. 4.119 gives:

$$\frac{\Delta F}{F_o} = \frac{\alpha}{h_o} aN \sin(iN\omega t + \theta_i) \qquad 1<i<\infty \qquad [4.120]$$

where θ_i is a phase angle of no physical significance in this analysis. Equation 4.120 states that only small amplitude surface geometry errors with a wavelength equal to $2\pi/iN$ on the rotating surface produce axial motion error. The frequency of the motion error is $iN\omega$ and the peak force amplitude is:

$$\Delta F = \alpha\, aNF_o/h_o \qquad [4.121]$$

The corresponding peak amplitude is $\Delta F/K$, where K is the bearing stiffness. Equation 4.121 gives an approximate relationship between small amplitude surface geometry errors and motion error for an orifice fed annular thrust bearing subject to a concentric load and based upon the surface errors regulating the mass flow rate of air through the bearing. Assuming the latter uniformly affects the complete pressure field in the bearing, then Eq. 4.121 gives an upper bound estimate of motion error. The effects of significant static pressure variations in the bearing produced by, for example, offset axial load, misalignment, distortion or variation in orifice flow characteristics, have not been considered. Significant scatter in determining the coefficient of discharge for apparently identical orifices is reported by [9]. For a given pressure ratio across the orifices, the coefficients varied by more than 10%. The main effect this will have is to increase the overall amplitude of motion errors generating frequencies, harmonic of basic rotational speed. This can be illustrated by considering a single orifice to have a different coefficient of discharge to the others used in the bearing. For this case, Eq 4.119 can be rewritten as:

$$\frac{\Delta F}{F_o} = \alpha \Sigma_1^N \left[\frac{a}{h_o}.\sin i(\omega t + 2\pi k/N) + \frac{b}{h_o}.\sin(2\pi jk/N)\right] + (\alpha_1 - \alpha)$$

$$\left[\frac{a}{h_o}.\sin i(\omega t + 2\pi k/N) + \frac{b}{h_o}.\sin(2\pi jk/N)\right] \qquad 2 < i < \infty, 2 < j < \infty \qquad [4.122]$$

where now $(\alpha_1 - \alpha)$ represents the difference in averaging factor between that of the normal orifices and the odd orifice. Thus, in addition to producing force fluctuations of frequency $Ni\omega$, the bearing now produces force fluctuation frequencies of $i\omega$ in addition.

It is interesting to note the benefit to motion error obtained by introducing pockets into the orifice fed bearing. They reduce the restriction to air flow at entry to the bearing film and reduce the sensitivity of the bearing to small changes in bearing gap arising from surface geometry errors: the larger the pocket diameter, the greater the potential for the bearing to average the effects of surface errors to a low level. It follows that other forms of compensation giving a more uniformly distributed flow of air into or out of the bearing have a greater potential to achieve low motion errors. Porous bearings, slot fed bearings and grooved or recessed bearings, for example, all allow air to be fed in uniformly around the circumference, generating a high averaging effect and if manufactured to the same standard as an orifice-fed bearing, should produce motion errors substantially lower.

4.13.3 Approximate motion error produced by journal bearings

The assumptions and approximations applied to annular thrust bearings in the preceding section may also be applied to journal bearings. Consider the case of a journal bearing subject to a central radial load for which the gap h_r is a function of angular position:

$$h_r = h_o(1 + \varepsilon.\cos\theta)$$

where the eccentricity ε is a small fraction of the concentric gap h_o. For small amplitude surface errors superimposed on the rotating surface of amplitude $a.\sin i(\omega t + \theta)$, the bearing gap is:

$$h_o(1 + \varepsilon.\cos\theta) + a.\sin i(\omega t + \theta_k) = {}_o(1 + \varepsilon.\cos\theta) + \Delta h \qquad [4.123]$$

For an orifice-fed bearing, motion error is assumed to be produced by Δh regulating the mass flow rate of air at entry to the bearing gap as for the annular thrust bearing considered in the previous section. Equation 4.116 giving the pressure perturbation for a bearing of constant gap h_o is modified to take account of gap varying with angular position:

$$\frac{\Delta P}{P_o} = \frac{\Delta h(1 - P_a^2/P_o^2)}{2h_r[1 + 16d_r^2\, h_r^2/K_d C_d^2 d_o^4]} \qquad [4.124]$$

This is approximated to:

$$\frac{\Delta P}{P_o} = \frac{\Delta h(1 - P_a^2/P_o^2) - (1 - \varepsilon.\cos\theta)}{2h_o[1 + 16d_r^2\, h_o^2/K_d C_d^2 d_o^4]} = \frac{\Delta h(1 - \varepsilon.\cos\theta)\alpha}{h_o} \qquad [4.125]$$

The force fluctuation local to a single orifice is:

$$\frac{\Delta F}{F_o} = \alpha\left[\frac{a}{h_o}.\sin i(\omega t + 2\pi k/N) + \frac{b}{h_o}.\sin(2\pi jk/N)\right]$$

$$(1 - \varepsilon.\cos\theta) \qquad 2 < i < \infty, 2 < j < \infty \qquad [4.126]$$

The net force generated by the bearing in the Y direction is:

$$\frac{\Delta F}{F_o} = \alpha \Sigma_1^N \left[\frac{a}{h_o} . \sin i(\omega t + 2\pi k/N) + \frac{b}{h_o} . \sin(2\pi jk/N) \right]$$

$$(1-\varepsilon . \cos 2\pi k/N) \cos 2\pi k/N \qquad 2<i<\infty, 2<j<\infty \qquad [4.127]$$

Considering just the rotating surface, Eq. 4.127 may be rewritten as:

$$\frac{\Delta F}{F_o} = \frac{\alpha a}{2h_o} \Sigma_1^N [\sin(i\omega t + 2\pi k(I+1)/N) + \sin(i\omega t + 2\pi jk(i-1)/N)] -$$

$$\frac{\alpha \varepsilon a}{4h_o} \Sigma_1^N [\sin(i\omega t + 2\pi k(I+2)/N) + \sin(i\omega t + 2\pi jk(I-2)/N)$$

$$+2 \sin(i\omega t + 2\pi jki/N)] \qquad 2<i<\infty \qquad [4.128]$$

Each of the terms in the summation may be simplified using [26], as for Eq. 4.119. This leads to terms of the form:

$$\sin(i\omega t + \theta').[\sin(x\pi)/\sin(x\pi/N)]$$

where θ' is a phase angle and x takes the value: $i-2$, $i-1$, i, $i+1$ or $i+2$. When x equals an integer multiple of N, then $\sin(x\pi)/\sin(x\pi/N) = N$ or otherwise zero. Thus Eq. 4.128 may be rewritten as:

$$\frac{\Delta F}{F_o} = \frac{\alpha a N}{2h_o} [\sin((jN+1)\omega t + \theta_{i+1}) + \sin((jN-1)\omega t + \theta_{i-1}] - \frac{\alpha \varepsilon a N}{4h_o}$$

$$[\sin((jN+2)\omega t + \theta_{i+2}) + \sin((jN-2)\omega t + \theta_{i-2}) + 2\sin(jN\omega t + \theta_i)] \quad 2<i<\infty \quad [4.129]$$

Equation 4.129 gives the net force fluctuations in a radial (Y) direction for a journal bearing subject to a light central radial load. The force fluctuations occur at frequencies harmonic of the bearings rotational speed times the number of orifices, $jN\omega$ and $jN\omega \pm 1$ and $jN\omega \pm 2$. Each frequency is generated by a different surface wavelength, $2\pi/(jN)$, $2\pi/(jN \pm 1)$, $2\pi/(jN \pm 2)$. The amplitude of motion error may be found by dividing ΔF by the bearing's radial stiffness as determined from a static analysis.

Equation 4.129 is an approximation used to explain the mechanism by which externally pressurised journal air bearings generate motion error. In general, all harmonics of basic rotational speed may be generated, particularly if significant static pressure variations in the bearing are produced, for example, by offset radial load, misalignment, distortion or variation in orifice flow characteristics. As for the annular thrust bearing, the orifice compensated aerostatic bearing has less potential to produce low motion error than, for example, porous, slot fed or micro-recessed bearing types.

4.13.4 Asynchronous motion error

Aerostatic bearings in normal operation produce exceptionally low levels of asynchronous motion error, typically less than 10 nm peak–to-peak and on the most accurate bearings lower than 1 nm. For aerostatic bearings, there is even less published work on this subject than on synchronous motion error, so the following discussion is based on the author's experience developing air spindles for noise testing ball bearings.

Considering an externally pressurised air bearing in isolation, any asynchronous motion error must either arise from random occurrences in the bearing or the air supply. Even for factory air lines, pressure pulsations are generally negligible and the standards of air cleanliness required for air bearings are easily met with low cost filters. So this is seldom an issue.

Random occurrences within the bearing can be generated by or associated with the air flow through it. For example, air exhausting from the bearing edges is often observed to create an audible sound, best described as a 'hiss'. Hiss approximates 'white noise' and contains a broad spectrum of frequencies that in this case is produced by small fluctuations in air pressure. It is not unreasonable to expect such fluctuations to create relative movement between the bearing surfaces, albeit at high frequencies and low amplitudes. Spindles required for noise testing ball bearings must have low vibration levels over a 50 to 10 kHz bandwidth and since bearing vibration is measured in terms of velocity, amplitudes at high frequencies must be extremely low; ideally at 10 kHz, below 1 nm.

An indication of the effect of air flow on motion error may be obtained by monitoring the broadband vibration level occurring on a static bearing for increasing air supply pressure. The simple arrangement shown in Figure 4.33 consists of a single journal bearing supported on a shaft that is vibration isolated from its surroundings. An accelerometer placed on the journal records the journal's radial component of vibration level. The spectrum of the vibration is obtained and rms acceleration converted to displacement at dominant frequencies in the 0 to 1 kHz range. The vibration measurement, in terms of displacement, is shown plotted against air supply pressure in Figure 4.34. Displacement is observed to increase from a background noise level of 0.024 nm with no air supply almost linearly to 2.5 nm at a pressure of 5 Bar. Since the bearing is static,

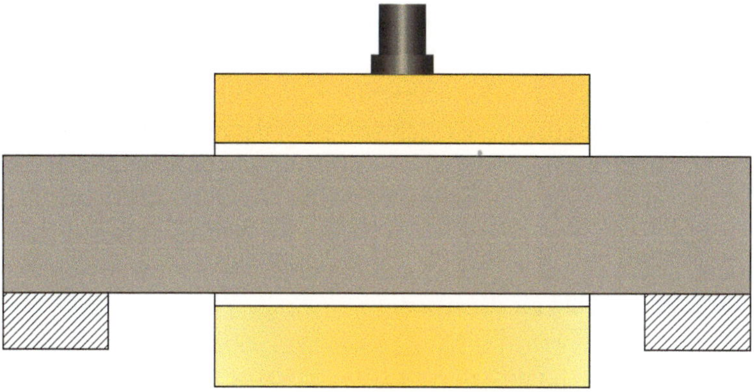

Figure 4.33 Experimental arrangement for monitoring asynchronous vibration on aerostatic journal bearing

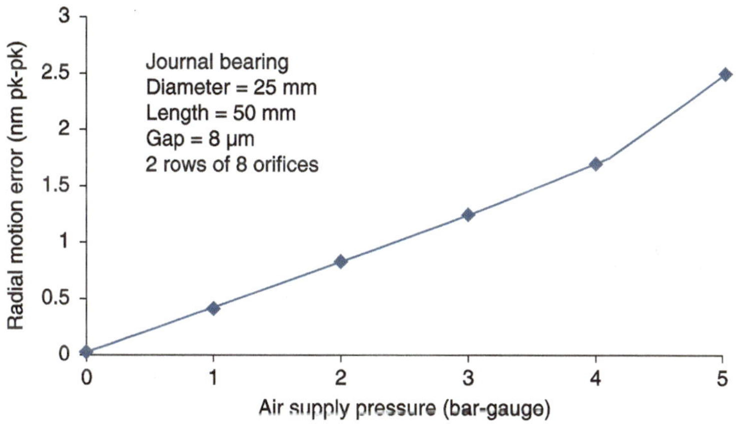

Figure 4.34 Air bearing vibration level versus supply pressure

there is no relationship to bearing speed, but instead the vibration is random, exciting structural resonances associated with the bearing and adjacent components.

The causes of asynchronous motion error are not clear from the experiments and it is speculated that localised turbulence within the bearing created in and around features such as orifices and pockets are responsible. However, whether this is correct or not, it follows that minimising the air flow rate through an aerostatic bearing reduces asynchronous motion error. One implication is that bearings with reduced gap size have the potential for low asynchronous motion error.

4.14 Thermal characteristics

Aerostatic bearings are characterised by low friction and power loss, nonetheless in ultra precision machinery even temperature rises as low as 2 or 3 °C can create unwanted thermal distortions, so it is still important to consider this aspect of performance. In this section, a simple calculation of friction torque and power loss is presented, a thermal model suitable for structural analysis proposed and the power loss associated with the supply of pressurised air discussed.

4.14.1 *Torque and power loss*

The total power loss occurring in an aerostatic bearing has two components, one due to relative motion between its surfaces shearing the air film, the other due to air flow through the bearing. The latter occurs even if the bearing is stationary and does not normally generate a significant torque. For simplicity, the two mechanisms are considered to be independent and that frictional torque arises only from shearing of the air film within the bearing where the film is assumed to obey the Newtonian law:

$$\tau = \eta.\ dU/dh \qquad [4.130]$$

In the above equation, the film thickness is h, U is the relative velocity of the bearing surfaces, η is the viscosity of air and τ is the shear stress acting across the film (Figure 4.35). Bearing torque is then found by integrating Eq. 4.130 over the bearing area.

Annular thrust bearings

Consider a bearing of bore diameter D_i and outside diameter D_o, for which one surface is rotating at a constant annular speed of ω (Figure 4.36). The torque generated by an annular element of width dR at radius R is:

$$dT_b = R.\ \tau.2\pi RdR = 2\pi R^2.dR.\eta.\ dU/dh$$

where:

$$dU/dh = \omega R/h$$

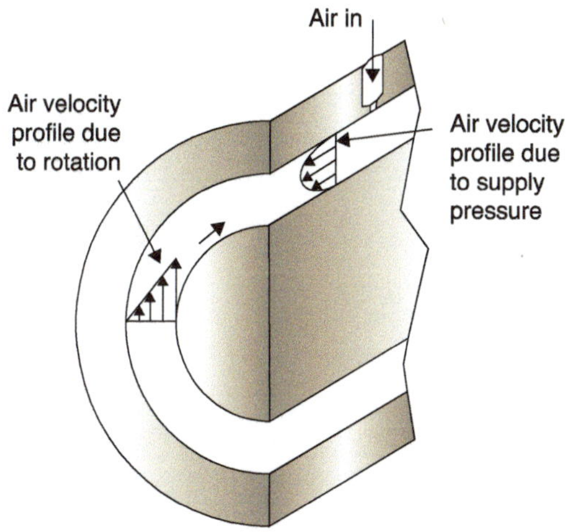

Figure 4.35 Air velocity profiles due to rotation and supply pressure

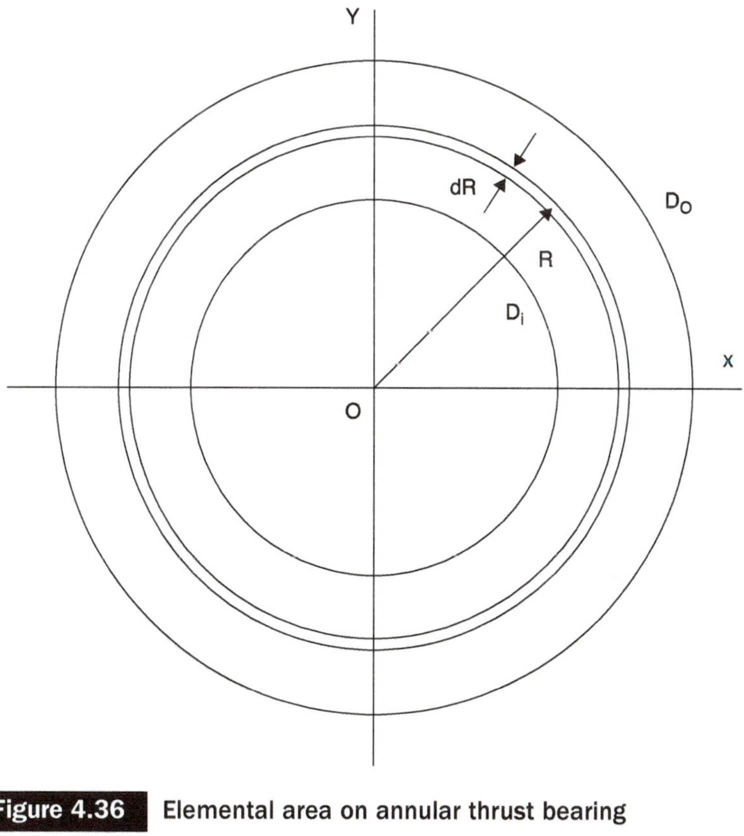

Figure 4.36 Elemental area on annular thrust bearing

Integrating over the bearing area:

$$T_b = \frac{\pi\eta\omega}{\partial h}\int_{R_i}^{R_o} R^3\, dR$$

$$T_b = \pi\eta\omega[D_o^4 - D_i^4]/32\,h \text{ where } D_o = 2R_o \text{ and } D_i = 2R_i \qquad [4.131]$$

The power dissipated by bearing rotation is:

$$W = T_b\omega = \pi\eta\omega^2[D_o^4 - D_i^4]/32\,h \qquad [4.132]$$

For annular thrust bearings, note how sensitive bearing power loss is to diameter. In ultra precision spindles using the double-sided thrust and two-journal bearing arrangement shown in Figure 4.1(a), the external diameter of the thrust bearing required to operate at high speeds is sometimes a compromise between a large diameter giving high static stiffness and a lower diameter limiting power losses.

Journal bearings

For a journal bearing of diameter D and length L, an annular surface element of width dZ has a length πD and if one of the bearing surfaces is rotating at constant angular speed ω, then the torque generated by the element is (Figure 4.37):

$$dT_b = (D/2).(\eta dU/dh).\pi D.dZ$$

where $dU/dh = \omega D/(2h)$. Integrating from –L/2 to L/2 gives the total friction torque of:

$$T_b = \pi\omega\eta L D^3/(4\,h) \qquad [4.133]$$

The power dissipated by bearing rotation is:

$$W = \pi\omega^2\eta L D^3/(4\,h) \qquad [4.134]$$

Air supply losses

The energy contained in compressed air supplied to an aerostatic bearing is dissipated in the bearing as it loses pressure, flowing through the bearing. The power loss is the product of supply pressure P_s and is given by:

$$W_p = P_a V_a \ln(P_s/P_a) \qquad [4.135]$$

where P_s is the supply pressure, P_a is ambient pressure and V_a is the volume flow rate is at ambient pressure.

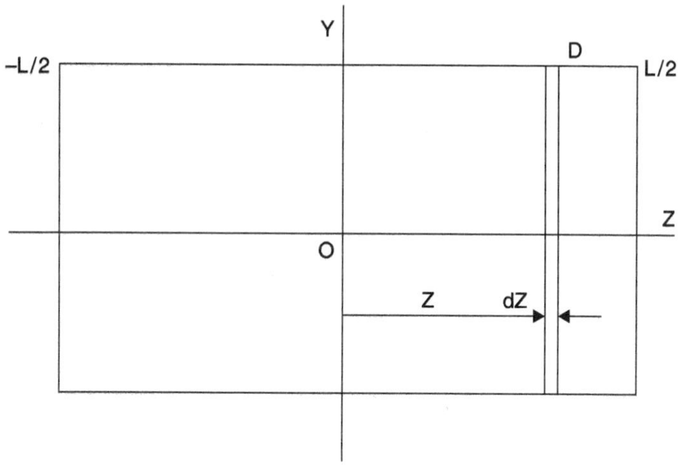

Figure 4.37 Elemental area on journal bearing

Air supply energy losses per bearing tend to be low and are not normally a significant source of heat generation. Furthermore, the energy dissipation is compensated for by air flow through the bearing removing some of the heat generated. Of more concern to some air bearing users are the costs associated with supplying the compressed air in the first place.

4.14.2 Thermal model

For the purpose of estimating temperature rise and thermal distortion in structures of which air bearings are part, the heat removed by the air exhausting from the bearing may be neglected and the power loss associated with the bearing assumed to be dissipated equally in the housing and rotor parts of the structure.

4.15 Reliability

There is no fundamental limitation to the life of an aerostatic bearing. Used within its designed operating conditions, the bearing surfaces are completely separated by a film of air, so there is no wear and life can be infinite. However, in practice, bearing life is limited by abuse of one form or another. The low operating clearances means that aerostatic bearings are sensitive to the quality of the air supply and an inherently low load carrying capacity means that they are intolerant of accidental loads or indeed temporary loss of air supply pressure.

Air bearing manufacturers make clear recommendations on the pressure and the quality of the air to be used with their products. Essentially the air must be free from solid or liquid contaminants that may bridge the operating gap, and dry to avoid a build up of oxide or other films on the bearing surfaces. The quality of compressed air is defined in the specification ISO 8573-1, where permissible particle sizes, water and oil content are defined in terms of different classes. Table 4.1 gives the first five classes.

For air bearings class 1 or 2 particle size limits, class 1 oil and class 5 water limits are normally recommended. This specification is easily met by ensuring an 'oil free' compressor is used to generate pressurised air and the air is passed through a standard filter set before reaching the bearing.

Bearing materials are chosen on the basis of dimensional stability, resistance to corrosion and resistance to damage or wear occurring when accidentally applied loads are sufficiently high to cause the bearing surfaces to contact. The 'crash proof' spindle is an ideal concept that some air bearing manufacturers have tried to develop, with varying degrees of success but is by no means universally applicable. Traditionally, wear or damage resistance is provided by using a hard material such as a martensitic stainless steel, hardened and tempered for one bearing surface and a low shear strength material such as lead bronze for the other. Under sliding contact, the lead acts as a boundary lubricant, minimising local heat generation and wear rates. The use of dissimilar materials also provides a degree of 'scuff' resistance, enabling the bearing geometry to be retained during and after overload.

Table 4.1 **Compressed air quality limits defined in ISO 8573-1**

Class	Particle Size (μm) and Number off /m^3			Water	Oil
	0.1–0.5	0.5–1.0	1.0–5.0	Dewpoint (C)	mg/m^3
1	100	1	0	−70	0.01
2	10k	1k	10	−40	0.1
3	–	10k	50	−20	1
4	–	–	1k	−3	5
5	–	–	2k	7	–

A wide range of surface coatings are available that can offer some improvement to crash resistance over and above the traditional materials. Thin film pvd or cvd coatings, such as diamond-like carbon, provide exceptional hardness and low friction and can be used in conjunction with other hard or semi-hard materials to provide both low heat generation and low wear rates.

4.16 References

1. Rowe, B.W. (2012), *Hydrostatic, Aerostatic and Hybrid Bearing Design*, UK: Butterworth–Heinemann.
2. Devitt, D. (2004), 'Analysis of pressurised porous air bearings', *20th Annual Workshop on Mathematical Problems in Industry*, University of Delaware, 21–25 June.
3. Arneson, H.E.G. (1966), 'Hydrostatic bearing structure', US Patent No. 3305282, 29 March.
4. Reynolds, O. (1886), 'On the theory of lubrication and its application to Mr Beauchamp Tower's experiments, including an experimental determination of the viscosity of olive oil', *Philosophical Transactions of the Royal Society*, **177**: 157–234.
5. Harrison, W.J. (1912–1925) 'The hydrodynamic theory of lubrication with special reference to air as a lubricant', *Transactions of the Cambridge Philosophical Society*, **XXII**: 6–54.
6. Hamrock, B.J. (1994), *Fundamentals of Fluid Film Lubrication*, New York: McGraw Hill.
7. Powell, J.W. (1970), *Design of Aerostatic Bearings*, London: Machining Publishing Co. Ltd.
8. Dudgeon, E.H. and Lowe, I.R.G. (1965) 'Theoretical analysis of hydrostatic gas journal bearings', *National Research Council of Canada*, Mech. Engineers Report No. MT-54.
9. Pink, E.G. and Stout, K.J. (1979), 'Orifice restrictor losses in journal bearings', *Proceedings of the Institute of Mechanical Engineers*, **193**: 47–52.
10. Stout, K.J. and Sweeney, F. (1984), 'Design of aerostatic flat pad bearings using pocketed orifice restrictors', *Tribology International*, **17**(4): 191–8.
11. Tawfik, M. and Stout, K.J. (1989), 'The design of high efficiency flat pad aerostatic bearings using laminar restrictions', *Tribology International*, **22**(4): 273–81.
12. Vohr, J.H. (1969), 'A study of inherent restrictor characteristics for hydrostatic gas bearings', *University of Southampton Gas Bearing Symposium*, Southampton.
13. McCabe, J.T., Elrod, H.G. and Carfagno, S.P. (1969), 'Entrance effects in gas bearings, Franklin Institute report', No. FIRL-I-C2429-1.
14. Gargiulo, E.P. (1979), 'Porous wall gas lubricated journal bearings: theoretical investigation', *Transactions of the ASME, Journal of Lubrication Technology*, **101**: 4587–65.

15. Gargiulo, E.P. (1979), 'Porous wall gas lubricated journal bearings: experimental investigation', *Transactions of the ASME, Journal of Lubrication Technology*, **101**: 4667–73.
16. Blondeel, E., Snoeys, R. and Devrieze, L. (1980), 'Dynamic stability of externally pressurised gas bearings', *Journal of Lubrication Technology*, **102**: 5117–19
17. Roblee, J.W. and Mote, C.D. (1990), 'Design of externally pressurised gas bearings for stiffness and damping', *Tribology International*, **23(5)**: 3337–45.
18. Mishra, A.K. (1998), 'Analysis of pneumatic instability of an aerostatic rectangular thrust bearing with offset load', *Wear*, **122**: 17–22.
19. Chen, M.F. and Lin, Y.T. (2002), 'Static behaviour and dynamic stability analysis of grooved rectangular aerostatic thrust bearings by modified resistance network method', *Tribology International*, **35**: 3297–38.
20. Lund, J.W. (1967), 'A theoretical analysis of whirl instability and pneumatic hammer for a rigid rotor in pressurised gas journal bearings', *Transactions of the ASME Journal of Lubrication Technology*, **89(F)**: 1547–66.
21. Rao, N.S. and Majumdar, B.C. (1979), 'Analysis of pneumatic instability of externally pressurised porous gas journal bearings', *Transactions of the ASME Journal of Lubrication Technology*, **101**: 487–53.
22. Yoshimoto, S., Anno, Y. and Ohashi, T. (1988), 'Stability of a rigid rotor supported by aerostatic journal bearings with circular slot restrictors (on the double-row admission bearing)', *Transactions of the ASME*, **110**: 2287–34.
23. Miyatake, M., Yoshimoto, S. and Sato, J. (2006), 'Whirling instability of a rotor supported by aerostatic bearings with a surface restricted layer', *Proceedings of the Institute of Mechanical Engineers, Journal of Engineering Tribology*, **220(J)**: 95–103.
24. Timoshenko, S., Young, D.H. and Weaver, W. (1974), *Vibration Problems in Engineering*, 4th edition, New York: John Wiley & Sons Inc.
25. Larson, R.H. and Richardson, H.H. (1962), 'A preliminary study of whirl instability for pressurised gas bearings', *Transactions of the ASME, Journal of Fluids Engineering*, **84(4)**: 5117–20.
26. Dwight, H.B. (1968), *Tables of Integrals and other Mathematical Data*, New York: The Macmillan Company.

4.17 Notation

a,b – Amplitude of surface geometry error
C_d – Coefficient of discharge of orifice
C_{orf} – Coefficient of discharge of orifice with pocket
D_i – Bore diameter
D_o – Outside diameter
D_p – Pitch circle diameter
d_o – Orifice diameter
d_r – Pocket diameter
F – Load or force

F_c – Load capacity
H – h/h_o
h – Bearing gap
h_o – Static equilibrium bearing gap
K – Static stiffness
K_d – Entrance loss coefficient
K_p – Permeability of porous media
L – Journal bearing length
M – Moment
m – Mass flow rate
N – Number of orifices
P – p^2
P_a – Atmospheric pressure
P_c – Pocket edge pressure
P_{dyn} – Dynamic pressure
P_e – Land entry pressure
P_o – Static equilibrium pressure
P_p – Pocket pressure
P_s – Supply pressure
P* – Critical pressure
P′ – p/P_a
p – Pressure at any point in the bearing gap
Q – Volume flow rate
R – Gas constant
Re – Reynolds number
r – Radius
T – Temperature
T_b – Bearing torque
t – Time
u,v – Surface velocities
V_a – Air flow rate at ambient pressure
W – Power loss
W_p – Pumping power loss
X,Y,Z – Cartesian coordinates
x – Wavelength factor
x_m – Peak amplitude

Symbols

$\underline{\alpha}$ – Averaging factor
γ – ratio of specific heats for air
γ_d – Damping ratio
Δh – Gap increment
Δp – Pressure increment
Δ_a – Logarithmic decrement
ε – Eccentricity
ξ – Efficiency factor

ψ – Tilt
η – Dynamic viscosity
θ – Angle of rotation
τ – Shear stress
ω – Angular speed
ρ – Density

5

Hydrodynamic herringbone and spiral grooved bearings

DOI: 10.1533/9780857092182.307

Abstract: Hydrodynamic herringbone and spiral grooved bearings are already widely used in computer disc drives and portable equipment, where they provide accurate running and high damping in a compact envelope and at low cost. Innovations such as liquid metal or ferrofluid lubricants have also helped this type of bearing to find an increasing number of applications where ease of sealing, high thermal conductivity or operation in a vacuum is important. The bearings operate with clearances in the 5 to 100 µm range, so most bearings do not demand unusually precise manufacturing or assembly techniques. But they are not available as standalone components from bearing companies, and equipment manufacturers generally design and manufacture their own bearings to suit the specific requirements of their products. In this chapter, a means of calculating the load capacity, static stiffness, damping and power loss of grooved annular thrust and cylindrical journal bearings is presented and the factors influencing motion error explained.

Key words: Self-acting bearings, hydrodynamic bearings, oil lubricated bearings, herringbone grooved bearings, spiral grooved bearings, chevron grooved bearings, fluid film bearings, high speed bearings, high stability bearings, whirl free bearings, low motion errors.

5.1 Introduction

'Self-acting bearings' is the name given to bearings that generate a lubricant film from relative motion between two surfaces dragging fluid into some form of wedge. When this happens, pressure builds up in the film, enabling it to support load. Hydrodynamic herringbone and spiral grooved bearings are a form of self-acting bearing that use a liquid, normally oil, as a lubricant and generate stable and accurate rotation. Such bearings are therefore highly portable, unlike hydrostatic and aerostatic types, have a low cost of ownership and are suitable for use in domestic, office, clean room and quiet environments, where compressors or high pressure power supplies are not practical or at least undesirable.

There are many types of hydrodynamic bearings but not all are suitable for precision engineering applications. In fact, most hydrodynamic cylindrical journal bearings rely on an eccentricity between opposing surfaces to create a wedge and hence pressure build-up in the fluid film. This eccentricity is not only a source of inaccuracy in its own right but the non-axi-symmetric pressure distribution it creates gives the bearing poor dynamic stability and the potential to generate high motion errors arising from any geometrical imperfections on its surfaces. The instability that conventional designs of self-acting cylindrical journal bearings are prone to and is unacceptable in high precision applications, is known as half speed whirl. Because these types of bearing rely on relative surface motion to create a lubricant film, any motion of the rotor's mass centre adds to or subtracts from the surface motion due to rotation and therefore affects the lubricant film thickness.

An orbital motion of the rotor's mass centre at a frequency of approximately half the rotational speed and in the same direction dramatically reduces film thickness and may lead to it collapsing altogether. Even if the film does not collapse, half speed whirl is readily excited by random events and results in unwanted sub-synchronous motion error that is not strictly synchronised with rotor speed. It is for this reason that attention is restricted to a particular class of self-acting bearing known as herringbone (journal) and spiral (thrust) grooved bearings. The essential feature of these bearings is that grooves machined into one of the bearing surfaces are used to pump fluid in an axial direction on a journal bearing or in a radial direction on a thrust bearing, creating an axi-symmetric pressure distribution, reducing the tendency to whirl and eliminating eccentricity or swash errors. The improvement in motion error and dynamic stability achieved is substantial; furthermore, the groove patterns can be arranged to act as mechanical seals aiding lubricant retention and minimising leakage in hydrodynamic bearing types.

In not requiring any auxiliary equipment or a sophisticated sealing arrangement, hydrodynamic herringbone and spiral grooved bearings simplify equipment design making this type of bearing very cost-effective. The static stiffness and load carrying capacity of the bearings is low compared to that of ball or hydrostatic bearings, but in having a fluid film of large area the damping capacity is high. The axi-symmetric pressure profile generated by this type of bearing also gives it the ability to average out bearing surface imperfections and hence the potential for low motion errors.

Herringbone and spiral grooved bearings may also use air as a lubricant, in which case they are referred to as aerodynamic bearings (Chapter 6). The differences in performance between hydrodynamic and aerodynamic bearing types are due to the differences between the characteristics of the lubricants they use. Oil has a relatively high viscosity and so enables the hydrodynamic bearing to support higher loads and generate higher damping, but at the expense of higher friction. Oil is also incompressible, which gives the hydrodynamic bearing the ability to respond rapidly to dynamic loads enhancing its stability. Air has a low viscosity so the aerodynamic bearing has low friction making it more suitable for use at high speeds, but the low viscosity of air coupled with its compressibility severely limits the bearing's load capacity. It also results in a small time delay in the bearing responding to dynamic loads. If the bearing speed is sufficiently high, this delay may lead to an unstable bearing motion.

In this chapter, a means of calculating the load capacity, static stiffness, damping and power loss occurring in hydrodynamic herringbone and spiral grooved bearings is described. Motion error arising from bearing surface imperfections is also considered and an approximate relationship between the amplitude and wavelength of surface imperfections and frequency and amplitude of motion derived. The analysis focuses on flat annular and cylindrical journal bearing geometries.

5.2 Bearing types

Grooves may be machined into tapered, spherical, cylindrical and flat annular bearing geometries. Figure 5.1 shows some examples of groove geometries typical of annular thrust bearings used on hard disc

Figure 5.1 Chevron grooved annular thrust bearings, 6 mm outside diameter

drives. Inward and outward pumping spiral grooves are arranged to form chevrons that pump lubricant towards a designed pitch circle diameter. The pumping characteristics of the inner and outer grooves are matched so that there is no net flow of lubricant and pressure is a maximum at the transition diameter. The bearings in Figure 5.1 are experimental bearings with different numbers of grooves and different groove angles used to determine the effect of groove design on load carrying capacity.

Figure 5.2 shows herringbone grooves machined into a shaft, again typical of a hard disc drive bearing. In this case the grooves are asymmetrical, so that there is a net flow of lubricant into a cavity formed between two bearings. This design is intended to minimise the risk of oil leakage.

An example of a spherical grooved bearing is shown in Figure 5.3, and in this case the axis of rotation is normal to the radial plane containing the circular groove, so that the spiral grooves pump lubricant into a central zone, where the surface is smooth and where the pressure is a maximum. The complete bearing extends over half of the ball's surface, thus giving it radial and axial load carrying capacity with respect to the axis of rotation.

Figure 5.2 Herringbone grooved bearing on 4 mm diameter shaft

Figure 5.3 Spherical grooved bearing 8 mm diameter

Spiral grooved thrust bearings take one of three forms (Figure 5.4), inward pumping, outward pumping or chevron grooved. With inward and outward pumping bearings, the lubricant is pumped towards a plain sealing land and pressure is a maximum at the groove–seal transition diameter. General practice is to optimise the transition diameter to give maximum load carrying capacity. If the sealing land width is too low, lubricant leakage rates are high and drain away some of the pressure generated by the grooves, whereas if it is too high, the grooved land is

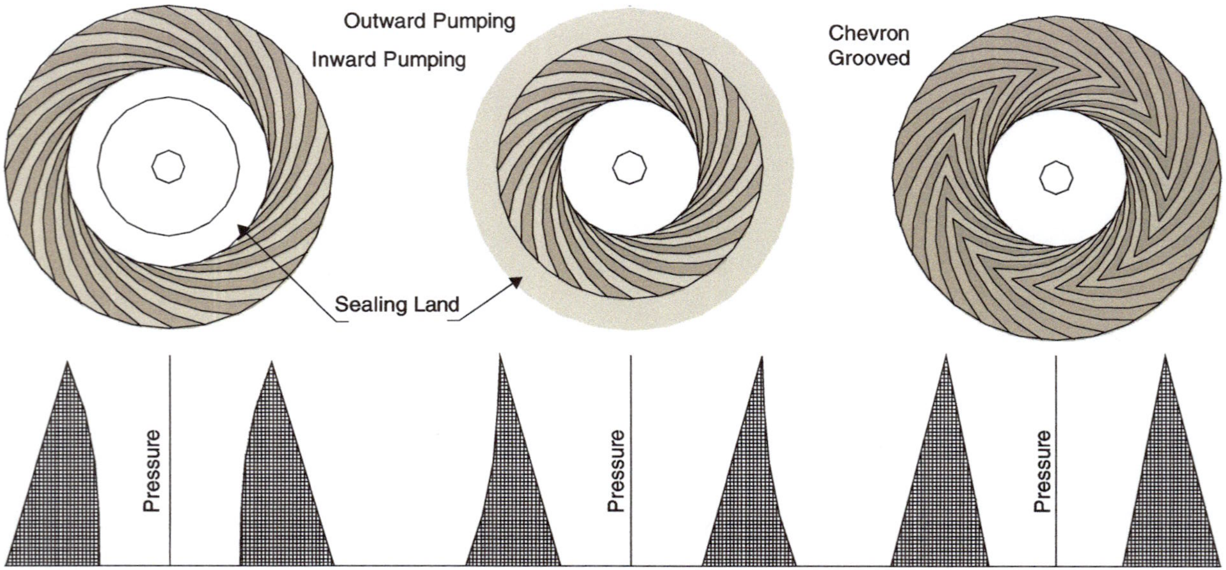

Figure 5.4 Inward pumping, outward pumping and chevron grooved thrust bearings and pressure profiles

narrow and does not have the capacity to generate a high pressure. Inward and outward pumping bearings thus create a net flow of lubricant, albeit the mass flow rate is very low.

For the chevron-grooved bearing, inward and outward groove designs are matched to generate the same pressure at the transition diameter; furthermore, the transition diameter is usually optimised to give maximum load carrying capacity. Thus the bearing has reasonable load carrying capacity and no net flow of lubricant. In this bearing design, the grooves act as mechanical seals helping to simplify the overall sealing arrangement of the bearing cavity. Furthermore, the bearing can be used with a magnetic oil [1], in which case no other form of sealing may be necessary.

Journal bearings are either partially grooved or fully grooved (Figure 5.5). The plain central land on the partially grooved bearing acts as a conventional hydrodynamic or aerodynamic journal bearing and adds to load carrying capacity, but at the expense of reduced dynamic stability. However, the fully grooved bearing has excellent stability and is the preferred choice where high precision rotation is required.

With all bearing types, the grooves may be placed on the stationary or rotating surface. Load capacities are similar and there is generally some improvement in stability in placing the grooves on the rotating component, but at the expense of motion error. The grooves create a pressure ripple superimposed on the static pressure profile and unless this is perfectly

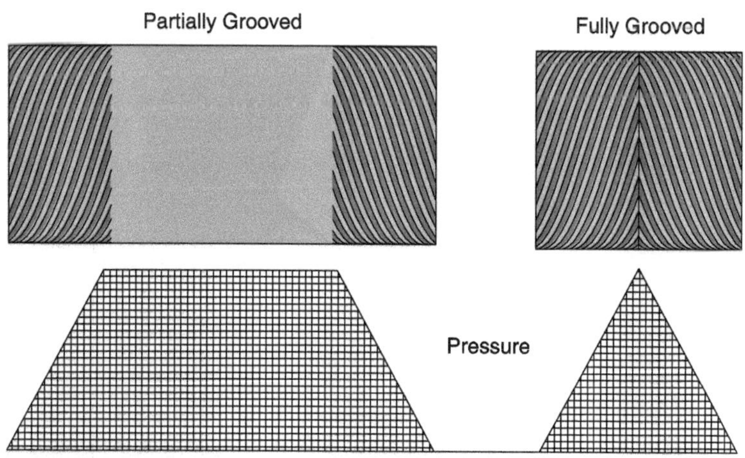

Figure 5.5 Partially grooved and fully grooved journal bearings

axi-symmetric, then rotating grooves will create motion error at frequencies harmonic of rotational speed.

5.3 Operating principles

Grooved bearings may be regarded as special cases of the stepped bearing first demonstrated by Lord Raleigh in 1918. The latter simply consists of two parallel surfaces, one of which is recessed to create regions of high and low gap (Figure 5.6). The edge of the step is arranged perpendicular to the direction of relative motion, so that fluid filling the space between the surfaces is dragged from a region of high to low gap. Under these conditions, continuity of mass flow rate dictates a change in velocity profile and hence a build up of pressure as the fluid passes over the edge of the step. In the stepped bearing, pressure at entry to and exit from the pad is at ambient and the pressure gradients in the two regions can be shown to be constant [22], so that the pressure profile is triangular with maximum pressure occurring at the step edge. Herringbone and spiral grooved bearings not only have a succession of steps but the step edges are arranged to be at some angle to the direction of relative motion. This leads to pressure gradients both in the direction of relative motion and normal to it, so that the local triangular pressure profile associated with each step becomes superimposed on a smooth profile (Figure 5.6).

The mechanism of pressure generation holds for all practical groove profiles. In developing analyses for predicting bearing performance, it is mathematically convenient to consider rectangular groove cross-sections and to consider narrow groove and ridge widths. Theoretically this also gives the maximum load carrying capacity, but in practice the grooves have to be manufactured and practical manufacturing techniques generate a range of forms where corners may be raised, groove sides inclined and the bottom of the grooves may be profiled. Even circular arc groove cross-sections have been found to work well, but prediction of their performance at the design stage requires published analyses to be factored.

5.4 Past developments

Herringbone and spiral grooved bearings and the analytical procedures for predicting their performance have evolved over more than half a

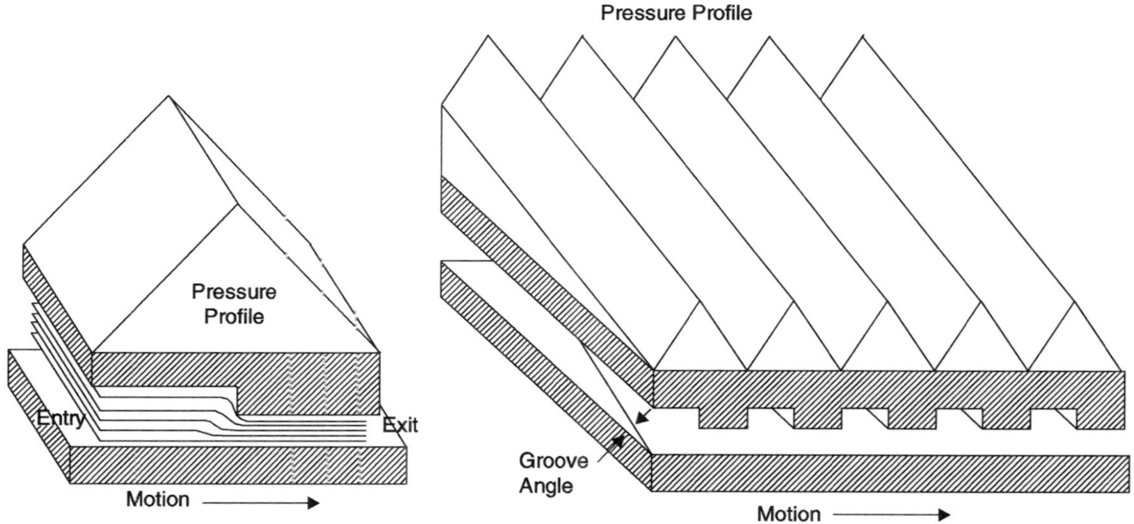

Figure 5.6 Cross-section through and pressure profiles for stepped and grooved bearings

century. A first satisfactory analysis of grooved bearing performance may be traced back to 1949, when Whipple [2] presented a basic model for predicting the pressure profile within a parallel sided thrust plate containing straight sided grooves. This analysis has since been improved in accuracy and extended to a wide range of bearing types and operating conditions by a number of researchers. Much of the early work, performed in the 1960s and 1970s, was driven by the need for low motion error bearings in recording equipment such as video recorders. Later, from about 1990 onwards, computer disc drives became the focus for grooved bearing development, as it was clear that disc capacity could be considerably increased by switching from ball to fluid film bearings having much lower levels of asynchronous motion error. In both of these fields of application, operating conditions of load and speed, coupled with the need for high damping, were better suited to hydrodynamic bearings than aerodynamic bearings and they were therefore the focus of attention.

The early development work examined all aspects of bearing performance necessary to produce a reliable, self-contained hydrodynamic bearing assembly and analytical procedures for predicting its performance. Muidermans improved upon existing analyses largely attributed to Whipple, by accounting more accurately for groove end effects and extended the analyses to other bearing geometries [3]. Elrod [4] and Booy [5] also considered groove-end effects, Reinhout [6] and Hirs [7] investigated bearing stability, whilst Chow and Vohr [8] examined turbulent operation of grooved bearings. Reinhout [9] and Remmers [10] both considered the use of grease as a lubricant. Thermal effects were investigated by Pan and Sternlicht [11] and by Bootsma [12], whilst Volman [13] examined the effect of spring preloaded bearing arrangements. These developments not only led to improvements in bearing performance but also to accurate and computationally efficient analytical procedures for predicting performance. In 1975, Bootsma [14] published, in the form of a PhD Thesis, a comprehensive analysis of the load capacity and dynamic stability of liquid lubricated grooved bearings having spherical, tapered, flat annular or cylindrical forms. The analysis was supported by experimental work, with fully and partially lubricated bearings also taking into account sealing characteristics. Bootsma [15] later followed this work by examining the conditions for which leakage free operation of herringbone grooved journal bearings could be obtained.

Publication of computer disc drive developments began around 1995. These focused on the development of disc drive motor units with integral hydrodynamic bearings and determined optimum bearing designs and arrangements, largely applying the earlier work described above rather than making fundamental bearing developments. In the disc drive application, the dynamic characteristics of grooved hydrodynamic bearings were of most interest. Leuthold [16] considered the dynamic requirements of a grooved bearing suitable for hard disc drive motors experiencing high inertial loads, whilst Ku [17] compared the dynamic characteristics of hydrodynamic and ball bearing types, concluding that the former had much higher damping levels and were therefore superior in the computer disc drive application. Overall system performance was also of interest and Finite Element based models developed for simulating both the static and dynamic characteristics of a complete thrust and journal bearing arrangement [18].

Aerodynamic bearing developments were performed in parallel with those of hydrodynamic bearings, but were related to the requirements of high speed turbo machinery. Aerodynamic bearings are discussed in Chapter 6, but as the analytical models developed for predicting aerodynamic bearing performance are more general and are equally applicable to hydrodynamic bearings, they are favoured here.

5.5 Load–deflection analysis of annular thrust bearings

The analysis of annular thrust bearings largely follows that of Malanoski and Pan [19]. They considered the static and dynamic performance characteristics of a spiral grooved annular thrust bearing subject to a concentrically applied axial load. In this section, the static analysis developed in [19] is used to describe the axial load–deflection characteristics of chevron grooved bearings and of inward or outward pumping spiral grooved bearings with sealing lands. The analysis is used to determine bearing load capacity and static stiffness; however, attention is restricted to concentrically loaded bearings operating with a constant gap and fully flooded with lubricant.

5.5.1 Mass flux in annular thrust bearings

The starting point for the analysis is to derive an expression for the mass flux of lubricant through an annular grooved land. Once this is known,

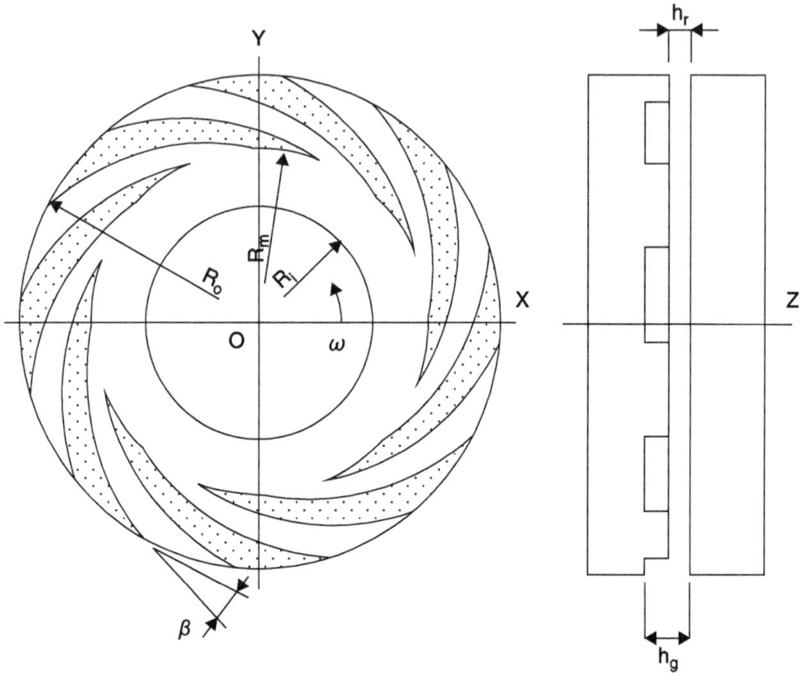

Figure 5.7 **Notation used for spiral thrust bearing analysis**

pressure gradients and hence pressure distributions can be determined in terms of bearing geometry and operating conditions. A sealing land may be considered as a special case of a grooved bearing with zero groove depth, thus the bearing first considered is an inward pumping spiral groove bearing with a sealing land rotating at constant speed ω (Figure 5.7). For this case, there is a net radial mass flow of lubricant towards the bearing centre. The outside radius of the bearing is R_o, the bore radius is R_i and the groove to seal transition radius is R_m. The clearance under a land, h_r and under a groove, h_g are both constant, as is the spiral groove angle, β.

The groove width is a_g and the land width is a_r, where the ratio $a_g/(a_g + a_r) = \alpha$ is also a constant. For mathematical convenience, the analysis is restricted to infinitesimally narrow groove and land widths.

To determine the load capacity and static stiffness of the bearing requires the smoothed pressure profile to be defined in terms of the geometrical parameters above, lubricant properties and bearing speed.

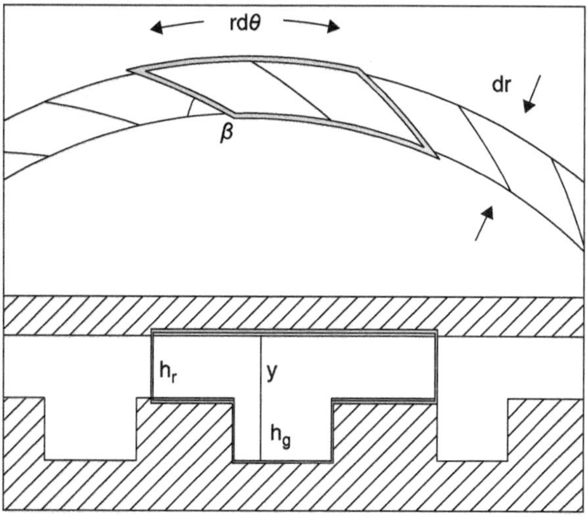

Figure 5.8 **Element of groove-ridge pair of width dr and length rdθ**

The starting point is to consider fluid flow in an element of a single groove–ridge pair.

An element of a groove–ridge pair is shown in Figure 5.8. The smoothed pressure profile has a radial gradient dp/dr and for a Newtonian fluid subject to the usual assumptions of flow in a narrow gap, generates a velocity profile:

$$U = y\left(\frac{y-h}{2\eta}\right)\frac{dp}{dr}$$

where y is the coordinate normal to the bearing surface. The radial mass flow rate due to a radial pressure gradient in the element is the sum of mass fluxes in the groove and in the ridge:

$$d\psi = d\psi_g + d\psi_r$$

$$d\psi = \int_{\theta}^{\theta+d\theta}[\rho r d\theta]\int_0^h U dy = \frac{-rh_g^3}{12\eta}\int_{\theta}^{\theta+\alpha d\theta}\left[\rho\frac{dp_g}{dr}\right]d\theta \frac{-rh_r^3}{12\eta}\int_{\theta+\alpha d\theta}^{\theta+d\theta}\left[\rho\frac{dp_r}{dr}\right]d\theta$$

As the groove and land widths are infinitesimally small, the integrals may be simplified:

$$d\psi_g = \frac{-rh_g^3}{12\eta}\rho\alpha d\theta\frac{dp_g}{dr} \quad \text{and} \quad d\psi_r = \frac{-rh_r^3}{12\eta}\rho(1-\alpha)d\theta\frac{dp_r}{dr} \qquad [5.1]$$

$$d\psi=\frac{-r\rho d\theta}{12\eta}\left[\alpha h_g^3\frac{dp}{dr}+h_r^3(1-\alpha)\frac{dp_r}{dr}\right]$$

where subscripts r and g refer to ridge and groove respectively. The pressure gradients dp_g/dr and dp_r/dr are found by considering continuity of mass flow normal to a groove edge boundary and by equating the local pressure rise across a groove to the local pressure rise across a ridge.

These equations are more conveniently expressed in terms of coordinates parallel and normal to the groove edges, which are γ and ε respectively. The pressure gradients dp/dr and $dp/rd\theta$ may be rewritten as:

$$\frac{dp}{dr}=\frac{dp}{d\varepsilon}\cos\beta+\frac{dp}{d\gamma}\sin\beta$$

$$\frac{1dp}{rd\theta}=\frac{-dp}{d\varepsilon}\sin\beta+\frac{dp}{d\gamma}\cos\beta \quad [5.2]$$

Mass flow normal to the groove–ridge boundary is driven by the surface velocity of the bearing and the pressure gradient. If the grooved surface rotates at angular speed ω_1 and the plain surface at speed ω_2, then for an element of width $d\gamma$ in the groove the mass flow rate is:

$$dm_g = d\gamma\rho\left\{\int_0^{hg} y\left(\frac{y-h_g}{2\eta}\right)\frac{dp}{d\varepsilon}.dy-\left(\frac{\omega_1-\omega_2}{2}\right)h_g r.\sin\beta\right\}$$

$$= d\gamma\rho\left\{\frac{-h_g^3}{12\eta}\frac{dp_g}{d\varepsilon}-\left(\frac{\omega_1-\omega_2}{2}\right)h_g r.\sin\beta\right\} \quad [5.3]$$

where $[(\omega_1-\omega_2)/2]r\sin\beta$ is the average speed of the fluid in the groove normal to the groove boundary relative to the plain bearing surface. Equation 5.3 must equal the elemental mass flow rate across a ridge:

$$dm_r=d\gamma\rho\left\{\frac{-h_r^3}{12\eta}\frac{dp_r}{d\varepsilon}-\left(\frac{\omega_1-\omega_2}{2}\right)h_r r.\sin\beta\right\} \quad [5.4]$$

Equating Eqs 5.3 and 5.4 and simplifying gives:

$$h_r^3\frac{dp}{d\varepsilon}-h_g^3\frac{dp}{d\varepsilon}=6\eta(\omega_1-\omega_2)(h_g-h_r)r.\sin\beta \quad [5.5]$$

Consider the pressure gradients across a groove–ridge pair (Figure 5.9). If the groove width is infinitesimally small, then the pressure rise across the groove in a direction normal to the groove edge is:

$$dp_g = \frac{dp_g}{d\varepsilon}(\alpha r d\theta \sin\beta) \text{ and across the ridge is } dp_r = \frac{dp_r}{d\varepsilon}(1-\alpha) r d\theta \sin\beta$$

These create an increase dp in the smoothed pressure profile such that:

$$dp = dp_g + dp_r \qquad \text{where} \qquad dp = \frac{dp}{d\gamma} r d\theta \cos\beta$$

Thus:

$$\alpha \frac{dp_g}{d\varepsilon} + (1-\alpha)\frac{dp_r}{d\varepsilon} = \frac{dp}{d\gamma}\,\mathrm{ctn}\beta \qquad [5.6]$$

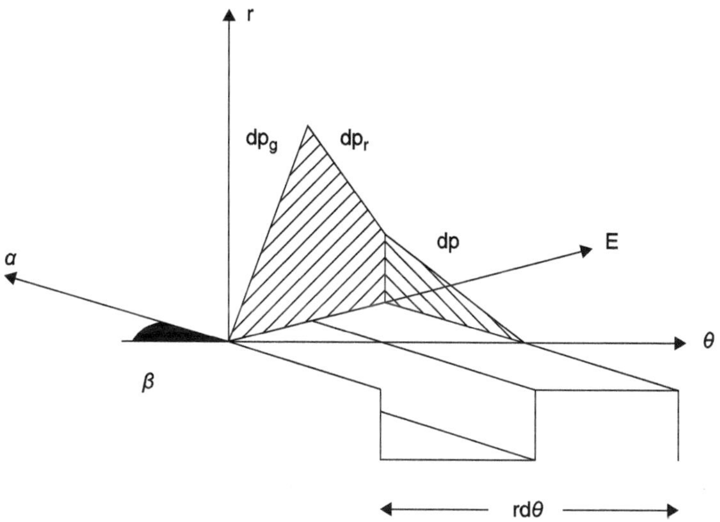

Figure 5.9 Local pressure equilibrium across a groove–ridge pair

Equations 5.5 and 5.6 enable $dp_g/d\varepsilon$ and $dp_r/d\varepsilon$ to be found:

$$\frac{dp_g}{d\varepsilon} = \left[-6\eta r(1-\alpha)(\omega_1-\omega_2)(h_g-h_r)\sin\beta + h_r^3 \frac{dp}{d\eta}\mathrm{ctn}\beta\right] \Big/ \left[\alpha h_r^3 + (1-\alpha)h_g^3\right] \quad [5.7]$$

$$\frac{dp_r}{d\varepsilon} = \left[6\eta r\alpha(\omega_1-\omega_2)(h_g-h_r)\sin\beta + h_g^3 \frac{dp}{d\eta}\mathrm{ctn}\beta\right] \Big/ \left[\alpha h_r^3 + (1-\alpha)h_g^3\right] \qquad [5.8]$$

Applying the coordinate transformation Eq. 5.2 to dp/dr enables Eqs 5.7 and 5.8 to be substituted into Eq. 5.1 to give the total flux density:

$$\frac{d\psi}{d\theta}=-\frac{\rho}{12\eta}\left\{\frac{h_g^3h_r^3+\alpha(1-\alpha)\sin^2\beta(h_g^3-h_r^3)^2 r}{(\alpha h_r^3+(1-\alpha)h_g^3)}\frac{dp}{dr}\right.$$

$$\left.\frac{-r^2\sin\beta\cos\beta.\alpha(1-\alpha)(h_g^3-h_r^3)(h_g-h_r)6\eta(\omega_1-\omega_2)}{(\alpha h_r^3+(1-\alpha)h_g^3)}\right\} \qquad [5.9]$$

Consider dynamic equilibrium of mass flow in an element of a groove–ridge pair (Figure 5.8). The mass of the element is:

$$dm=\rho dr.rd\theta[\alpha h_g+(1-\alpha)h_r]$$

and the rate of change of mass is:

$$\frac{dm}{dt}=rdr\, d\theta\frac{d}{dt}\{\rho[\alpha h_g+(1-\alpha)h_r]\}$$

The difference in flux entering and leaving the element is:

$$d\theta\left[d\psi/d\theta+dr\frac{d}{dr}(d\psi/d\theta)\right]-d\theta\, d\psi/d\theta$$

For equilibrium:

$$dr.rd\theta\frac{d}{dt}\{\rho[\alpha h_g+(1-\alpha)h_r]\}+d\theta\left[\frac{d\psi}{d\theta}+dr\frac{d}{dr}\left(\frac{d\psi}{d\theta}\right)\right]-d\theta\frac{d\psi}{d\theta}=0$$

Simplifying:

$$\frac{d}{dt}\{\rho[\alpha h_g+(1-\alpha)h_r]\}+\frac{1}{r}\frac{d}{dr}\left(\frac{d\psi}{d\theta}\right)=0 \qquad [5.10]$$

Equations 5.9 and 5.10 describe both the steady state and dynamic behaviour of the bearing. They can be applied to hydrodynamic and aerodynamic bearings with inward or outward pumping grooves and to sealing lands to determine the smoothed pressure profiles and hence the bearing's load capacity and static stiffness. They can also be used to determine the bearing's response to a time varying load and hence its damping capacity and/or degree of stability.

5.5.2 Pressure distribution in annular thrust bearings

For oil lubricated annular thrust bearings operating with constant gap, exact analytical expressions for the smoothed pressure profile and load carrying capacity can be derived by ignoring the time dependent term in Eq. 5.10 and integrating the remaining flux density term:

$$\frac{1}{R}\frac{d}{dr}\left(\frac{d\psi}{d\theta}\right)=0 \qquad [5.11]$$

For hydrodynamic bearings, the fluid density ρ is constant and Eq. 5.9 can be rewritten in the form:

$$\frac{d\psi}{d\theta}=\frac{-\rho}{12\eta}\left\{G_1 r\frac{dp}{dr}-G_2 r^2 6\eta(\omega_1-\omega_2)\right\} \qquad [5.12]$$

where:

$$G_1=\frac{h_r^3\left[H^3+\alpha(1-\alpha)\sin^2\beta(H^3-1)^2\right]}{(\alpha+(1-\alpha)H^3)} \quad \text{and}$$

$$G_2=\frac{h_r\sin\beta\cos\beta.\alpha(1-\alpha)(H^3-1)(H-1)}{(\alpha+(1-\alpha)H^3)}$$

and $H=h_g/h_r$ are constants that depend only on bearing geometry. For sealing lands, $\alpha=0$ and $h_g=h_r$, so the constants can be simplified to:

$$G_{1s}=h_r^3 \quad \text{and} \quad G_{2s}=0$$

Thus for a grooved land substituting Eq. 5.12 into Eq. 5.11 and simplifying gives:

$$\frac{\rho}{12\eta}[G_1 r(dp/dr)-G_2 r^2 6\eta(\omega_1-\omega_2)]=c_1 \qquad [5.13]$$

and for a sealing land:

$$\frac{\rho h_r^3}{12\eta}\frac{rdp}{dr}=c_2 \qquad [5.14]$$

Equations 5.13 and 5.14 may be integrated and the constants of integration found from boundary conditions. For the sealing land of an inward pumping bearing, $P=P_a$ at radius r_i, and $p=P_m$ at radius r_m. Integrating Eq. 5.14 and eliminating the constants gives the pressure at any radius:

$$\frac{P-P_a}{P_m-P_a}=\frac{\ln(r/r_i)}{\ln(r_m/r_i)} \qquad [5.15]$$

For the grooved land, $p=P_a$ at radius r_o and $P=P_m$ at radius r_m. Integrating Eq. 5.14 and eliminating the constants gives:

$$P-P_a=\left(P_m-P_a\right)\frac{\ln(r/r_o)}{\ln(r_m/r_o)}+3\eta\frac{G_2}{G_1}(\omega_1-\omega_2)\left[(r^2-r_i^2)-(r_m^2-r_i^2)\frac{\ln(r/r_o)}{\ln(r_m/r_o)}\right] \quad [5.16]$$

The pressure profile for the seal is only dependent on boundary conditions, whereas for the grooved land it depends on groove geometry, speed of rotation, fluid viscosity as well as boundary conditions. The pressure P_m at the seal–grooved land interface is unknown and is determined from continuity of mass flow. Equating flux densities for the seal and grooved lands at the interface, r_m, provides the necessary conditions:

$$\left.\frac{d\psi}{d\theta}\right|_{r_m}^{\text{seal}}=\left.\frac{d\psi}{d\theta}\right|_{r_m}^{\text{grooved land}}$$

Differentiating Eqs 5.15 and 5.16 gives the pressure gradients for the seal and grooved land, for the seal:

$$\frac{dp}{dr}=\frac{(P_m-P_a)}{r\ln(r_m/r_i)} \quad [5.17]$$

and for the grooved land:

$$\frac{dp}{dr}=\frac{(P_m-P_a)}{r\ln(r_m/r_o)}+3\eta\frac{G_2}{G_1}(\omega_1-\omega_2)\left[2r-\frac{(r_m^2-r_o^2)}{r\ln(r_m/r_o)}\right] \quad [5.18]$$

Substituting these into Eq. 5.12 and equating seal and grooved land flux densities at $r=r_m$ gives:

$$P_m-P_a=-3\eta\frac{G_2(\omega_1-\omega_2)(r_m^2-r_o^2)\ln(r_m/r_i)}{\left[h_r^3\ln(r_m/r_o)-G_1\ln(r_m/r_i)\right]} \quad [5.19]$$

Equation 5.19 together with Eqs 5.15 and 5.16 defines the pressure distribution across the bearing. The load supported by the bearing is determined by integrating Eqs 5.15 and 5.16 with respect to radius. For a grooved land of outer radius r_o:

$$F=\frac{\pi}{2}(P_m-P_a)\left[2r_m^2-\frac{(r_m^2-r_o^2)}{\ln(r_m/r_o)}\right]+\frac{3}{2}\pi\eta(\omega_2-\omega_1)\frac{G_2}{G_1}(r_m^2-r_o^2)\left[\frac{r_m^2-r_o^2}{\ln(r_m/r_o)}-(r_m^2+r_o^2)\right] \quad [5.20]$$

The preferred method of determining static stiffness is to obtain bearing load as a function of gap and to find the change in load produced by a small increment in gap. In obtaining the load–gap relationship, the degree of non-linearity of the bearing and hence the sensitivity of stiffness to gap error is evident. Bearing load capacity is defined at a practical minimum ridge gap.

The above analysis is based on an inward pumping grooved bearing with a sealing land, but it can be applied to an outward pumping bearing with sealing land simply by interchanging the boundary conditions. For the grooved land of an outward pumping bearing, $P=P_a$ at $r=r_o$ and $P=P_m$ at $r=r_m$. For the sealing land, $P=P_a$ at $r=r_o$ and $P=P_m$ at $r=r_m$.

An advantage of a chevron grooved bearing is that the net flow of lubricant through the bearing can be designed to be zero and hence simplify sealing arrangements. Inboard and outboard lands are therefore designed to have zero mass flux and the same pressure, P_m at the transition radius. From Eq. 5.12, zero flux gives:

$$\frac{dp}{dr}=\frac{G_2}{G_1}6\eta(\omega_1-\omega_2)r$$

Integrating and applying to inboard and outboard lands respectively gives:

$$P-P_a=\frac{G_{2i}}{G_{1i}}6\eta(\omega_1-\omega_2)(r-r_i)$$
$$P-P_a=\frac{G_{2o}}{G_{1o}}6\eta(\omega_1-\omega_2)(r-r_o) \qquad [5.21]$$

The design procedure amounts to finding the value of r, which gives the same pressure rise across both lands. For ease of manufacture, it is normal practice to keep the same groove depths and widths for both inboard and outboard lands. However, groove angles can differ on an optimised bearing design.

5.6 Annular thrust bearing design examples

Consider an annular thrust bearing having an outside diameter of 10 mm and a bore of 4 mm required to operate at 10 000 rpm with a ridge clearance of 6 μm and lubricated with oil of 20 cSt viscosity. Find the optimum designs of inward pumping, outward pumping and chevron grooved bearing are based on a maximum load capacity criterion. For this study a single-sided thrust bearing is considered. The resulting optimum groove geometry and transition diameter also apply to a double-sided bearing design, but the magnitude of the maximum load supported will differ.

5.6.1 Inward pumping bearings

The objective is to find the groove depth, angle, width ratio and grooved land–seal land transition diameter that gives maximum load for the operating gap, speed and oil viscosity specified. The optimum bearing design is readily found from a parametric study using Eq. 5.19. As the relationship between groove parameter and bearing load is well behaved, guessing at an initial bearing design and then finding the optimum value of each groove parameter in turn leads to a rapid iteration to a final bearing design.

The initial bearing design has a groove depth of 12 μm, width ratio of 0.5, angle of 20 degrees and a groove–seal transition diameter of 7 mm, generating a maximum pressure of 14.7 Bar supporting a load of 14.2 N. Varying groove depth between 4 and 24 μm, Figure 5.10 shows a groove depth of 14 μm to give maximum bearing pressure but the gradient of the pressure–groove depth curve is low either side of the optimum, so groove depth can depart from the optimum by ± 2 μm (±14%), reducing pressure by less than 3%.

The optimum groove angle is found to be 20 degrees (Figure 5.11), but it can vary between 17 and 23 degrees, (±15%), reducing pressure by less than 1.5%.

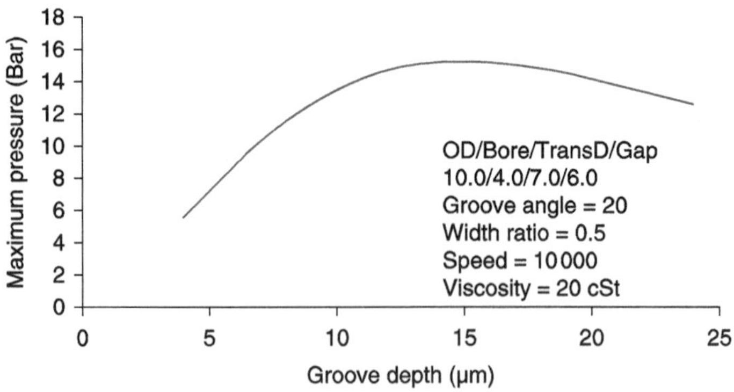

Figure 5.10 Effect of groove depth on maximum bearing pressure

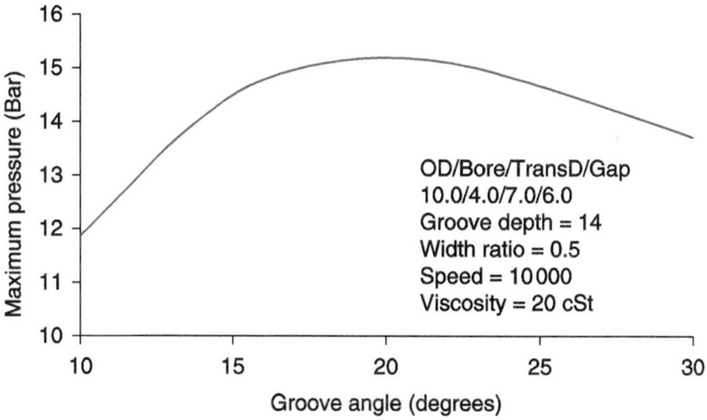

Figure 5.11 The effect of groove angle on bearing pressure

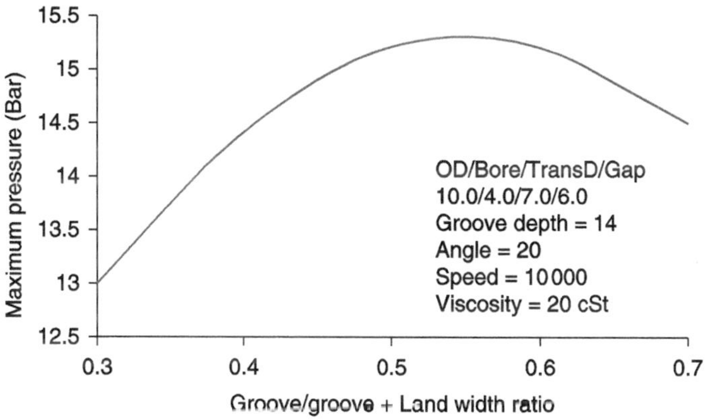

Figure 5.12 Effect of groove width ratio on bearing pressure

Similarly, groove width ratio can be varied by ±15% with only a small effect on pressure (Figure 5.12). The optimum groove width ratio is 0.5. Optimising groove depth, angle and width has increased maximum bearing pressure from 14.2 to 15.3 Bar.

Finally optimising groove–seal diameter shows an optimum value of 5.9 mm (Figure 5.13), increasing maximum bearing pressure to 17.1 Bar. The optimum bearing design supports 35.1 N compared to 14.1 N for the initially guessed value. A second iteration of the groove variables shows a small effect on maximum bearing pressure and load capacity. The final bearing design is as shown in Table 5.1.

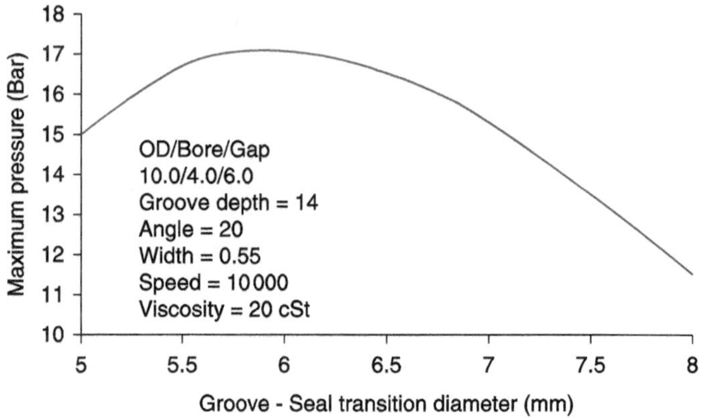

Figure 5.13 Effect of groove–seal transition diameter on bearing pressure

Table 5.1 Optimum inward pumping thrust bearing design

Inward pumping bearing	
Outside diameter (mm)	10
Bore (mm)	4
Groove–seal transition diameter (mm)	5.6
Clearance (μm)	6
Groove depth (μm)	16
Groove angle (°)	20
Groove width ratio	0.6
Speed (rpm)	10 000
Oil viscosity (cSt)	20
Maximum bearing pressure (bar)	17.5
Load (N)	40.5

5.6.2 Outward pumping bearings

The outward pumping bearing follows the same design procedure. The optimum design is given in Table 5.2.

Table 5.2 Optimum outward pumping thrust bearing design

Outward pumping bearing	
Outside diameter (mm)	10
Bore (mm)	4
Groove–seal transition diameter (mm)	8.1
Clearance (μm)	6
Groove depth (μm)	17
Groove angle (°)	19
Groove width ratio	0.6
Speed (rpm)	10000
Oil viscosity (cSt)	20
Maximum bearing pressure (bar)	10.4
Load (N)	25.6

5.6.3 Chevron grooved bearings

For the chevron grooved bearing, the mass flux is zero and for ease of manufacture, widths and depths of the inward and outward pumping grooves are kept the same. The design process is again iterative, guessing at an initial set of groove parameters, finding the optimum inward–outward groove transition diameter and adjusting groove parameters until a maximum bearing load is found. Figure 5.14 shows the maximum

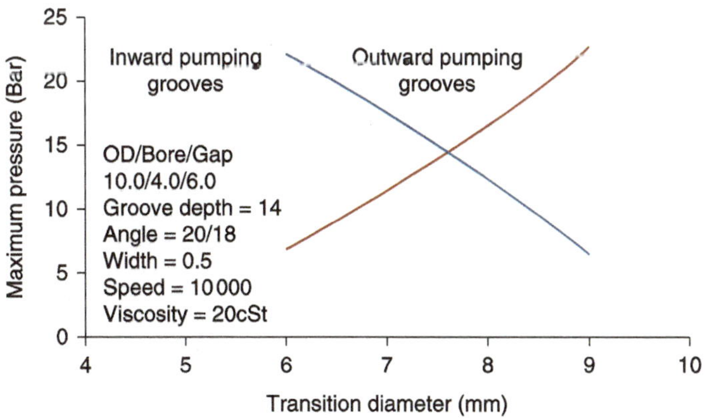

Figure 5.14 Optimum transition diameter for chevron grooved bearing

Table 5.3 Optimum chevron grooved bearing design

Chevron grooved bearing	
Outside diameter (mm)	10
Bore (mm)	4
Groove–seal transition diameter (mm)	7.6
Clearance (μm)	6
Groove depth (μm)	14
Groove angle outward pumping (°)	18
Groove angle outward pumping (°)	20
Groove width ratio	0.5
Speed (rpm)	10 000
Oil viscosity (cSt)	20
Maximum bearing pressure (bar)	14.6
Load (N)	48.1

pressure–transition diameter relationship for the inward and outward pumping grooves subject to zero mass flux. The diameter at which the curves intersect is where the two parts of the bearing are in equilibrium, i.e. the maximum pressures match, a necessary condition for zero mass flux. The optimum bearing design is summarised in Table 5.3.

5.6.4 Optimum design of chevron grooved bearings

Of the three bearing designs, the chevron grooved bearing supports the highest load, 48.1 N. The inward pumping bearing supports slightly less load, 40.5 N, whilst the outward pumping bearing has a relatively low load capacity of 25.6 N. With grooves extending all the way across the chevron grooved bearing, its average clearance is higher than the other two bearings, giving it the advantage of a lower friction loss; furthermore, its grooves act as mechanical seals making the provision and retention of lubricant much simpler. Thus for a majority of applications it is the preferred design.

5.6.5 Static stiffness characteristics of chevron grooved bearings

The preceding examples derived inward pumping, outward pumping and chevron grooved bearing designs that generated a maximum load carrying capacity for a given operating gap, speed and lubricant viscosity. Static stiffness is derived from the slope of the load–gap curve for a single-sided bearing or load–deflection curve for a double-sided bearing. In this section, attention is focused on the optimised design of chevron groove bearing given in Table 5.3.

The axial load–deflection/gap characteristics of single- and double-sided bearings are shown in Figure 5.15. For a single-sided bearing, the load–gap relationship is highly non-linear with load generating capacity reducing almost exponentially with increasing gap. However, when arranged as a double-sided bearing, the load–deflection relationship is near linear. In this example, load capacity is defined at a minimum gap of 2 μm at which the single-sided bearing supports a slightly higher load of 177.6 N compared to 163.1 N for the double-sided bearing.

Stiffness of the single-sided bearing decreases from 44.1 N/μm at 2 μm through 17.3 N/μm at 6 μm to 2 N/μm at 12 μm gap. The decrease in stiffness with gap is large, but the stiffness/load ratio changes only slightly. Thus when used as a flotation pad to support a dead weight, the single-sided bearing results in an axial resonant frequency that is almost independent of the weight supported.

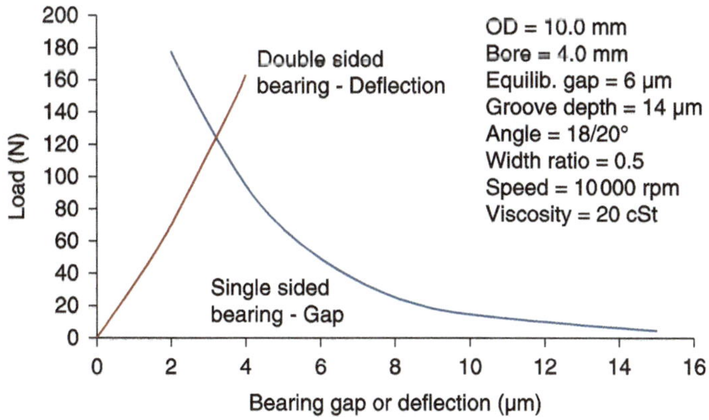

Figure 5.15 Axial load–deflection characteristics of a chevron grooved bearing

In contrast, the double-sided bearing maintains a high stiffness for all deflections. Under no load, at the equilibrium gap of 6 μm, it has a static stiffness of 34.6 N/μm, but this is also the minimum stiffness.

Figure 5.15 assumes oil viscosity to remain constant at 20 cSt throughout the range of gaps considered. For the single-sided bearing, heat generation and hence temperature increases with reducing gap and the associated reduction in oil viscosity reduces the slope of the load–deflection curve. For the double-sided bearing, the one variation in heat generation with load is much lower, as a reduction in gap on one side of the bearing is offset by an increase in gap on the other.

For a given bearing geometry and operating gap, inspection of Eq. 5.19 reveals bearing pressure, and therefore load carrying capacity and static stiffness are directly proportional to speed of rotation and oil viscosity.

5.7 Radial load–deflection analysis of herringbone grooved journal bearings

The analysis of a grooved cylindrical journal bearing follows that of [21]. They considered the general case of a bearing operating with a compressible lubricant subject to small static and dynamic eccentricities and derived the differential equations describing the bearing's radial load–deflection and whirl characteristics. Oil lubrication is considered to be a special case of zero compressibility for which lubricant density remains constant [22]. As for the grooved thrust bearing, the analysis is restricted to bearings with narrow grooves, fully flooded with lubricant. Radial load is assumed to act through the centre of the bearing for which the gap is independent of axial position.

The bearing considered is fully grooved with the two sets of grooves symmetrically disposed about a central radial plane (Figure 5.16). The bearing is of radius R, length L and the grooves may be placed on the stationary or rotating component. In Figure 5.16, they are shown on the shaft that rotates at constant speed ω_1. For this situation, the grooves pump towards the bearing centre, but because of symmetry there is no net flux of fluid through the bearing. However, there is build up of fluid pressure along the length of the bearing, reaching a maximum at its centre. As for the thrust bearing, the pressure profile consists of a smoothed profile with a 'saw tooth' profile superimposed (Figure 5.17). The latter is due to a local pressure variation at a groove–ridge pair and

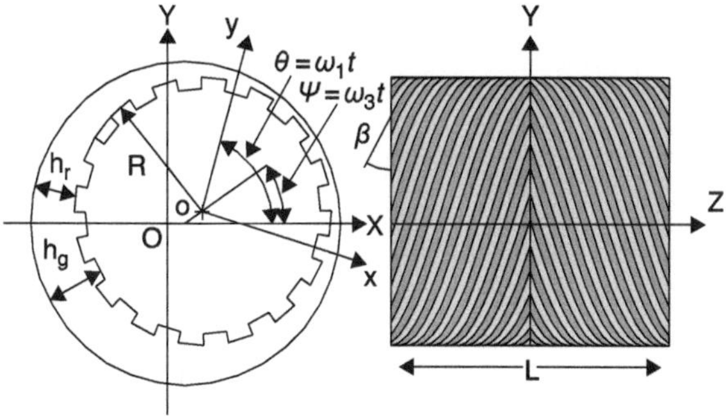

Figure 5.16 Notation for herringbone grooved journal bearing

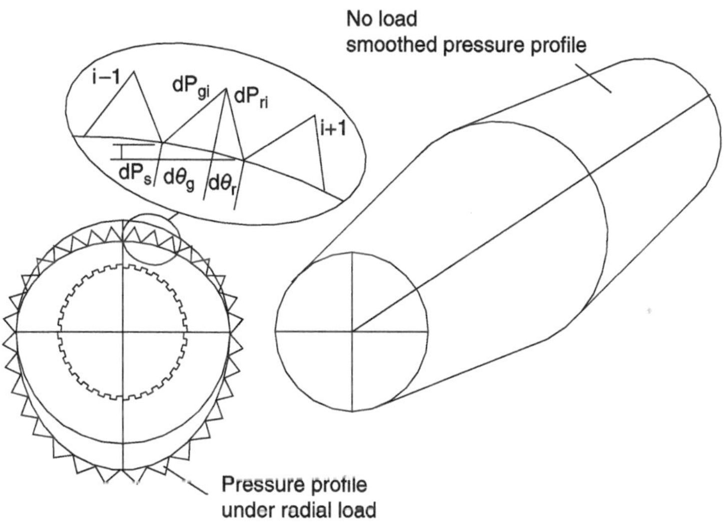

Figure 5.17 Pressure profiles in grooved journal bearing

for the case of narrow groove and ridge widths the effect of the local 'saw tooth' pressure gradients on load capacity are neglected.

Under a centrally applied radial load the clearance under a ridge, h_r is only a function of angular position and is given by:

$$h_r = C(1 - \varepsilon_r \cdot \cos\theta)$$

where ε_r is the eccentricity ratio, e/C and C is the concentric bearing gap. Groove depth is constant and the ratio of groove depth to ridge clearance under no load is denoted by $H_o C$, so the local clearance over the groove,

$h_g = h_r + H_o C$, is also a function of θ. This generates smoothed and saw tooth pressure profiles that are both a function of angular and axial position:

$$P_s = P_s(\theta, z) \quad \text{and} \quad dP = dP(\theta, z)$$

The groove width is a_g and the ridge width is a_r, where the ratio $a_g/(a_g + a_r) = \alpha$, and is constant. Groove helix angle, β is also constant.

To determine the load capacity and static stiffness of the bearing requires the smoothed pressure profile to be defined in terms of the geometrical parameters above, lubricant properties and bearing speed. The starting point is to consider fluid flow local to a single groove–ridge pair.

The mass flux in a narrow gap can be derived from Reynolds equation [24]. Applying this to groove and ridge regions in a cylindrical gap:

$$M_{\theta r} = -\frac{\rho h_r^3}{12\eta}\frac{1}{R}\frac{dP_r}{d\theta} + \frac{\rho h_r}{2}(U+V) \qquad [5.22]$$

$$M_{zr} = -\frac{\rho h_r^3}{12\eta}\frac{1}{R}\frac{dP_r}{dz} \qquad [5.23]$$

$$M_{\theta g} = -\frac{\rho h_g^3}{12\eta}\frac{1}{R}\frac{dP_g}{d\theta} + \frac{\rho h_r}{2}(U+V) \qquad [5.24]$$

$$M_{zg} = -\frac{\rho h_g^3}{12\eta}\frac{1}{R}\frac{dP_g}{dz} \qquad [5.25]$$

where U and V are the linear speeds of the smooth and grooved surfaces respectively and the local saw tooth pressure gradients $dP_r/d\theta$, dP_r/dz, $dP_g/d\theta$ and dP_g/dz are constant for a single groove–ridge pair (Figure 5.17). In order to derive the differential equations describing the smoothed pressure profile, it is necessary to define the smoothed pressure profile gradients in terms of the local saw tooth pressure gradients. This is done by considering equilibrium of pressures local to a groove–ridge pair. Referring to Figure 5.17, the pressure change across a single groove–ridge pair in an angular direction is:

$$dP_s = \frac{dP_g}{d\theta} d\theta_g + \frac{dP_r}{d\theta} d\theta_r$$

where:

$$d\theta_g = \alpha . d\theta \text{ and } d\theta_r = (1 - \alpha)\, d\theta$$

Thus:

$$\frac{dP_s}{d\theta}=\frac{\alpha.dP_g}{d\theta}+(1-\alpha)\frac{dP_r}{d\theta} \qquad [5.26]$$

Similarly, considering pressure gradients in an axial direction:

$$\frac{dP_s}{dz}=\frac{.dP_g}{dz}+(1-\alpha)\frac{dP_r}{dz} \qquad [5.27]$$

The local pressure gradients are determined from continuity of mass flux normal to a groove ridge interface:

$$M_{\theta g}\sin\beta - M_{zg}\cos\beta = M_{\theta r}\sin\beta - M_{zr}\cos\beta \qquad [5.28]$$

and equilibrium of pressure gradients parallel to the groove–ridge interface in the groove and ridge regions:

$$\frac{1}{R}\cos\beta\frac{dP_g}{d\theta}+\sin\beta\frac{dP}{dz}=\frac{1}{R}\cos\beta\frac{dP_r}{d\theta}+\sin\beta\frac{dP_r}{dz} \qquad [5.29]$$

Equations 5.26 to 5.29 may be used to obtain expressions for the local pressure gradients and mass fluxes in terms of the smoothed pressure profile gradients:

$$\frac{dP_g}{d\theta}=\left[\left\{h_r^3+(1-\alpha)(h_g^3-h_r^3)\cos^2\beta\right\}\frac{dP_s}{d\theta}+(1-\alpha)(h_g^3-h_r^3)R\cos\beta.\sin\beta\frac{dP_z}{dz}\right.$$
$$\left.+(1-\alpha)6\eta(U-V)(h_g-h_r)R\sin^2\beta\right]\Big/\left[\alpha h_r^3+(1-\alpha)h_g^3\right] \qquad [5.30]$$

$$\frac{dP_g}{dz}=\left[(1-\alpha)(h_g^3-h_r^3)\cos\beta.\sin\beta\}\frac{dP_s}{Rd\theta}+\{\alpha h_r^3+(1-\alpha)h_g^3\right.$$
$$\left.-(1-\alpha)(h_g^3-h_r^3)\cos^2\beta\}\frac{dP_z}{dz}-(1-\alpha)6\eta(U-V)(h_g-h_r)\cos\beta\sin\beta\right]\Big/ \qquad [5.31]$$
$$\left[\alpha h_r^3+(1-\alpha)h_g^3\right]$$

$$\frac{dP_r}{d\theta}=\left[\{h_g^3-\alpha(h_g^3-h_r^3)\cos^2\beta\}\frac{dP_s}{d\theta}-\alpha(h_g^3-h_r^3)R\cos\beta.\sin\beta\frac{dP_z}{dz}\right.$$
$$\left.-\alpha 6\eta(U-V)(h_g-h_r)R\sin^2\beta\right]\Big/\left[\alpha h_r^3+(1-\alpha)h_g^3\right] \qquad [5.32]$$

$$\frac{dP_g}{dz}=-\alpha(h_g^3-h_r^3)\cos\beta.\sin\beta\}\frac{dP_s}{Rd\theta}+\{\alpha h_r^3+(1-\alpha)h_g^3+\alpha(h_g^3-h_r^3)\cos^2\beta\}$$
$$\frac{dP_z}{dz}+\alpha 6\eta(U-V)(h_g-h_r)\cos\beta\sin\beta]/[\alpha h_r^3+(1-\alpha)h_g^3] \qquad [5.33]$$

$$M_{\theta r} = -\rho h_r^3\Big[\{h_g^3-\alpha(h_g^3-h_r^3)\cos^2\beta\}\frac{dP_s}{Rd\theta}-\alpha(h_g^3-h_r^3)\cos\beta.\sin\beta\frac{dP_s}{dz}$$
$$-\alpha 6\eta(U-V)(h_g-h_r)\sin^2\beta\Big]\Big/\Big\{12\eta[\alpha h_r^3+(1-\alpha)h_g^3]\Big\}$$
$$+\rho h_r(U+V)/2 \qquad [5.34]$$

$$M_{zr} = -\rho h_r^3\Big[-\alpha(h_g^3-h_r^3)\cos\beta\frac{dP_s}{Rd\theta}+\Big\{\alpha h_r^3+(1-\alpha)h_g^3+\alpha(h_g^3-h_r^3)\cos^2\beta\Big\}\frac{.dP_s}{dz}$$
$$+\alpha 6\eta(U-V)(h_g-h_r)\cos\beta\sin\beta\Big]\Big/\Big\{12\eta[\alpha h_r^3+(1-\alpha)h_g^3]\Big\} \qquad [5.35]$$

$$M_{\theta g} = -\rho h_g^3\Big[\{h_r^3+(1-\alpha)(h_g^3-h_r^3)\cos^2\beta\}\frac{dP_s}{Rd\theta}+(1-\alpha)(h_g^3-h_r^3)\cos\beta.\sin\beta$$
$$\frac{dP_s}{dz}+(1-\alpha)6\eta(U-V)(h_g-h_r)\sin^2\beta\Big]\Big/\Big\{12\eta[\alpha h_r^3+(1-\alpha)h_g^3]\Big\}$$
$$+\rho h_g(U+V)/2 \qquad [5.36]$$

$$M_{zg} = -\rho h_g^3\Big[(1-\alpha)(h_g^3-h_r^3)\cos\beta.\sin\beta\frac{dP_s}{Rd\theta}+\{\alpha h_r^3+(1-\alpha)h_g^3$$
$$+(1-\alpha)(h_g^3-h_r^3)\cos^2\beta\}.\frac{dP_s}{dz}-(1-\alpha)6\eta(U-V)(h_g-h_r)$$
$$\cos\beta\sin\beta\Big]\Big/\Big\{12\eta[\alpha h_r^3+(1-\alpha)h_g^3]\Big\} \qquad [5.37]$$

So far equations have been derived giving the local pressure gradients and mass fluxes in terms of the smoothed pressure profile gradients. The final step in the analysis is to apply the principle of mass conservation to an elemental control volume over a single groove–ridge pair.

It is convenient to consider an elemental control volume bounded by two radial surfaces S_q, S_{q+dq}, and two surfaces parallel to the groove–ridge edges, S_e, S_{e+de} (Figure 5.18). The grooved surface and hence the control volume are moving at speed V. Two coordinate systems are defined. The orthogonal system $OZ\theta$ is such that OZ is fixed and is parallel to the bearing's axis and $O\theta$ is moving in the direction of motion of the groove–ridge element considered. The coordinate system Oqe is skewed and is moving such that Oq remains parallel to the edge of the groove ridge element and Oe is parallel to $O\theta$. The coordinates are related by the transforms:

$$e=\theta-\frac{Z}{R}\cot\beta \qquad q=\frac{Z}{\sin\beta} \qquad [5.38]$$

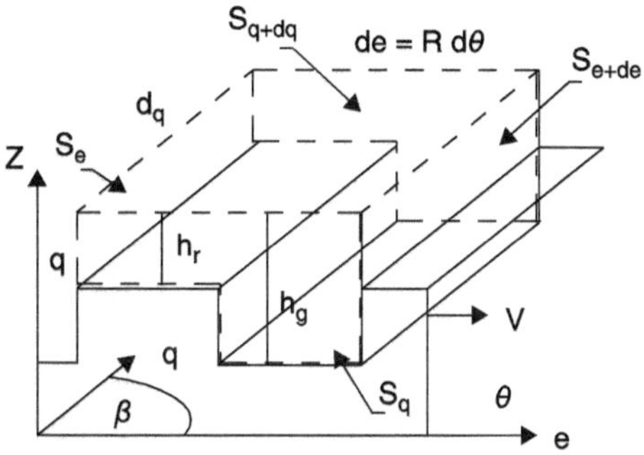

Figure 5.18 **Elemental control volume over single groove–ridge pair**

$$Z = q.\sin\beta \qquad \theta = e + \frac{q}{R}\cos\beta \tag{5.39}$$

Conservation of mass applied to the elemental control volume requires that the net mass flow rate of fluid into the element must equal the rate of change of mass of the element. As the control volume is moving at speed V the rate of change of mass is:

$$\frac{\partial}{\partial t}(\Delta M) + \frac{V}{R}\frac{\partial}{\partial \theta}(\Delta M) \tag{5.40}$$

where:

$$\Delta M = \rho[\alpha h_g + (1-\alpha)h_r]Rd\theta dZ = \rho[\alpha h_g + (1-\alpha)h_r]\sin\beta dedq \tag{5.41}$$

Mass flows into the element through the four boundary surfaces. Through S_e:

$$[M_{\theta r}\sin\beta - M_{zr}\cos\beta - \rho V h_r \sin\beta]dq\,|_e \tag{5.42}$$

Through S_{e+de}:

$$[M_{\theta r}\sin\beta - M_{zr}\cos\beta - \rho V h_r \sin\beta]dq\,|_{e+de} \tag{5.43}$$

Through S_q, assuming a narrow groove–ridge pair (width rate):

$$R[M_{zr}(1-\alpha)de + M_{zg}\alpha de]\,|_q \tag{5.44}$$

Through S_{q+dq}:

$$R[M_{zr}(1-\alpha)de + M_{zg}\alpha de]\,|_{q+dq} \tag{5.45}$$

Thus for equilibrium:

$$\frac{d}{de}[M_{\theta r}\sin\beta - M_{zr}\cos\beta - \rho V h_r \sin\beta]dqde + \frac{d}{dq}R[M_{zr}(1-\alpha)+M_{zg}\alpha]dedq +$$
$$\left\{\frac{\partial}{\partial t}+\frac{V}{R}\frac{\partial}{\partial\theta}\right\}\{\rho[\alpha h_g+(1-\alpha)h_r]\,R\sin\beta dedq\}=0 \qquad [5.46]$$

Transforming to z, θ coordinates where from Eq. 5.45:

$$\frac{\partial}{\partial e}=\frac{\partial}{\partial\theta} \quad \text{and} \quad \frac{\partial}{\partial q}=\frac{\partial}{\partial\theta}\cos\beta+\frac{\partial}{\partial Z}\sin\beta$$

This gives:

$$\frac{1\partial}{R\partial\theta}[M_{\theta r}\sin\beta+M_{zg}\alpha\cos\beta-M_{zr}\alpha\cos\beta-\rho V h_r\sin\beta]+$$
$$\frac{\partial}{\partial Z}[M_{zr}(1-\alpha)+M_{zg}\alpha]+\left\{\frac{\partial}{\partial t}+\frac{V\partial}{R\partial\theta}\right\}\{\rho[\alpha h_g+(1-\alpha)h_r]\sin\beta\}=0 \qquad [5.47]$$

Together with Eqs 5.34 to 5.37 for mass flux, Eq. 5.47 enables the smoothed pressure profile to be obtained. A solution to these equations is derived for the case of a bearing operating under steady whirl at a constant speed, ω_3. A rotating frame of reference oxy is defined for which the axes are fixed in the shaft rotating at speed ω_1. The shaft is whirling at speed ω_3 and amplitude ε (Figure 5.16). It is convenient to define the coordinate, ψ such that:

$$\psi=\theta-\omega_3 t \qquad [5.48]$$

Ridge and groove gaps are therefore given by:

$$h_r=C+\varepsilon_r\cos\psi \quad \text{and} \quad h_g=CH_o+\varepsilon_r\cos\psi \qquad [5.49]$$

where C is the no load bearing gap and H_o is the ratio h_g/h_r, also under no load. As:

$$\frac{\partial}{\partial\psi}=\frac{\partial}{\partial\theta} \quad \text{and} \quad \frac{\partial h_g}{\partial t}=\frac{\partial h_g}{\partial\psi}\frac{\partial\psi}{\partial t}$$

$$V=R\omega_1 \quad \text{and} \quad U=R\omega_2$$

Equation 5.46 can be rewritten as:

$$\frac{1\partial}{R\partial\psi}[M_{\theta r}\sin\beta+M_{zg}\alpha\cos\beta-M_{zr}\alpha\cos\beta-\rho R\omega_1 h_r\sin\beta]+$$
$$\frac{\partial}{\partial Z}[M_{zr}(1-\alpha)+M_{zg}\alpha]+(\omega_1-\omega_3)\frac{\partial}{\partial\psi}\{\rho[\alpha h_g+(1-\alpha)h_r]\sin\beta\}=0 \quad [5.50]$$

Equation 5.50 together with the Eqs 5.34 to 5.37 for mass flux, gives the smoothed pressure profile for a grooved cylindrical journal operating under the condition of steady whirl and may be applied to liquid and gas lubricated bearings. For the case of zero whirl speed, these equations define the radial load–deflection characteristics of the bearing. An analytical solution to Eq. 5.49 is only possible for the special case of zero radial deflection, where axial symmetry exists and the smoothed pressure profile is therefore a function of axial position only. To determine the whirl and radial load–deflection characteristics requires a numerical solution and to find the whirl response over a range of speeds and radial loads requires lengthy computation times. An approximate solution for the case of a bearing subject to a small radial deflection, Δ, where the smoothed pressure profile can be considered to consist of a small dynamic component superimposed on a static concentric pressure profile:

$$P(\psi,Z)=P_o(Z)+\Delta P_1(\psi,Z)$$

considerably reduces computation time. This effectively linearises the solution, but is accurate for the small deflections of most interest in precision bearing applications and is the preferred method of solution here. The mass fluxes corresponding to the pressure profile of Eq. 5.50 are also assumed to take the form:

$$\begin{aligned} M_{\theta r}&=M_{\theta ro}+\Delta M_{\theta r1}\\ M_{zr}&=M_{zro}+\Delta M_{zr1}\\ M_{\theta g}&=M_{\theta go}+\Delta M_{\theta g1}\\ M_{zg}&=M_{zgo}+\Delta M_{zg1} \end{aligned} \quad [5.51]$$

These can be substituted into Eq. 5.49 and the static and dynamic terms separated to give equations for P_o and P_1. For P_o, terms containing $\partial P/\partial\psi$ are zero, so Eq. 5.57 can be simply reduced to:

$$\partial[\alpha M_{zgo}+(1-\alpha)M_{zro}]=0$$

Since the analysis is restricted to symmetrical bearing designs for which the net axial mass flux is zero, it follows that:

$$\alpha M_{zgo} + (1-\alpha)M_{zro} = 0 \qquad [5.52]$$

Substituting for M_{zgo} and M_{zro}, using Eqs 5.35 and 5.37 and simplifying gives:

$$\frac{dP_o}{dZ} = \frac{\alpha(1-\alpha)6\eta(U-V)(h_g-h_r)\left(h_g^3-h_r^3\right)\sin\beta\cos\beta}{\alpha(1-\alpha)\left(h_g^3-h_r^3\right)\sin^2\beta + h_g^3h_r^3} \qquad [5.53]$$

Integrating, applying the boundary condition $P_o = P_a$ at $Z = 0$ and putting $H_o = h_g/h_r$ gives:

$$P_o = \frac{\left[\alpha(1-\alpha)6\eta(U-V)(H_o-1)\left(H_o^3-1\right)\sin\beta\cos\beta\right]}{C^2\left[\alpha(1-\alpha)\left(H_o^3-1\right)\sin^2\beta + H_o^3\right]} z + P_a \qquad [5.54]$$

Note that Eq. 5.55 is independent of fluid density ρ and therefore applies to both oil and air lubricated bearings. For hydrodynamic bearings, ρ is constant so it is convenient to put:

$$\begin{aligned} M'_{\theta r} &= M_{\theta r}\rho \\ M'_{\theta g} &= M_{\theta g}\rho \\ M'_{zr} &= M_{zr}\rho \\ M'_{zg} &= M_{zg}\rho \end{aligned} \qquad [5.55]$$

and to rearrange Eq. 5.50, cancelling ρ:

$$\frac{1\partial}{R\partial\psi}[M'_{\theta r}\sin\beta + (M'_{zg} - M'_{zr})\alpha\cos\beta] + \frac{\partial}{\partial z}[\alpha M'_{zg} + (1-\alpha)M'_{zr}]\sin\beta + \omega_3 e\sin\psi\cos\beta = 0 \qquad [5.56]$$

where:

$$\frac{\partial h_g}{\partial\psi} = \frac{\partial h_r}{\partial\psi} = -e\sin\psi$$

Substituting for M'_θ, M'_{zg} and M'_{zr}, using Eqs 5.34 to 5.37 and 5.55 and simplifying gives:

$$C_1\frac{\partial^2 P_1}{\partial\psi^2} + C_2\frac{\partial^2 P_1}{\partial\psi\partial Z} + C_3\frac{\partial P_1}{\partial\psi} + C_4\frac{\partial P_1}{\partial Z} + C_5\frac{\partial^2 P_1}{\partial Z^2} + C_6 = 0 \qquad [5.57]$$

where:

$$C_1 = -\frac{\left[h_r^3h_g^3 + \alpha(1-\alpha)(h_g^3-h_r^3)^2\cos^2\beta\right]}{R^2\left[\alpha h_r^3 + (1-\alpha)h_g^3\right]} \qquad [5.58]$$

$$C_2=-\frac{\left[2\alpha(1-\alpha)(h_g^3-h_r^3)^2\cos\beta\sin\beta\right]}{R\left[\alpha h_r^3+(1-\alpha)h_g^3\right]} \quad [5.59]$$

$$C_3=\frac{3\left[h_r^3h_g^3+\alpha(1-\alpha)(h_g^3-h_r^3)^2\cos^2\beta\right]+\left[3(h_r^2h_g^3+h_r^3h_g^2)+6\alpha(1-\alpha)(h_g^3-h_r^3)(h_g^3-h_r^3)\cos^2\beta e\sin\psi\right.}{R\left[\alpha h_r^3+(1-\alpha)h_g^3\right]} \quad [5.60]$$

$$C_4=-\frac{3\alpha\cos\beta\sin\beta)\left(h_g^3-h_r^3\right)\left[2\left(h_g^2-h_r^2\right)-\left(h_g^3-h_r^3\right)\right]e\sin\psi}{R\left[\alpha h_r^3+(1-\alpha)h_g^3\right]} \quad [5.61]$$

$$C_5=-\frac{\left[h_r^3h_g^3+\alpha(1-\alpha)(h_g^3-h_r^3)^2\sin^2\beta\right]}{\left[\alpha h_r^3+(1-\alpha)h_g^3\right]} \quad [5.62]$$

$$C_6=6\eta e\sin\psi(2\omega_3-\omega_1-\omega_2)+18\eta\alpha(1-\alpha)(\omega_1-\omega_2)$$
$$\frac{\sin^2\beta(h_g^3-h_r^3)^2h_g^2h_r^2e\sin\psi}{[\alpha h_r^3+(1-\alpha)h_g^3]^2} \quad [5.63]$$

and $h_r = C + e\cos\psi$ and $h_g = CH_o + e\cos\psi$ are functions of ψ. Equation 5.57 may be solved for the pressure distribution P_1 using relaxation techniques subject to the boundary conditions:

$$P_1(\psi,Z) = P_a \text{ at } Z=0$$

and if the bearing is fully grooved:

$$\alpha M_{zg1} + (1-\alpha) M_{zr1} = 0 \text{ at } Z = L/2$$

The latter is the condition for zero mass flux at the centre of the bearing. Integrating pressure over the bearing area in OXZ and OYZ planes gives the radial load components:

$$F_x=2\int_0^{L/2}\int_0^{2\pi} P_1(Z,\psi)\cos\psi \, dZ \, d\psi \quad [5.64]$$

$$F_y=2\int_0^{L/2}\int_0^{2\pi} P_1(Z,\psi)\sin\psi \, dZ \, d\psi \quad [5.65]$$

The magnitude of the load is:

$$W = (F_x^2 + F_y^2)^{1/2} \quad [5.66]$$

and the attitude angle is:

$$\phi=\mathrm{Tan}^{-1}(F_y/F_x) \quad [5.67]$$

5.8 Load capacity and static stiffness characteristics of herringbone grooved journal bearings

The analysis presented above is based on small radial deflections and effectively linearises the relationship between radial load and deflection. As a result, the analysis predicts the load capacity of the bearing to be simply the product of static stiffness and maximum permissible deflection. Conventional plain hydrodynamic journal bearings exhibit a non linear load–deflection characteristic [22], which takes the form:

$$W \propto 12\pi e/[(2+e^2)(1-e^2)^{1/2}]$$

where W is load and e is eccentricity. Thus, for small eccentricities, load is approximately proportional to eccentricity but for large eccentricities, the bearing exhibits a load hardening characteristic and neglecting practical limitations, load capacity tends to infinity as ε_r approaches unity. Thus for most bearing designs, the linearised analysis provides a lower bound estimate of load capacity. An exception might be the unusual situation where bearings with deep grooves operate at eccentricities close to unity for which groove depth is poorly matched to the local clearance. However, taking account of the bearing's surface geometry, as well as heat generation at low operating gaps, a practical maximum eccentricity at which to define load capacity is somewhat less at 0.5 to 0.8.

For all fluid film journal bearings the application of a radial load results in a component of deflection in line with the load and a component normal to it. The ratio of the two components defines the attitude angle, as given by Eq. 5.66. Considering the bearing to be a machine element, it is convenient to define static stiffness in terms of the components:

$$[K] = \begin{bmatrix} K_{xx} & K_{xy} \\ K_{yx} & K_{yy} \end{bmatrix} \qquad [5.68]$$

In general, for a bearing subject to a static radial load, the four stiffness coefficients are all different in magnitude; however, for a bearing operating with zero static eccentricity, $K_{xx} = K_{yy}$ and $K_{xy} = -K_{yx}$. In which case, K_{xx} and K_{xy} may be expressed in terms of total bearing load, W, eccentricity, e, operating gap, C and attitude angle, ϕ, using Eqs 5.64 and 5.66:

$$K_{xx} = \frac{W}{eC}\cos\phi \quad \text{and} \quad K_{xy} = \frac{W}{eC}\sin\phi \qquad [5.69]$$

A limitation to the analysis presented in Section 5.8 is that it does not consider tilt deflections, so angular stiffness components are neglected. As on fully grooved bearings, most part of the lift occurs at or close to the centre of the bearing, so it is reasonable to expect angular stiffness to be small.

The attitude angle is created by viscous shearing forces within the lubricant film. When the shaft and journal centres are coincident, the bearing gap is constant and viscous shearing simply creates a net torque. Introducing a small eccentricity between shaft and journal centres by applying an external radial load increases the viscous shear stresses on one side of the bearing and reduces them on the other, so creating a net radial force that has a component perpendicular to the applied load. For a plain journal bearing operating at or close to zero eccentricity, this viscous shear force can be much larger than the applied load, creating an attitude angle close to 90 degrees. The example considered in Figure 5.19 shows the attitude angle of a plain journal bearing to be −88.7 degrees at a speed of 8000 rpm. Adding grooves has the effect of reducing attitude angle, in this case increasing groove depth to a maximum of 50 μm, about twice the bearing gap, reduces attitude angle to −18.9 degrees when the grooves are rotating with respect to the applied load vector.

Static stiffness is influenced by groove design. Simply adding grooves to a plain shaft or journal increases the average clearance of the bearing and reduces static stiffness. Figure 5.20 is an example of the effect of groove depth on stiffness. The 'minimum gap' curve corresponds to adding grooves to a plain shaft, the minimum clearance of the bearing

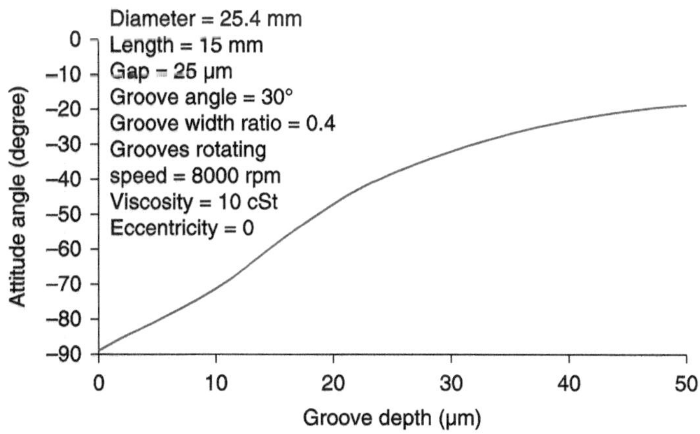

Figure 5.19 Effect of groove depth on attitude angle

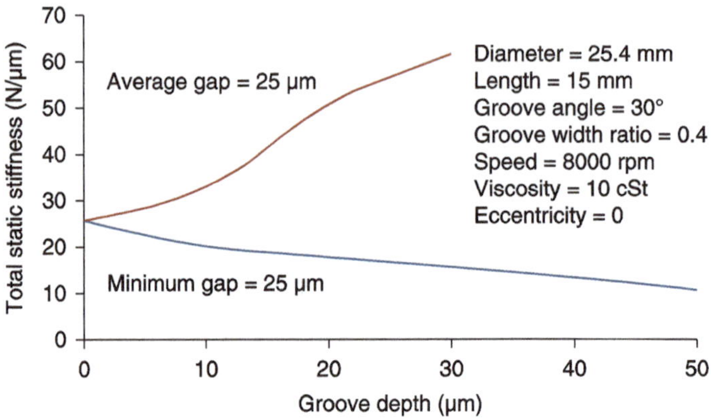

Figure 5.20 **Effect of groove depth on static stiffness**

being maintained as constant, in this case at 25 μm. Static stiffness gradually reduces as groove volume is increased, but the loss in performance is accompanied by a reduction in friction and power loss. Conversely adding grooves to a plain shaft so as to maintain a constant average gap and hence constant power loss, increases static stiffness but this means the minimum gap over the ridges of the bearing is reduced to compensate for groove volume, requiring a higher level of manufacturing precision. Thus whether or not a grooved bearing has a higher static stiffness than a plain bearing depends on the basis of comparison. If the same minimum gap and level of precision is the criterion then the plain bearing has an advantage, but if it is the ratio of static stiffness to power loss that is important then the grooved bearing offers a better performance.

Groove width has a similar effect on static stiffness to that of groove depth (Figure 5.21). Compared on the basis of minimum gap, then increasing groove volume by increasing width reduces static stiffness relative to that of a plain bearing. However, compared on the basis of friction and power loss, increasing groove width increases static stiffness.

Groove angle is independent of groove volume and for a given groove width and depth exhibits an optimum value. As for the annular thrust bearing, the groove angle has a small but significant effect on static stiffness, the optimum value is close to 20 degrees, but angles of between 15 and 30 degrees vary static stiffness by only 6% (Figure 5.22). This characteristic is remarkably consistent across a wide range of grooved bearing designs and operating conditions.

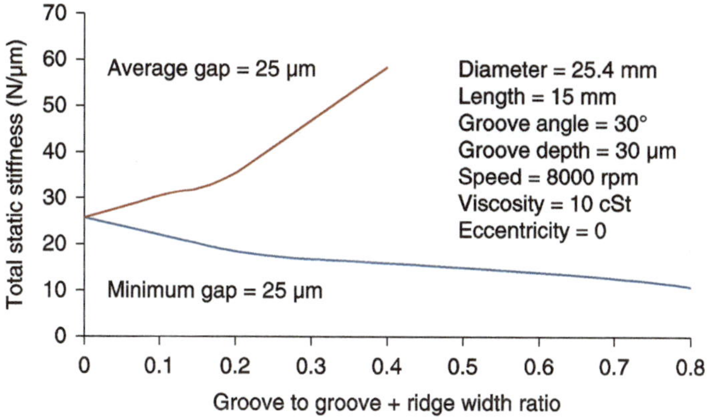

Figure 5.21 Effect of groove width on static stiffness

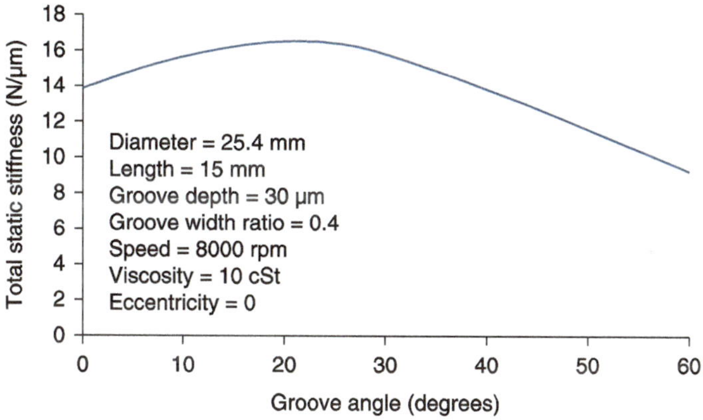

Figure 5.22 Effect of groove angle on static stiffness

5.9 Damping and dynamic stability

In this section, the dynamic aspects of grooved bearing performance are considered.

Although, in principle, oil lubricated annular thrust bearings may exhibit a 'swash' motion type of instability, they are generally used with journal bearings that constrain them to rotate about a fixed axis. Thus in practice, grooved annular thrust bearings are not only well behaved but can also exhibit a high damping capacity. For this type of bearing, attention is restricted to its damping characteristics and an expression for damping coefficient is derived for the case of an annular thrust bearing subject to a dynamic, concentrically applied, axial load. The expression is used to determine the damping characteristics of the chevron grooved bearing of Table 5.3.

However, oil lubricated journal bearings may exhibit two types of instability. Half speed whirl, as explained in the introduction, arises from an interaction between an orbital motion of the rotating component and the lubricant film. A forward orbital motion at a frequency of half the rotational speed dramatically reduces the lubricant entrainment velocity in the loaded zone of the bearing and hence substantially reduces its load carrying capacity. Half speed whirl creates sub-synchronous motion error and is unwanted in precision bearing applications. Journal bearings are also susceptible to out-of-balance loads that create a whirling motion at the speed of rotation. For oil lubricated bearings, load capacity increases linearly with speed but out-of-balance forces generated by a rotor of mass M increase with speed squared. If speed continues to increase, a critical value will be reached, above which the bearing cannot support the load. In this section, Eq. 5.58 is used to examine the effect of groove design on half speed whirl and to derive dynamic coefficients for defining a whirl stability criterion for rotor–bearing systems.

5.9.1 Damping in annular thrust bearings

Equation 5.10 for dynamic equilibrium of mass flux and Eq. 5.12 for flux density are the starting point for deriving damping forces in oil lubricated thrust bearings. Damping forces are derived for the case of a small simple harmonic perturbation to the bearing clearance superimposed on the much larger static equilibrium clearance:

$$h_r = h_{ro} + \Delta h.\sin\Omega t \quad \text{and} \quad h_g = h_{go} + \Delta h.\sin\Omega t \qquad [5.70]$$

This is assumed to generate a perturbation to the equilibrium pressure distribution:

$$P = P_o + \Delta p\,\sin(\Omega t - \phi) \qquad [5.71]$$

where the perturbed pressure has in phase, $\Delta p.\sin\Omega t.\cos\phi$ and quadrature, $\Delta p.\cos\Omega t.\sin\phi$ components relative to the phase of Δh, and the perturbed pressures are small compared to the static equilibrium pressure, P_o. Substituting for h_r and h_g in Eq. 5.12 and using Eq. 5.10 gives:

$$\Delta h\omega.\cos\omega t=\frac{1}{12\eta}\left[G_1\frac{\partial^2 P}{\partial r^2}+\frac{G_1}{r}\frac{\partial P}{\partial r}-12\eta(\omega_1-\omega_2)G_2\right] \quad [5.72]$$

The parameters G_1 and G_2 are functions of bearing gap and groove geometry and are hence time dependent. Thus they may also be written in terms of a static equilibrium value and a much smaller perturbed quantity:

$$G_1=G_{1o}+\Delta g_1 \quad \text{and} \quad G_2=G_{2o}+\Delta g_2 \quad [5.73]$$

Equation 5.73 may be substituted into Eq. 5.72 and the static terms and the first-order terms separated to give two equations, one describing static equilibrium, i.e. a restatement of Eq. 5.13, the other dynamic equilibrium. Here only the dynamic equilibrium is given:

$$12\eta.\Delta h\omega\cos\omega t=\left[G_{1o}\left(\frac{\partial^2 p}{\partial r^2}+\frac{1\partial p}{r\partial r}\right)+\Delta g_1\left(\frac{\partial^2 P_o}{\partial r^2}+\frac{1\partial P_o}{r\partial r}\right)-12\eta(\omega_1-\omega_2)\Delta g_2\right] \quad [5.74]$$

where p is given by Eq. 5.71. Δg_1 and Δg_2 can be derived by putting $h_r=h_{ro}+\Delta h.\sin\Omega t$ and $h_g=h_{go}+\Delta h.\sin\Omega t$ into Eq. 5.12 and separating out the first-order terms. The result is:

$$\Delta g_1=3h_r^2H^3(1-\alpha)\frac{[H^3-(H^3-1)\alpha\sin^2\beta]}{[\alpha+(1-\alpha)H^3]^2}\Delta h.\sin\Omega t \quad [5.75]$$

$$\Delta g_2=\frac{\alpha(1-\alpha)\sin\beta\cos\beta[\alpha-(1-\alpha)H^6+2(2-\alpha)H^3-3H^4]\Delta h.\sin\Omega t}{[\alpha+(1-\alpha)H^3]^2} \quad [5.76]$$

In Eq. 5.74, G_{1o}, $\partial^2P_o/\partial r^2$ and $\partial P_o/\partial r$ are either constant or a function of radius, and Δg_1 and Δg_2 are time-dependent variables containing $\sin\Omega t$. The pressure perturbation p is a function of radius and time. In-phase and quadrature components of p can be found by separating out the cos and sin terms of Eq. 5.74 using 5.71. In phase:

$$G_{1o}\left(\frac{\partial^2 p}{\partial r^2}+\frac{1\partial p}{r\partial r}\right)\cos\varphi=12\eta(\omega_1-\omega_2)\Delta g_2-\Delta g_1\left(\frac{\partial^2 P_o}{\partial r^2}+\frac{1\partial P_o}{r\partial r}\right) \quad [5.77]$$

Quadrature:

$$\frac{\partial^2 p}{\partial r^2}+\frac{1\partial p}{r\partial r}=\frac{12\eta.\Delta h\Omega}{G_{1o}\sin\phi} \quad [5.78]$$

Note that the derivative terms of P_o in Eq. 5.77 may be found from Eq. 5.16 to give:

$$\frac{\partial^2 P_o}{\partial r^2}+\frac{1\partial P_o}{r\partial r}=\frac{12\eta G_2(\omega_1-\omega_2)}{G_1}$$

So that for the in phase pressure:

$$\frac{\partial^2 p}{\partial r^2}+\frac{1\partial p}{r\partial r}=\frac{12\eta(\omega_1-\omega_2)}{G_{1o}\cos\varphi}\left[\Delta g_2-\Delta g_1\frac{G_2}{G_1}\right] \qquad [5.79]$$

Equations 5.78 and 5.79 apply to both grooved and plain sealing lands. For a sealing land $h_g=h_r$ and $\alpha=0$, so that G_1, G_2, Δg_1 and Δg_2 may be simplified:

$$G_1=h_r^3 \text{ and } G_2=0$$
$$\Delta g_1=3h_r^2 \text{ and } \Delta g_2=0$$

Thus for the plain sealing land, the in-phase and quadrature components of pressure are given by:

$$\frac{\partial^2 p}{\partial r^2}+\frac{1\partial p}{r\partial r}=0 \qquad [5.80]$$

and:

$$\frac{\partial^2 p}{\partial r^2}+\frac{1\partial p}{r\partial r}=\frac{12\eta.\Delta h\Omega}{h_r^3} \qquad [5.81]$$

Equation 5.81 takes the same form as Eq. 5.11; describing the static equilibrium pressure and the pressure distribution across the land can therefore be found from Eq. 5.15. The damping force, of most interest here, is determined by integrating Eq. 5.78 and imposing boundary conditions. Rearranging Eq. 5.78 gives:

$$\frac{\partial}{\partial r}\left(\frac{r\partial p}{\partial r}\right)=\frac{12\eta r.\Delta h\Omega}{G_{10}\sin\varphi}$$

and integrating twice with respect to r gives:

$$p_q=\frac{3\eta r^2\Delta h\Omega+A_1\ln(r)+A_2}{G_{10}} \qquad [5.82]$$

A_1 and A_2 are constants determined from the boundary conditions at the bore and outside diameter of the bearing land and $p_q=p\sin\phi$, the quadrature component of p. For an inward pumping bearing design with the seal around the bore, the perturbation in pressure $p=0$ at $r=r_i$ and $p=p_m$ at $r=r_m$. Solving for A_1 and A_2 gives the pressure distribution as a function of p_m and r:

$$p_q = \frac{3\eta\Delta h\Omega}{h_r^3}\left[\left(r^2 - r_i^2\right) - \left(r_m^2 - r_i^2\right)\frac{\ln(r/r_i)}{\ln(r_m/r_i)}\right] + p_m \frac{\ln(r/r_i)}{\ln(r_m/r_i)} \qquad [5.83]$$

r_m is the transition radius between seal and grooved lands and the pressure p_m at this point is found by considering equilibrium of mass flux for both parts of the bearing. The solution to Eq. 5.78 for a grooved land is the same as for a sealing land, except that h_r^3 is replaced by G_{1o}. Thus for an outward pumping land:

$$p_q = \frac{3\eta\Delta h}{G_{1o}}\Omega\left[\left(r^2 - r_i^2\right) - \left(r_m^2 - r_i^2\right)\frac{\ln(r/r_i)}{\ln(r_m/r_i)}\right] + p_m \frac{\ln(r/r_i)}{\ln(r_m/r_i)} \qquad [5.84]$$

and for an inward pumping land:

$$p_q = \frac{3\eta\Delta h}{G'_{1o}}\Omega\left[\left(r^2 - r_o^2\right) - \left(r_m^2 - r_o^2\right)\frac{\ln(r/r_o)}{\ln(r_m/r_o)}\right] + p_m \frac{\ln(r/r_o)}{\ln(r_m/r_o)} \qquad [5.85]$$

where r_o is the radius at the outside edge of the land for which the perturbation pressure is also zero and G'_{1o} is the groove parameter for the outer land. For an inward pumping bearing with sealing land, p_m is determined by equating the perturbation to mass flux density at the boundary between the two regions. Considering a perturbation to Eq. 5.12:

$$\frac{\partial}{\partial\theta}(\psi + \Delta\psi) = \frac{-\rho}{12\eta}\left[r(G_1 + g_1)\frac{\partial}{\partial r}(P_o + p) - (G_2 + g_2)r^2 6\eta(\omega_2 - \omega_1)\right]$$

which gives at the boundary, $r = r_m$:

$$\frac{\partial}{\partial\theta}(\Delta\psi) = \frac{-\rho}{12\eta}\left[r_m\left(G_1\frac{\partial p}{\partial r} + g_1\frac{\partial P_o}{\partial r}\right) - g_2 r_m^2 6\eta(\omega_2 - \omega_1)\right] \qquad [5.86]$$

Since the perturbation to flux density for the inner and outer bearing lands must be the same at the boundary:

$$\left[\left(G_1\frac{\partial p}{\partial r} + g_1\frac{\partial P_o}{\partial r}\right) - g_2 r_m 6\eta(\omega_2 - \omega_1)\right]_{\text{inner}} = \left[\left(G_1\frac{\partial p}{\partial r} + g_1\frac{\partial P_o}{\partial r}\right) - g_2 r_m 6\eta(\omega_2 - \omega_1)\right]_{\text{outer}} \qquad [5.87]$$

For an inward pumping bearing with a sealing land:

$$G_1\frac{\partial p_g}{\partial r}+g_1\frac{\partial P_{go}}{\partial r}-g_2 r_m 6\eta\left(\omega_2-\omega_1\right)=h_{ro}^3\frac{\partial p_s}{\partial r}+3h_{ro}^2\frac{\partial P}{\partial r} \qquad [5.88]$$

The pressure gradients are found from Eqs 5.15, 5.16, 5.83 and 5.85. Substituting these into Eq. 5.88 and simplifying gives:

$$\begin{aligned} p_m = &\left\{6\eta r_m^2\left(\omega_2-\omega_1\right)\left(g_2'-g_2+g_1'\frac{G_2'}{G_1'}-g_1\frac{G_2}{G_1}\right)\right. \\ &+(P_m-P_a)\left(\frac{g_1'}{\ln\left(r_o/r_m\right)}-\frac{g_1}{\ln\left(r_i/r_m\right)}\right) \\ &\left.+3\eta\Delta h\Omega\left[\left(\frac{r_m^2-r_o^2}{\ln\left(r_o/r_m\right)}\right)-\left(\frac{r_m^2-r_i^2}{\ln\left(r_i/r_m\right)}\right)\right]\right\}\Bigg/\left[\frac{G_1'-G_1}{\ln(r_o/r_m)\ln\left(r_i/r_m\right)}\right] \end{aligned} \qquad [5.89]$$

where ′ denotes the outer land of the bearing. For a sealing land, $G_2=g_2=0$ and $G_1=h_r^3$ and $g_1=3h_r^2$. Equations 5.83 and 5.85 together with Eq. 5.89 define the perturbed pressure distribution within an inward pumping grooved thrust bearing subject to a small dynamic deflection $\Delta h.\sin\omega t$. The dynamic force acting on the bearing is found by integrating Eqs 5.83 and 5.85. For an outward pumping grooved land:

$$\begin{aligned} F_d = &\frac{\pi}{2}p_m\left[2r_m^2-\frac{r_m^2-r_i^2}{\ln\left(r_m/r_i\right)}\right] \\ &+\frac{3\pi\eta}{2G_1}\left(\omega_2-\omega_1\right)\Delta h\Omega\left(r_m^2-r_i^2\right)\left[\frac{r_m^2-r_i^2}{\ln\left(r_m/r_i\right)}-\left(r_m^2+r_i^2\right)\right] \end{aligned} \qquad [5.90]$$

For an inward pumping grooved land, replace r_i by r_m and r_m by r_o and for a sealing land, replace G_1 by h_r^3. Equation 5.90 expresses damping in terms of force, and dividing left- and right-hand sides of the equation by $\Delta h\Omega$ gives the damping coefficient in terms of force per unit velocity of approach of the bearing surfaces, i.e. in SI units N/(m/s).

5.9.2 Annular thrust bearing damping characteristics

Examples of how damping coefficients vary with bearing parameters is given here for the bearing design considered in Section 5.7.4. Damping coefficient increases with oil viscosity and bearing area but decreases with gap and rotational speed. For the chevron grooved bearing defined in Table 5.3, the damping force as a function of bearing gap is shown in Figure 5.23. Damping coefficient is expressed in N/(mm/s) for gaps in the range 2 to 20 μm and it can be seen that for this type of bearing damping, levels decay exponentially with increasing gap.

The decrease in damping coefficient with rotational speed is shown in Figure 5.24, where it can be seen that within a practical range of speeds for the bearing considered, i.e. 1 to 20 krpm, the relationship is near linear.

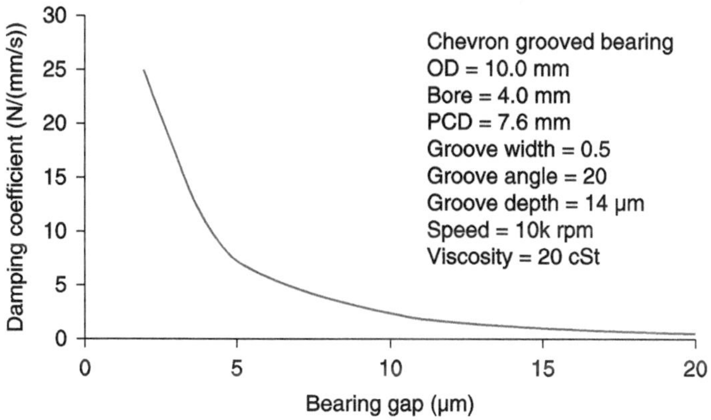

Figure 5.23 Example of the effect of gap on damping coefficient for a chevron grooved bearing

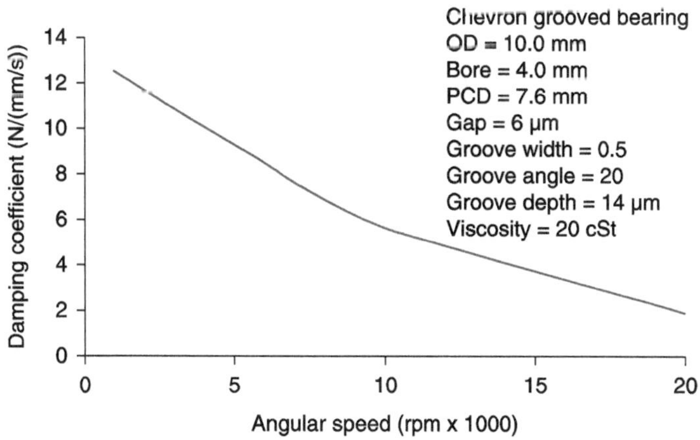

Figure 5.24 Example of the effect of angular speed on damping coefficient

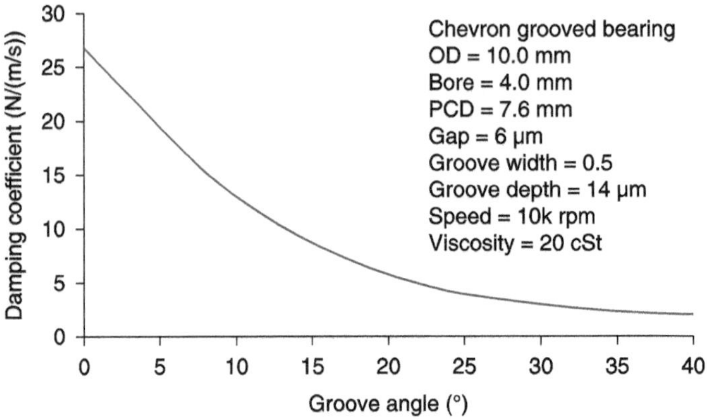

Figure 5.25 **Example of the effect of groove angle on damping force**

Damping coefficient is also influenced by groove design. Increasing groove depth and groove width increases the average bearing gap and reduces the damping force in a manner illustrated by Figure 5.23. However, damping is also strongly dependent on groove angle, decreasing with increasing groove angle (Figure 5.25). For this bearing, the optimum groove angle was 20 degrees, based on a maximum static stiffness criterion. Reducing groove angle to 10 degrees more than doubles damping levels but reduces stiffness by only 15%.

5.9.3 Dynamic characteristics of herringbone grooved journal bearings

Equation 5.58 enables the whirl response of a single oil lubricated journal bearing to be determined and can be used to evaluate the half speed whirl and synchronous whirl performance of grooved bearings. It should be noted that in deriving Eq. 5.58, a steady state whirl equilibrium was assumed for which squeeze film effects occurring in the lubricant film were neglected. This enables a reasonable estimate of the forces generated by a given whirl amplitude and frequency to be made and can be used to determine the effective stiffness of the bearing in a rotor–bearing system in order to identify resonant frequencies. But in neglecting squeeze film effects, damping is assumed to be zero and response amplitudes may in theory become infinite under some running conditions.

Whirl response

A major advantage of herringbone grooved journal bearings compared to other types of fluid film journal bearings is their improvement in whirl response when operating at zero or low eccentricities. It is this aspect of their performance that makes them far more suitable for precision applications than all other types of self acting fluid film journal bearing.

[23] presents the results of an experimental study of the dynamic characteristics of a herringbone grooved journal bearing. The study makes a direct comparison between the dynamic response of plain and grooved journal bearings using an impulse response measuring technique. The whirl response of the two types of bearing is shown in Figure 5.26 and compared with theoretical predictions from the analysis given in Section 5.8.

In Figure 5.26, a whirl ratio of 1 corresponds to a whirl speed equal to the bearing's rotational speed of 7800 rpm. The dashed lines represent experimental measurements and the solid lines are theoretical predictions. In using an impulse response measuring technique, [23] was measuring the frequency response of the bearing–shaft system but taking the 0 to 130 Hz part of the response and normalising it with respect to the shaft's rotational speed of 7800 rpm (130 Hz) produces the response shown in Figure 5.26. Agreement between the theoretical whirl response and the measured frequency response is reasonable over the range of whirl speeds considered. However, of most importance is the difference

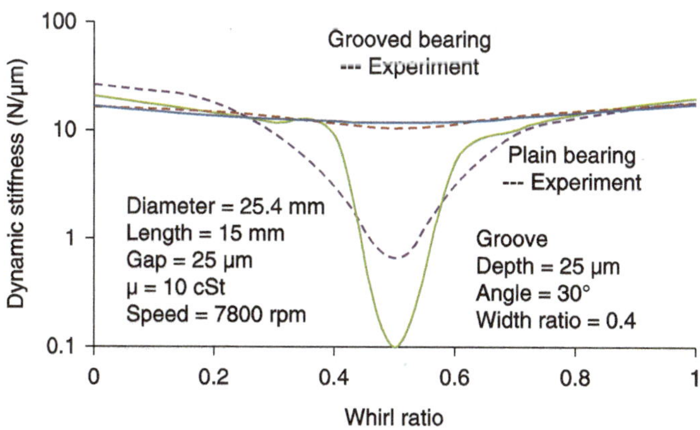

Figure 5.26 **Example of whirl response of plain and grooved journal bearings**

between the plain bearing and grooved bearing responses. At a whirl speed ratio of 0.5, the measured dynamic stiffness of the plain bearing reduces to 0.65 N/μm from a value of 26 N/μm at zero whirl speed. This compares to measured dynamic stiffness of 10.5 and 16.5 N/μm respectively for the grooved bearing. Thus the plain bearing operating at low eccentricities is extremely sensitive to any disturbing forces that are impulsive or contain frequencies close to half of its running speed and as such is prone to creating relatively large amplitude sub-synchronous motion errors.

For the grooved bearing, the minimum dynamic stiffness also occurs at half the rotational speed but is influenced by groove design. Groove parameters are often optimised with respect to a minimum dynamic stiffness criterion to achieve a bearing highly insensitive to whirl and sub-synchronous disturbing forces. Figures 5.27 to 5.29 show how minimum dynamic stiffness depends on groove depth, width and angle for the bearing design example considered in Figure 5.26. Each of the parameters exhibits a distinct optimum value. Groove depth is 1.2 times the bearing gap, groove width ratio is 0.5 times the groove plus ridge width and groove angle is 25 degrees. The dynamic stiffness–whirl speed characteristics of the optimised bearing design are shown in Figure 5.30.

The optimised bearing design has a minimum dynamic stiffness of 24.5 N/μm, only 18% lower than the stiffness of 29.9 N/μm at zero whirl speed (Figure 5.30). In this case, groove parameters were optimised at one specific operating condition but the optimum values of depth

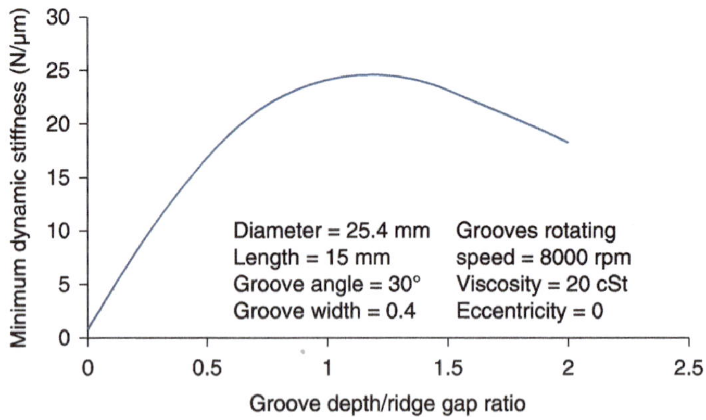

Figure 5.27 Example of the effect of groove depth on minimum dynamic stiffness

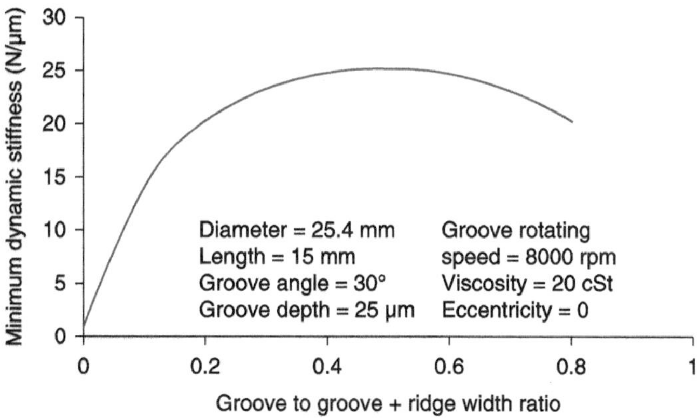

Figure 5.28 Example of the effect of groove width on minimum dynamic stiffness

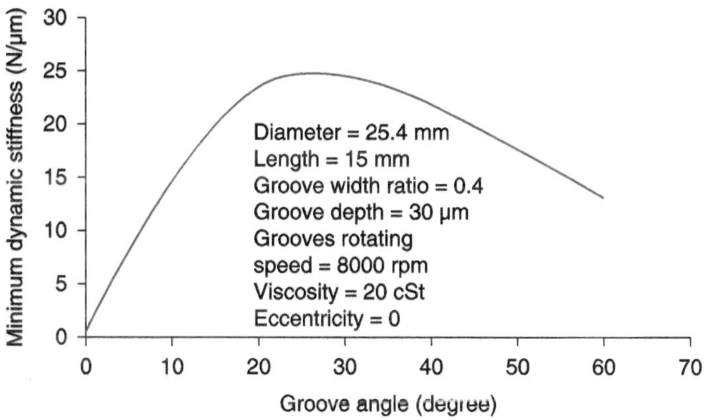

Figure 5.29 Example of the effect of groove angle on minimum dynamic stiffness

ratio = 1.2; width ratio = 0.5 and angle = 25 degrees are typical and remarkably insensitive to bearing speed, clearance, length/diameter ratio and whether or not the grooves are rotating with respect to the applied load vector.

However, the ratio of minimum dynamic stiffness to static stiffness is dependent on the bearing length to diameter ratio (Figure 5.31). Static stiffness and hence load capacity increase rapidly with bearing length, as increasing the latter not only increases bearing area but enables the grooves to achieve higher pressures in the centre of the bearing. Minimum

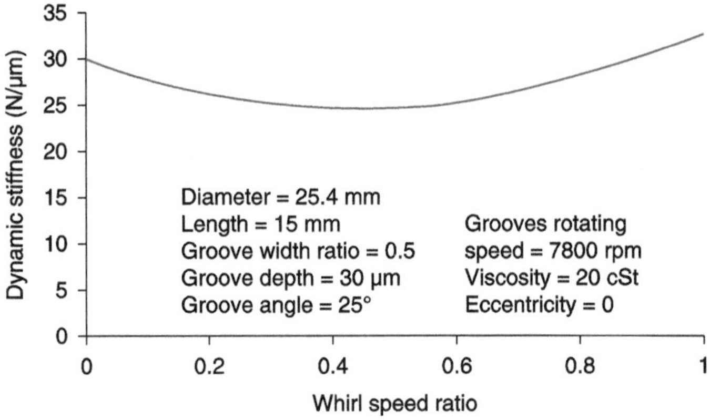

Figure 5.30 **Dynamic stiffness of optimised bearing design**

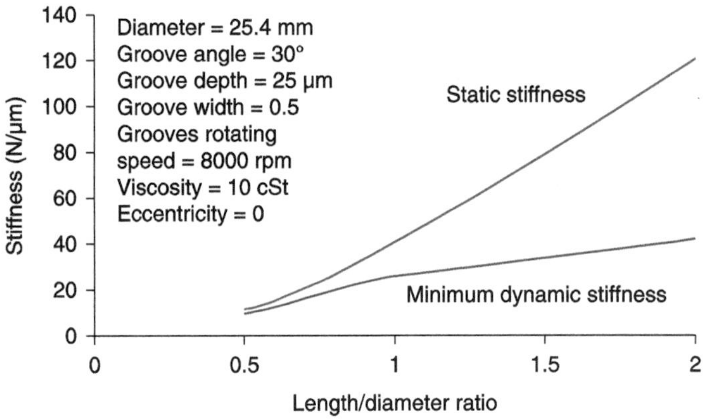

Figure 5.31 **Example of the effect of length on diameter ratio on bearing stiffness**

dynamic stiffness increases more slowly with bearing length, indicating that long bearings are more susceptible to half speed whirl. In practice, length/diameter ratios are normally in the range 0.5 to 2.

The analysis presented in the preceding section states that bearing load capacity and hence static stiffness is proportional to rotational speed and viscosity. In practice, increased speed creates higher friction levels, in turn increasing the bearing operating temperature and reducing oil viscosity. In Figure 5.32, static stiffness is measured as a function of speed and compared with two theoretical predictions, one assuming constant

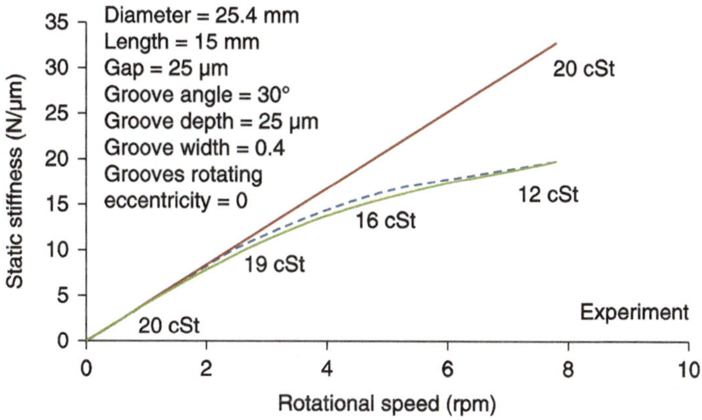

Figure 5.32 **Example of the effect of speed on static stiffness**

viscosity, the other correcting for temperature rise. Reasonable agreement is achieved for the latter. It follows that in estimating bearing performance at the design stage, it is important to know the oil viscosity at the required operating speed.

Dynamic coefficients of herringbone grooved journal bearings

The whirl analysis discussed in the preceding sections, although limited to concentric rotation, is invaluable in optimising the dynamic performance of individual journal bearings. But to determine how the bearing or bearings behave as part of a rotor–bearing system, stiffness and damping coefficients are more useful. Lund [24] reviewed the use of eight coefficients contained in stiffness [K] and damping [C] matrices for representing the dynamic response of fluid film journal bearings in 1987, and concluded that they were particularly useful in determining the unbalance response, damped natural frequencies and stability of rotors supported by fluid film bearings. He also concluded that the limitation of considering a perturbation or small amplitude dynamic response was of little practical importance in most rotor dynamic applications. Dynamic coefficients are an aspect of herringbone grooved bearings that is not fully developed and the following is an approximate approach, limited to concentric rotation for which coefficients are derived from the whirl analysis presented in Section 5.7. It should be stressed that damping coefficients derived in this way only account for the cross-coupling between motions in X and Y directions and do not actually contain a damping term.

The static stiffness of hydrodynamic journal bearings was discussed in Section 5.8 and because this type of bearing exhibits an attitude angle when subjected to a static radial load, it was necessary to describe its static stiffness by a set of four coefficients:

$$[K] = \begin{bmatrix} K_{xx} & K_{xy} \\ K_{yx} & K_{yy} \end{bmatrix}$$

where the subscripts x and y refer to orthogonal stiffness components contained in a radial plane and in the general case of a bearing subject to a significant radial load, all four stiffness components may differ in magnitude. The static force acting on the bearing is related to static x and y deflections by:

$$\begin{bmatrix} F_{ox} \\ F_{oy} \end{bmatrix} = \begin{bmatrix} K_{xx} & K_{xy} \\ K_{yx} & K_{yy} \end{bmatrix} \begin{bmatrix} x_o \\ y_o \end{bmatrix} \qquad [5.91]$$

If now the bearing is subject to a dynamic radial load, the attitude angle phenomenon causes it to respond in the form of an orbit. Furthermore, since the lubricant film thickness is now continually changing with time, squeeze film effects must be taken into account. Thus for an orbit of small amplitude, $\Delta e = (x^2 + y^2)^{1/2}$, the dynamic force is related to the amplitude and speed of orbit by:

$$\begin{bmatrix} F_{dx} \\ F_{dy} \end{bmatrix} = \begin{bmatrix} K_{xx} & K_{xy} \\ K_{yx} & K_{yy} \end{bmatrix} \begin{bmatrix} x \\ y \end{bmatrix} + \begin{bmatrix} C_{xx} & C_{xy} \\ C_{yx} & C_{yy} \end{bmatrix} \begin{bmatrix} dx/dt \\ dy/dt \end{bmatrix} \qquad [5.92]$$

where [K] contains the stiffness coefficients and [C] is the damping coefficients. The eight coefficients are those referred to by Lund in his review of [24]. The analysis of a grooved journal bearing presented in Eq. 5.7 considered a small amplitude perturbation superimposed upon a 'no load' or concentric rotation static equilibrium condition. The perturbation took the form of an orbit of constant amplitude, eC, and constant angular speed, $d\phi/dt = \omega_3$. Static stiffness coefficients K_{xx} and K_{xy} were determined from the bearing load, attitude angle and deflection for the special case of $d\phi/dt = 0$. The four stiffness coefficients used in Eq. 5.93 are for a bearing subject to a static radial deflection, and for concentric rotation it is assumed that $K_{xx} = K_{yy}$ and $K_{xy} = -K_{yx}$. Lund [24] determined the stiffness coefficients for a plain journal bearing for eccentricities of between 0.02 and 0.9 and showed substantial differences between the four coefficients for large eccentricities. At an eccentricity of 0.02, K_{xy} approximated $-K_{yx}$ but there was still a significant difference between K_{xx} and K_{yy}. The differences may be less for a grooved bearing than a plain bearing, as the attitude angle for the latter to approach 90 degrees at concentric rotation, whereas for a grooved bearing it is typically 50 to 60 degrees, but nonetheless this is a limitation of the present approach.

In the following, damping coefficients are determined for the case of concentric rotation. For damping, Lund [25] found $C_{xx} = C_{yy}$ and $C_{xy} = C_{yx}$ at an eccentricity of 0.02, so the assumption that this also applies to concentric rotation is reasonable. Rearranging Eq. 5.92 and using Eq. 5.91 gives:

$$\begin{bmatrix} C_{xx} & C_{xy} \\ C_{yx} & C_{yy} \end{bmatrix} \begin{bmatrix} dx/dt \\ dy/dt \end{bmatrix} = \begin{bmatrix} F_{dx} - F_{ox} \\ F_{dy} - F_{oy} \end{bmatrix} \qquad [5.93]$$

x and y in Eq. 5.93 are related to eccentricity e and attitude angle ϕ in Eq. 5.68 by:

$$x = e\cos\varphi \text{ and } y = e\sin\varphi \qquad [5.94]$$

and for steady state whirl, the derivatives are given by:

$$dx/dt = -e\sin\phi.(d\phi/dt) \text{ and } dy/dt = e\cos\phi.(d\phi/dt) \qquad [5.95]$$

where $d\phi/dt = \omega_3$, the whirl speed used in the analysis of Section 5.7. The forces F_{ox}, F_{oy}, F_{dx} and F_{dy} are also determined from the whirl analysis of Eq. 5.7. F_{ox} and F_{oy} are the forces corresponding to a static deflection eC of the bearing, i.e. for $\omega_3 = 0$. F_{dx} and F_{dy} are the dynamic forces determined for finite values of ω_3 at the same amplitude eC. Since, for concentric bearing operation $C_{xx} = C_{yy}$ and $C_{xy} = C_{yx}$, then substituting Eq. 5.95 into Eq.5.94 gives:

$$\begin{bmatrix} C_{xx} & C_{xy} \\ C_{xy} & C_{xx} \end{bmatrix} \begin{bmatrix} -e\omega_3\sin\phi \\ e\omega_3\cos\phi \end{bmatrix} = \begin{bmatrix} F_{dx} - F_{ox} \\ F_{dy} - F_{oy} \end{bmatrix} \qquad [5.96]$$

where the dynamic force components are given by:

$$F_{dx} = 2\int_o^{L/2}\int_o^{2\pi} P_1(Z,\psi)\cos\psi \, dZ \, d\psi \qquad [5.97]$$

$$F_{dy} = 2\int_o^{L/2}\int_o^{2\pi} P_1(Z,\psi)\sin\psi \, dZ \, d\psi \qquad [5.98]$$

for non-zero values of ω_3. Damping coefficients are determined from Eq. 5.97.

Example of damping coefficients

The 'no load' damping coefficients are dependent on whirl speed and bearing design. For the bearing considered in Section 5.9, the effect of groove depth on the coefficients C_{xx} and C_{yy} is shown in Figure 5.33 for a whirl speed ratio of 1. The damping coefficients for this condition largely influence the bearing's imbalance response. It can be observed that the cross-coupling coefficient C_{xy} reduces steadily with groove depth, whereas the direct coefficient C_{xx} exhibits a maximum value at a groove depth of 28 μm, close to the optimum value of 30 μm giving maximum bearing stiffness. A groove depth of zero represents a plain journal bearing for which it is observed that C_{xx} is low compared to that for the optimal design of grooved bearing.

The effect of groove angle on damping coefficient is shown in Figure 5.34. Again the direct damping coefficient is a maximum for the optimum groove angle based on a bearing stiffness criterion, whilst the cross-coupled coefficient reduces with increasing groove angle.

Instability

It is clear from Sections 5.8 and 5.9 that if, for whatever reason, a rotor takes up a whirling motion, it not only imposes a rotating load on

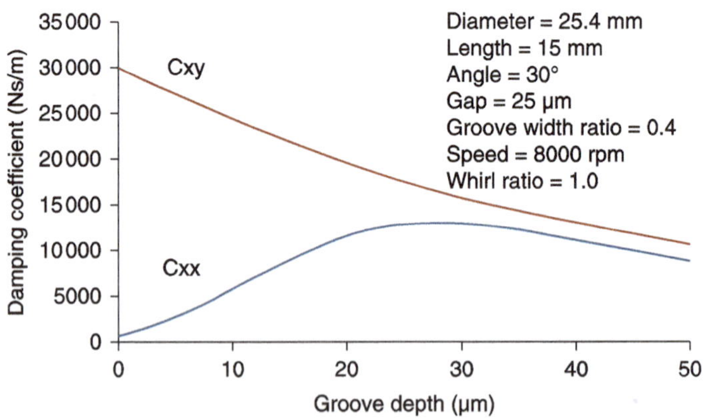

Figure 5.33 **Example of the effect of groove depth on damping coefficients**

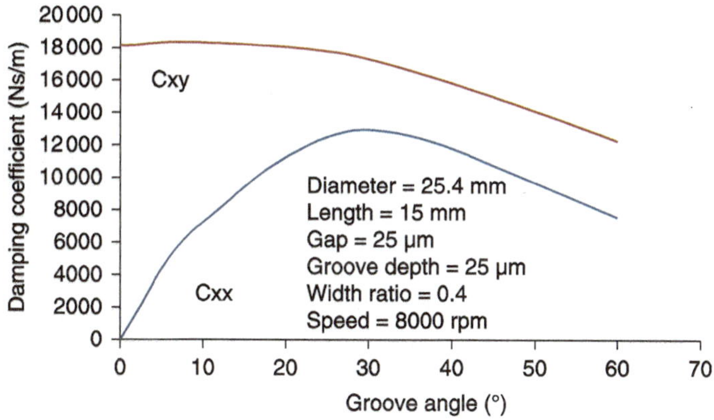

Figure 5.34 **Example of the effect of groove angle on damping coefficients**

the journal bearings but also interacts with the fluid film influencing the bearing's ability to support the load. The whirl motion is stable if its amplitude remains constant with time, at which condition whirl forces are within the bearing's dynamic load carrying capacity. However, if the whirl forces are increased, they can reach a threshold value that the bearing can no longer support, giving rise to an instability where the whirl amplitude continues to increase rapidly with time leading to seizure.

The stiffness and damping coefficients defined in Section 5.9 can be used to develop a criterion for instability. First, the bearing is assumed to support a mass M and to be whirling with small amplitudes x and y in a plane Oxy normal to the bearing's axis. The equations of motion are:

$$\begin{aligned} &Md^2x/dt^2 + C_{xx}\, dx/dt + K_{xx}x + C_{xy}dy/dt + K_{xy}y = 0 \\ &Md^2y/dt^2 + C_{yy}\, dy/dt + K_{yy}y + C_{yx}dx/dt + K_{yx}x = 0 \end{aligned} \qquad [5.99]$$

Note that for the mass M whirling about the bearing centre, $K_{xx} = K_{yy}$, $K_{xy} = -K_{yx}$ and $C_{xx} = C_{yy}$, $C_{xy} = C_{yx}$. The solution to Eq. 5.99 is assumed to take the form:

$$x = x\ \exp(\lambda t) \text{ and } y = y\ \exp(\lambda t) \qquad [5.100]$$

where λ is a complex number, $\lambda = \varsigma + i\omega$. If ς is positive, the solution is unstable, whereas if ς is negative, it is stable. ω is the whirl speed of the rotor. Substituting for x and y in Eq. 5.99 and using Eq. 5.100 results in:

$$\begin{bmatrix} (K_{xx} - M\omega^2 + i\omega C_{xx}) & (K_{xy} + i\omega C_{xy}) \\ (K_{yx} + i\omega C_{yy}) & (K_{yy} - M\omega^2 + i\omega C_{yy}) \end{bmatrix} \begin{bmatrix} x \\ y \end{bmatrix} = 0 \qquad [5.101]$$

A non-trivial solution to Eq. 5.101 can be found by forming the determinant and setting it equal to zero. Separating the real and imaginary terms of the determinant results in the following two equations:

$$K_{xx} C_{yy} + K_{yy} C_{xx} - K_{xy} C_{yx} - K_{yx} C_{xy} = M_c \omega_c^2 (C_{xx} + C_{yy}) \qquad [5.102]$$

$$(K_{xx} - M_c \omega_c^2)(K_{yy} - M_c \omega_c^2) - K_{xy} K_{yx} = \omega_c^2 (C_{xx} C_{yy} - C_{xy} C_{yx}) \qquad [5.103]$$

from which the critical mass $M = M_c$ and the associated whirl frequency $\omega = \omega_c$ can be determined. The coefficients K and C are defined by bearing design and conditions of operating speed and lubricant, and for any given conditions a stability boundary of mass versus frequency can be determined.

5.10 Motion error

Providing the bearing is not being used in an unstable part of its operating regime, then the main sources of motion error in self-acting bearings are geometrical errors of the critical load bearing surfaces and the dynamic pressure perturbation inherent to a groove–ridge pair. Surface geometry errors can take the form of out-of-flatness errors in annular thrust bearings; out-of-roundness in cylindrical journal bearings or groove geometry errors. All types of surface error and the pressure perturbations result in motion error synchronous with the speed of the rotating bearing component.

In this section, the effects of geometrical surface errors, groove geometry errors and dynamic pressure perturbations on bearing motion error are considered. For oil lubricated annular thrust and cylindrical journal bearings, the effect of amplitude and wavelength of each type of error on the magnitude and frequency of pressure fluctuations in the bearing are explained and the net effect on motion error discussed.

5.10.1 Bearing model

For the purpose of explaining how bearing surface imperfections create motion error, the simple structural model of Figure 5.35 is used.

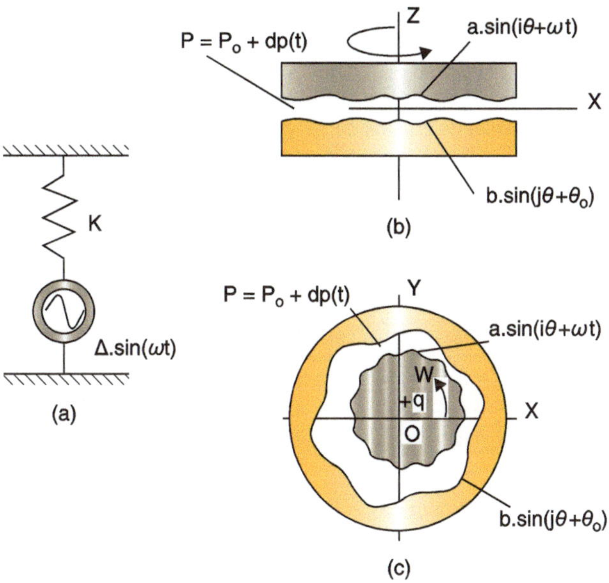

Figure 5.35 (a) Structural model; (b) thrust bearing; and (c) journal bearing with surface imperfections

As for the other types of bearing considered in this book, the structural model of the self-acting bearing, Figure 5.35(a) comprises a linear spring of stiffness K in series with a source of excitation $\Delta.\sin(\omega t)$, where Δ is the amplitude of the bearing's motion error arising from imperfections on its surfaces. Depending on the bearing arrangement and how the bearing is mounted, the excitation may generate a relative displacement between the shaft and housing and/or transmit a disturbing force. In a limiting case, the shaft and housing can be considered completely unrestrained, in which case the relative movement between the two will be $\Delta.\sin(\omega t)$. In another limiting case, both housing and shaft can be considered rigidly fixed in position and then the disturbance will generate a force between the two given by:

$$F = K.\Delta.\sin(\omega t) \qquad [5.104]$$

where K is the stiffness of the bearing. In determining the motion error of a self-acting bearing, it is convenient to consider the latter. The shaft and housing are rigidly fixed in position so that any imperfections on the bearing surfaces only create a fluctuation in the pressure within the fluid film. Integrating pressure over the bearing area gives the disturbing force,

F and the amplitude of the motion error, Δ is then simply given by F/K from Eq. 5.105.

Figures 5.35(b) and (c) show models of thrust and journal bearings to be considered. In each case, one of the surfaces is stationary and the other is rotating at a constant angular speed given by ω. For the thrust bearing, the two surfaces are concentric (Figure 5.35(b)). For the journal bearing, the stationary outer surface is fixed with its centre at O, but the shaft and hence the inner surface of the bearing rotates at constant angular speed ω about its centre at q.

Both surfaces of each bearing are assumed to contain small amplitude geometrical errors. In general, these will be a function of angular position θ and radius r in the case of a thrust bearing, or θ and axial position z in the case of a journal bearing. It is convenient to consider these geometrical errors to be simple harmonic in form and only vary in height in the direction of motion, i.e. for both thrust and journal bearings:

For the stationary surface:

$$b = b.\sin(j\theta + \theta_o) \qquad 2 < j < N$$

where b is the amplitude of the error, j is the number of waves per circumference and θ_o is a phase angle defining angular position with respect to the frame of reference. θ_o has no physical significance in this analysis and may be put equal to zero. Similarly, for the rotating surface:

$$a = a.\sin i(\theta' + \theta'_o) \qquad 2 < i < N$$

where a is amplitude, θ' is defined with respect to a frame rotating at angular speed ω, θ'_o is a phase angle defining the position of the rotating frame with respect to a fixed frame at any point in time and i is the number of waves/circumference. θ'_o may be expressed as a function of time, $\theta'_o \omega t$. Thus the above equations may be expressed as:

$$\begin{aligned} b &= b.\sin(j\theta) && 2 < j < N \\ a &= a.\sin i(\omega t + \theta') && 2 < i < N \end{aligned} \qquad [5.105]$$

for both annular and cylindrical surfaces. Fourier series can be used to describe more complex surface geometries. In fact, any cyclic geometry can be considered to be the sum of a number of simple harmonic terms. For example, for the stationary bearing surface:

$$b(\theta) = b_o + \Sigma_2^N b_{1m}\sin(m\theta) + \Sigma_2^N b_{2m}\cos(m\theta) \qquad [5.106]$$

In this analysis, only small amplitude surface geometry errors and small amplitude motion errors are considered where it is reasonable to assume that the relationship between the two is linear. Thus for the case of a complex surface geometry, the effect of individual wavelengths in Eq. 5.106 upon pressure in the bearing's fluid film can be considered separately and added linearly to determine the net effect.

Since the analysis is confined to small amplitude motion errors and surface imperfections, then the relationship between the two can be determined by considering a perturbation to the bearing's static equilibrium pressure distribution. For self-acting bearings, three sources of pressure perturbation are considered:

- geometrical errors on plain surfaces and ridge surfaces of grooved lands;
- groove geometry errors, namely variations in width and depth;
- the inherent pressure perturbation to the smoothed pressure profile generated by a groove–ridge pair.

Thus it is convenient to consider plain and grooved surfaces separately. For thrust bearings, the plain surface and sealing land are subject to out-of-flatness errors, whereas for the grooved land all three types of error listed above can occur. Similarly, for a cylindrical journal bearing, plain surfaces are only subject to out-of-roundness, whereas grooved lands may have all three types of error.

Plain surface geometry errors are considered as small departures from an otherwise geometrically perfect surface that result in a small change to the bearing clearance. For a bearing rotating at constant speed, the gap variation with position and time is given by:

$$h = h_r + \Delta h(\theta,t) \tag{5.107}$$

where h_r is the static equilibrium ridge clearance and is constant for a thrust bearing but may vary with position on a radially loaded journal bearing and Δh is the perturbed value of the gap that varies with position and time. It is assumed that h, as defined by Eq. 5.107, produces a pressure distribution in the bearing gap given by:

$$p = P_o + p(\theta,r,t) \tag{5.108}$$

where P_o is the static equilibrium pressure field and p is the time dependent perturbation in pressure produced by Δh.

5.10.2 Grooved annular thrust bearings

Annular thrust bearings may have one surface fully grooved as in a chevron grooved bearing or partially grooved as in an inward or outward pumping bearing. Furthermore, the grooved surface may be chosen to be stationary or rotating. Here, an inward pumping bearing design is considered.

With pressure perturbations, surface geometry and groove geometry errors, there are many different situations that need to be covered. Here the overlap between the different situations is minimised by confining attention to an inward pumping bearing design comprising an inboard sealing land with two plain opposing surfaces, together with an outboard pumping land where one of the opposing surfaces is grooved. By considering sealing lands and pumping lands separately, the analysis is easily adapted to the two other types of bearing.

Sealing land

The sealing land is assumed to comprise two nominally flat, parallel surfaces, each of which has a small amplitude geometry error superimposed upon it (Figure 5.36). For convenience, the geometry errors are simple harmonic in form, with an amplitude dependent only on angular position. For the stationary surface, the error is described by $b = b.\sin j\theta$, where j is the number of waves/circumference and θ is the angle relative to the OX axis of the coordinate system OXYZ. For the rotating surface, the geometry error is described by $a = a.\sin i(\omega t + \theta)$, where i is the number of waves/circumference, ω is the angular speed of the surface and θ is again a phase angle defined with respect to OX.

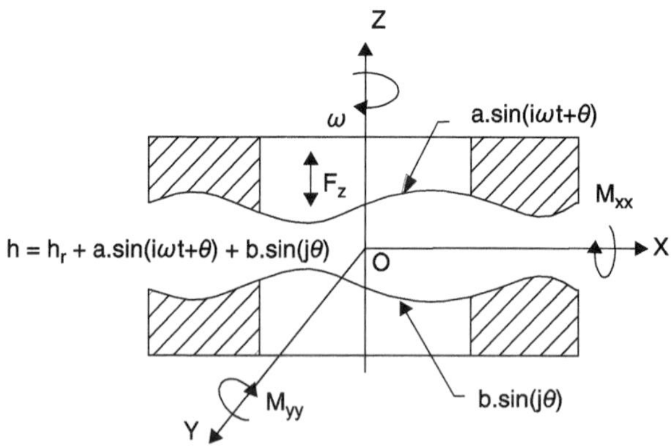

Figure 5.36 **Surface geometry errors on sealing land of annular thrust bearing**

The surface errors generate a bearing clearance that varies with angular position and time:

$$h = h_r + a.\sin i(\omega t + \theta) + b.\sin j\theta = h_r + \Delta h \qquad [5.109]$$

This is assumed to create a dynamic perturbation to the steady state equilibrium pressure in the bearing that has the potential to produce a linear motion error in the direction of the bearing axis OZ and angular motion errors M_{xx} about the axis OX and M_{yy} about OY.

Equation 5.14 describing the steady state pressure gradient in a sealing land is the starting point for the analysis. Putting $h = h_r + \Delta h$ and $P = P_o + p$ gives:

$$\frac{\rho}{12\eta}(h_r + \Delta h)^3 \frac{rd(P_o + p)}{dr} = C_2$$

Separating out the first-order terms results in:

$$\frac{dp}{dr} + \frac{3\Delta h}{h_r}\frac{dP_o}{dr} = 0 \qquad [5.110]$$

and integrating:

$$p + \frac{3\Delta h}{h_r}P_o = C \qquad [5.111]$$

where p is a function of r, θ and t and P_o is the steady state pressure distribution given by Eq. 5.15. The pressure perturbation $p = 0$ when $P_o = P_a$, therefore $C = 3\Delta h/h_r$ and:

$$p = \frac{-3\Delta h}{h_r}(P_o - P_a) \qquad [5.112]$$

Substituting for Δh using Eq.5.109 and $(P_o - P_a)$ using Eq. 5.15:

$$p = \frac{-3}{h_r}[a.\sin i\omega t + b.\sin j\theta](P_m - P_a)\frac{\ln(r/r_i)}{\ln(r_m/r_i)} \qquad [5.113]$$

Thus the effect of the surface geometry errors on sealing lands is to create a pressure perturbation proportional to the amplitude of the error, the steady state pressure distribution and inversely proportional to the clearance. If the error is on the stationary surface, the pressure perturbation is only a function of position but if it is on the rotating surface, it creates a time dependent pressure fluctuation harmonic of rotational speed. The net dynamic axial force and moments are found by integrating pressure over the area of the land. For axial force:

$$F_a = \frac{-3}{h_r}(P_m - P_a)\int_0^{2\pi}\int_{ri}^{rm}\left[a.\sin i(\omega t+\theta)+b.\sin j\theta\right]r\frac{\ln(r/r_i)}{\ln(r_m/r_i)}dr.d\theta \qquad [5.114]$$

However, the integrals of a.siniω and b.sinjθ over the interval 0 to 2π are both zero, so there is no net axial dynamic force. For moments:

$$M_{xx} = M_{yy} = \int_o^{2\pi}\int_{ri}^{rm} p.r.r\sin\theta dr.d\theta$$

$$M_{xx} = \frac{-3}{h_r}(P_m - P_a)\int_0^{2\pi}\left[a.\sin i(\omega t+\theta)+b.\sin j\theta\right]\sin\theta d\theta \qquad [5.115]$$

$$\int_{ri}^{rm} r^2\frac{\ln(r/r_i)}{\ln(r_m/r_i)}dr$$

Multiplying the sin terms and then integrating Eq. 5.110 with respect to θ gives:

$$M_{xx} = \frac{3\pi}{2h_r}(P_m - P_a)\left(\frac{3r_m^3\ln(r_m/r_i)-r_m^3+r_i^3}{9\ln(r_m/r_i)}\right)\left[a.\cos\omega t+\pi b\right] \qquad [5.116]$$

Note that sinjθsinθ integrates to $\pi/2$ for $j=1$ over the interval 0 to 2π and zero for all other values of j. This means that surface errors on the stationary land create a static moment, but do not create a dynamic moment variation. Similarly, sini$(\omega t+\theta)$sinθ integrates to $(\cos\omega t)/2$ for $i=1$ and zero for all other values of i, which means that a small misalignment or tilt of the rotating surface creates a net dynamic moment with a frequency equal to the speed of rotation.

Grooved land

Two sources of motion error are considered, pressure perturbations inherent in groove and ridge pairs and surface geometry errors. In the analysis of Section 5.6, the pumping land was assumed to have an infinite number of grooves so that the perturbation was infinitesimally small, a mathematically convenient assumption enabling an analytical expression for the static pressure distribution to be derived. In practice, pumping lands have a finite number of grooves and a pressure perturbation that is significant in magnitude compared to the magnitude of the overall 'smoothed' pressure profile. Here, an estimate of the magnitude of the pressure perturbation associated with a land having a finite number of grooves is first made using a simple analysis. The purpose is to identify the important variables and to assess their influence on motion error before considering the effect of surface geometry errors.

For estimating the magnitude of the pressure perturbation across a groove–ridge pair, the notation of Figure 5.9 is used. Normal to a groove boundary, the pressure gradients associated with the groove and ridge is assumed constant and the pressure rise across the groove is dp_g and the pressure drop across the ridge is dp_r. The latter is less than dp_g by an amount dp, which is due to the radial static pressure gradient across the land:

$$dp = dp_g - dp_r \qquad [5.117]$$

Considering flow normal to a groove–ridge boundary, then for continuity of mass flow:

$$\frac{-h_g^3}{12\eta}\frac{dp_g}{de} + \frac{Uh_g}{2} = \frac{-h_r^3 dp_r}{12\eta de} + \frac{Uh_r}{2} \qquad [5.118]$$

where $U = \omega r \sin\beta$ is the surface velocity normal to the groove–ridge boundary. If the land has N grooves/circumference, the groove to groove + ridge width ratio is α, then at radius r the groove and ridge widths are:

$$a_g = 2\pi\alpha r\sin\beta/N \text{ and } a_r = 2\pi(1-\alpha)r\sin\beta/N \qquad [5.119]$$

For small groove and ridge widths:

$$dp_g/de \approx dp_g/a_g$$

and using Eq. 5.117:

$$dp_r/de \approx -(dp_g - dp)/a_r \qquad [5.120]$$

These relationships enable Eq. 5.118 to be rearranged:

$$dp_g[(1-\alpha)h_g^{\,3} + \alpha h_r^{\,3}] = 12\pi\omega\eta r^2\alpha(1-\alpha)(h_g - h_r)\sin^2\beta/N - dp\alpha h_r^{\,3} \qquad [5.121]$$

To a first approximation, the pressure gradient across a grooved land is constant and for an inward pumping land is given by:

$$dp/dr \approx (P_m - P_a)/(r_o - r_m) \qquad [5.122]$$

so that dp in Eq. 5.122 is:

$$dp = 2\pi r(P_m - P_a)\cos\beta/[N(r_o - r_m)] \qquad [5.123]$$

Combining Eqs 5.123 and 5.121 gives:

$$dp_g[(1-\alpha)h_g^3+\alpha h_r^3]=12\pi\omega\eta r^2\sin^2\beta\alpha(1-\alpha)(h_g-h_r)/N \\ -2\pi\,\alpha\,h_r^3 r(P_m-P_a)\cos\beta/[N(r_o-r_m)] \quad [5.124]$$

which enables the pressure perturbation, p_g across a groove–ridge pair to be determined. Figure 5.37 is an example of the magnitude of the pressure perturbations produced by the bearing of Table 5.1, obtained using Eq. 5.124. This bearing generates a maximum static pressure of 17.5 Bar at the grooved land-sealing land transition diameter at 10 000 rpm with a 20 cSt lubricant. The magnitude of the perturbations is a maximum at the outside radius where the surface speed is highest. The magnitude is also inversely proportional to the number of grooves.

The perturbations have, at least to a first approximation, a triangular waveform and so at a constant radius the pressure profile in the bearing may be expected to take the form:

$$P=P_o+\Sigma_1^M p_j\cos jN\theta+\Sigma_1^M p_j\sin jN\theta \quad [5.125]$$

for the grooves on the stationary land, or:

$$P=P_o+\Sigma_1^M p_j\cos jN\omega t+\Sigma_1^M p_j\sin jN\omega t \quad [5.126]$$

for grooves on the rotating land. Whilst the perturbations are significant in magnitude, they theoretically integrate to zero over the interval 0 to 2π to give a zero axial dynamic force in the same way as the surface geometry errors on the sealing lands do, i.e. Eq. 5.114. Similarly, considering

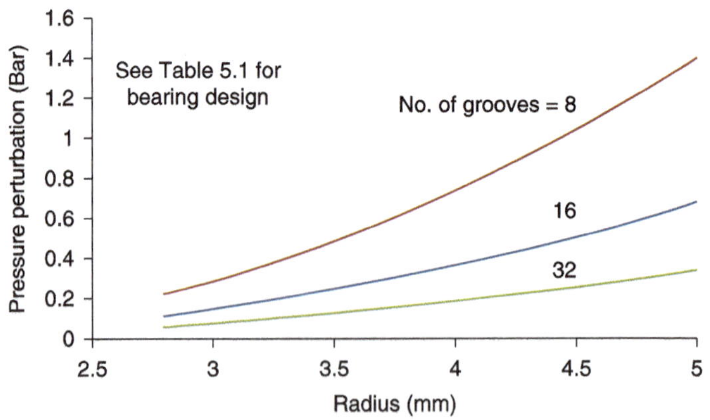

Figure 5.37 **Example of groove–ridge pressure perturbation for the bearing of Table 5.1**

moment equilibrium, grooves on the stationary surface create a net static moment and small misalignments of the rotating surface create a dynamic moment with a frequency equal to the rotating speed, i.e. as for Eq. 5.116. These deductions are based on the assumption that all grooves are perfectly equal in geometry and the bearing is subject to a concentrically applied axial load. In practice, manufacturing errors may result in groove-to-groove variations in depth, width and angle. Furthermore, depending on the manufacturing process, the variations may be local to a single groove or distributed over several grooves. But even so, the pressure perturbations are still described by Eqs 5.125 and 5.126, only now the set of coefficients, p_j, are different. When there are groove-to-groove variations, the pressure perturbations do not generally integrate to zero but instead integrate to a constant value over the interval 0 to 2π. Thus groove geometry variations on the stationary surface affect the net static axial force generated by the bearing and create a net static moment. Groove variations on the rotating surface affect the net static axial force generated by the bearing but more importantly also create a net moment that rotates with the surface to generate a tilt motion error at the basic rotational speed of the bearing.

A first estimate of the sensitivity of the bearing to groove geometry errors can be made using the static analysis of Section 5.5 by varying each of the groove parameters by $\pm 10\%$ and observing the effect on static axial force. Dividing the force by the number of grooves gives the approximate effect of a single groove. Figure 5.38 gives an example of the effect of groove parameter errors on axial force for the bearing

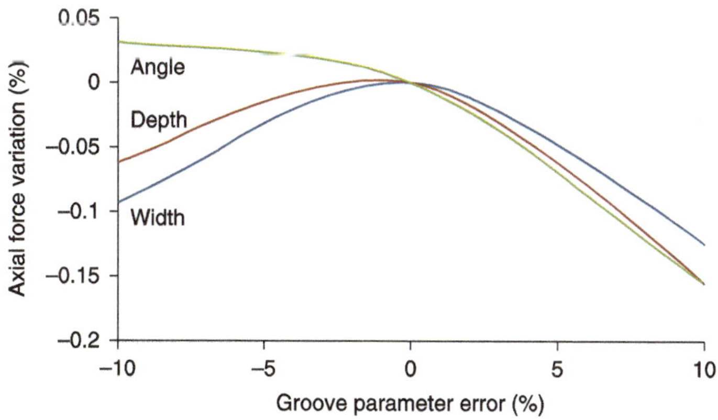

Figure 5.38 **Effect of an error to a single groove on axial force**

of Table 5.1 with eight grooves. This shows a 10% error in the width, depth or angle of a single groove to produce less than a 0.15% variation in axial force, suggesting the bearing to be highly insensitive to manufacturing errors. If the grooves are on the rotating surface, the net error in axial force rotates about the bearing axis setting up a rotating moment and hence the tilt motion error referred to earlier. However, this is also proportionally low in magnitude.

Other types of surface error, namely flatness errors on the grooved or opposing land, also create pressure fluctuations in the lubricant film. For errors small in amplitude compared to the bearing gap, a first approximation to their effect on axial force and moment is described by Eqs 5.114 and 5.116, and they add linearly to the effect of groove geometry errors.

The good averaging effect of the grooved bearing described in the preceding analysis applies to the situation where there is an axi-symmetric static pressure distribution in the lubricant film, surface geometry errors are small in amplitude compared to the bearing gap and the grooves are equi-spaced. An axi-symmetric static pressure distribution only occurs on a bearing subject to a concentric axial load. Under eccentric loading or misalignment, the static pressure distribution becomes a function of angular position and this interacts with both groove and surface geometry errors on the rotating surface to create motion errors that in general contain frequencies harmonic of rotational speed. The mechanism of motion error generation differs from that described above, in that now the bearing film has regions of high and low pressure and a given surface geometry error creates a larger fluctuation in the high pressure zone than in the low pressure zone. Eccentric loading and misalignment of grooved bearings are not considered in this book.

5.10.3 Grooved journal bearings

The arguments presented in Section 5.10.2 also apply to a journal bearing under concentric rotation. As the gap is constant along and around the bearing, the static pressure distribution is axi-symmetric and pressure perturbations produced by the grooves or surface geometry errors theoretically integrate to zero. However, under radial load, the pressure fluctuations produced by grooves or surface geometry errors are larger in the high pressure zone on the loaded side of the bearing than in the unloaded zone and result in motion error at frequencies harmonic of the speed of rotation.

The pressure fluctuations produced by grooves on journal bearings are given by Eq. 5.121, as for annular thrust bearings only r is the journal radius, which is constant and equal to D/2 and for a fully grooved bearing of length L with N grooves/circumference:

$$dp = 2\pi D(P_m - P_a)\cos\beta/[NL] \qquad [5.127]$$

where P_m is the static pressure at the centre of the bearing. Thus the perturbation dp_g is given by:

$$dp_g[(1-\alpha)h_g^3 + \alpha h_r^3] = 3\pi\omega\eta D^2\alpha(1-\alpha)(h_g - h_r)\sin^2\beta/N - 2\pi D(P_m - P_a)\ \alpha h_r^3\cos\beta/[NL] \qquad [5.128]$$

If under the action of a radial load the gap over a ridge is:

$$h_r = h_{ro}(1 - e\cos\theta) \text{ and hence } h_g = h_{go}(1 + e\cos\theta) \qquad [5.129]$$

then for small eccentricities substituting for h_g and h_r in Eq. 5.128 using Eq. 5.129 gives the pressure perturbation variation with eccentricity and angular position.

For the bearing studied in Section 5.8, Figures 5.39 and 5.40 give examples of the magnitude of the pressure perturbation and its dependence on eccentricity and number of grooves. Figure 5.39 shows the magnitude of the pressure perturbations in line with the radial displacement of the rotor on the loaded and unloaded sides of the bearing for the case of the bearing having eight grooves. The pressure perturbations are significant in magnitude, but it is the difference between the two curves that is

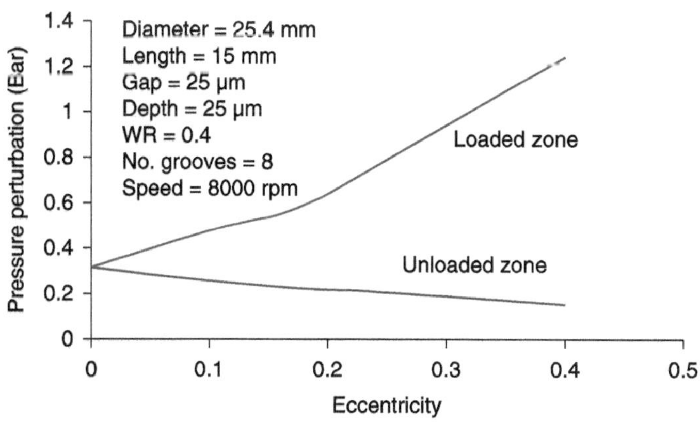

Figure 5.39 Example of pressure perturbation in groove journal bearing

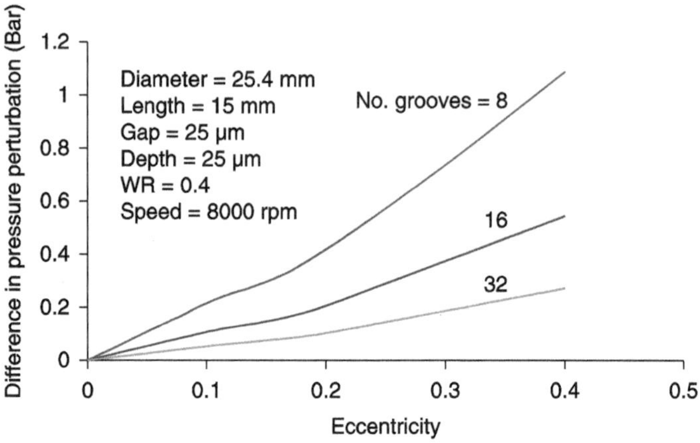

Figure 5.40 **Imbalance in pressure perturbations in grooved journal bearing**

important, as this does not integrate to zero with angular position (Figure 5.40). Thus if the grooves are placed on the rotating component and the bearing subjected to a radial load, a motion error is created that contains frequencies harmonic of N times the rotational speed.

Although the pressure perturbations are significant in magnitude, they only create a net dynamic force if the grooves are placed on the rotating surface and this is found by integrating the perturbations produced by all grooves over the total bearing area for different angular positions of the rotor. It is the force variation with position that creates motion error. Numerical integration of the pressure perturbations shown in Figure 5.39 result in a dynamic force variation of only 0.001 N for an eccentricity of 0.4, assuming the bearing has 8 identical grooves. The bearing stiffness is 41.3 N/µm, so the motion error at 0.024 nm is negligible.

Of more significance is the effect of groove geometry errors and how the pressure perturbations interact with other surface geometry errors to create bearing motion error. An error to the depth, width or angle of a single groove creates a significant pressure variation and if it is on the rotating surface, a significant change in bearing deflection as it passes through the load zone (Figure 5.41).

A sensitivity analysis of the bearing design used to generate Figure 5.39 shows that errors to groove depth, width or angle of up to 10% produce a small but significant change to the radial deflection of the bearing. In this example, the bearing has a relatively large nominal operational eccentricity of 0.4 corresponding to a radial deflection of 10 µm and a low number of grooves, 8, and as such is an extreme. Furthermore, the

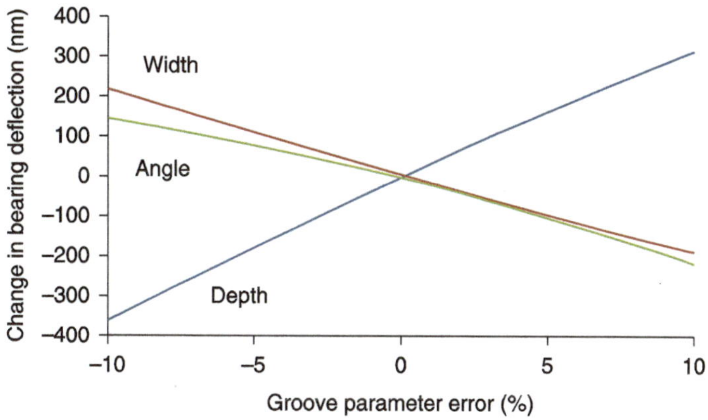

Figure 5.41 Example of the effect of groove parameter error on radial deflection of a journal bearing

error applies to all grooves whereas an error to a single groove may be expected to produce much lower deflections of the order of tens of nm.

5.11 Thermal characteristics

Self-acting bearings generate heat by viscous shearing of the lubricant in the bearing gap. The mechanism of heat generation is the same for both hydrodynamic and aerodynamic types and in both cases may have a significant effect on bearing performance.

For hydrodynamic bearings, even modest speeds may generate a significant power loss when lubricated by a viscous oil or grease and as the only means of heat removal is conduction through the shaft and housing, a significant rise in operating temperature may occur. Not only does this provide unwanted thermal distortions in precision bearing systems, but it also reduces the lubricant viscosity and hence the stiffness and load carrying capacity of the bearing itself. However, the fact that lubricant viscosity reduces with temperature does provide a degree of self regulation, limiting the maximum power loss and operating temperature of the bearing, therefore reducing the risk of it being thermally unstable. An accurate calculation of bearing stiffness and load capacity should nonetheless take into account the effect of temperature on lubricant viscosity. It follows that either the temperature must be known at the required operating conditions or it must be determined analytically or experimentally. The former requires the bearing's power loss as an input parameter to a numerical model simulating heat flow through the system. Thus the purpose of this section is to provide a simple calculation of bearing torque and power loss for annular thrust and cylindrical journal bearing types.

5.11.1 Lubricant flow in a stepped channel

Whilst the analysis of fluid flow presented in Sections 5.5 and 5.7 was based on the assumption of an infinite number of infinitesimally narrow grooves, in practice manufacturing techniques dictate a finite number of grooves of a width that is relatively large compared to the clearances used in grooved bearings. Thus the assumptions made here are that:

1. The groove profile is perfectly rectangular in the direction of relative surface motion (Figure 5.8).
2. The groove and ridge widths are large compared to groove and ridge clearances.
3. The lubricant is subject to a constant velocity gradient equal to U/h_g in the groove, and U/h_r in the ridge region.
4. The lubricant is Newtonian, i.e. the shear stress $\tau = \eta.U/h$.
5. Velocity gradients in the lubricant film in a direction normal to the relative surface motion are negligible.
6. The velocity gradients remain constant over the full width of the groove and the ridge.
7. The same velocity gradients are generated, no matter whether the grooves are on the moving or the stationary surface.

In addition, the usual assumptions of negligible fluid inertia, body forces and constant viscosity are made (Figure 5.42). Assumption 3 is an approximation for the purpose of estimating friction losses in the fluid

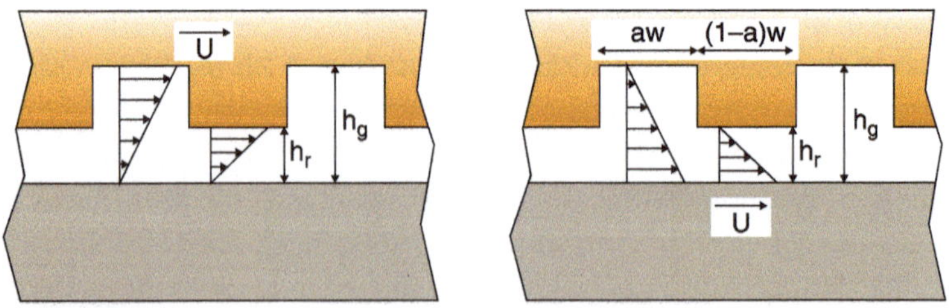

Figure 5.42 Assumed velocity gradients in lubricant film

film. The condition can only exist if there is a discontinuity in fluid flow at the groove ridge boundary.

5.11.2 Torque and power loss in annular thrust bearings

Subject to the above assumptions, consider a bearing of bore diameter D_i and outside diameter D_o, for which one surface has spiral grooves machined into it and is rotating at a constant angular speed of ω (Figure 5.43). The groove to groove + ridge width ratio is constant and equal to α, the ridge clearance is also constant and equal to h_r and the groove clearance is h_g. Shear stress in the lubricant in the ridge region is:

$$\tau_r = \eta U/h_r$$

and in the groove region:

$$\tau_g = \eta U/h_g$$

The torque generated by an annular element of width dR at radius R is:

$$dT = R[\alpha\tau_g + (1 - \alpha\tau_r)]2\pi R dR = 2\pi\eta U[\alpha/h_g + (1 - \alpha)/h_r]R^2 dR \qquad [5.130]$$

where the surface velocity $U = \omega R$. Substituting for U and expressing torque as an integral:

$$T = \pi\eta\omega[\alpha/h_g + (1 - \alpha)/h_r]/2 \int_{ei}^{eo} R^3 dR$$

which gives:

$$T = \pi\eta\omega[\alpha/h_g + (1 - \alpha)/h_r]\ [D_o^4 - D_i^4]/32 \qquad [5.131]$$

The power loss of the bearing is:

$$Q = \omega = \pi\eta\omega^2[\alpha/h_g + (1 - \alpha)/h_r]\ [D_o^4 - D_i^4]/32 \qquad [5.132]$$

Thus for an annular thrust bearing, power loss is sensitive to speed and diameter being proportional to rotational speed squared and diameter raised to the fourth power (Figure 5.43).

Equation 5.133 applies to a single-sided bearing, where h_r and h_g are the clearances under load and the grooves extend the full width of the bearing. For an inward pumping bearing comprising a grooved land and a sealing land, the total power loss is:

$$Q = \pi\eta\omega^2[\alpha/h_g + (1 - \alpha)/h_r]\ [D_o^4 - D_m^4]/32 + \pi\eta\omega^2\ [D_m^4 - D_i^4]/32h_r \qquad [5.133]$$

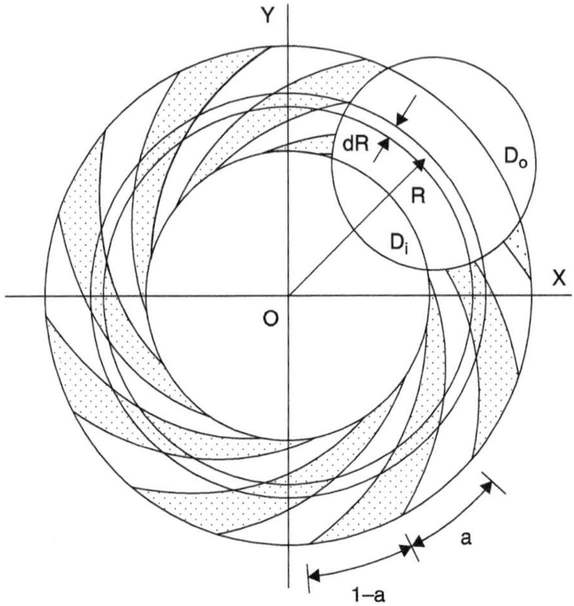

Figure 5.43 Elemental area on grooved annular thrust bearing

where D_m is the groove–seal transition diameter. Similar expressions may be written for an outward pumping bearing and a chevron grooved bearing where unusually different depths or widths may be used on inboard and outboard grooves.

For a symmetrical, double-sided bearing, where the no load clearance on each side of the bearing is h_r, the power loss is 2Q. The power loss remains at 2Q under the action of an externally applied axial load but a greater proportion of the loss is generated on the loaded side of the bearing. The resulting uneven temperature distribution can in some instances significantly 'bow' the thrust rotor, affecting bearing stiffness and load carrying capacity.

5.11.3 Torque and power loss in journal bearings

Consider a partially grooved cylindrical journal bearing of diameter D and length 2L (Figure 5.44). The grooved length of the bearing is $2(L - L_g)$, the groove width ratio is α and under no load the clearances over the ridge and groove are constant, equal to h_r and h_g, respectively. For a rotational speed of ω, the shear stress in the fluid in ridge and groove regions is:

$$\tau_r = \eta\omega D/2h_r \quad \text{and} \quad \tau_g = \eta\omega D/2h_g$$

In the grooved region of the bearing, an annular surface element of width dZ generates a torque:

$$dT_g = (D/2)\eta\omega D[\alpha/2h_g + (1-\alpha)/2h_r].\pi D.dZ \qquad [5.134]$$

Integrating from L_g to L gives a torque of:

$$T_g = \pi\omega\eta(L-L_g)D^3[\alpha/h_g + (1-\alpha)/h_r]/4$$

For the plain region of the bearing, the clearance is h_r and $\alpha = 0$, so the torque generated by half the region is:

$$T_p = \pi\omega\eta L_g D^3/4h_r$$

The total torque generated by the bearing is $2(T_p + T_g)$:

$$T = \pi\omega\eta D^3 \{(L-L_g)[\alpha/2h_g + (1-\alpha)/2h_r] + L_g D^3/2h_r\} \qquad [5.135]$$

and the power loss is:

$$Q = \pi\omega^2\eta D^3\{(L-L_g)[\alpha/2h_g + (1-\alpha)/2h_r] + L_g/2h_r\} \qquad [5.136]$$

The application of a radial load to the bearing will create a variation in bearing clearance with angular position, but the total torque and power loss of the bearing remains unchanged.

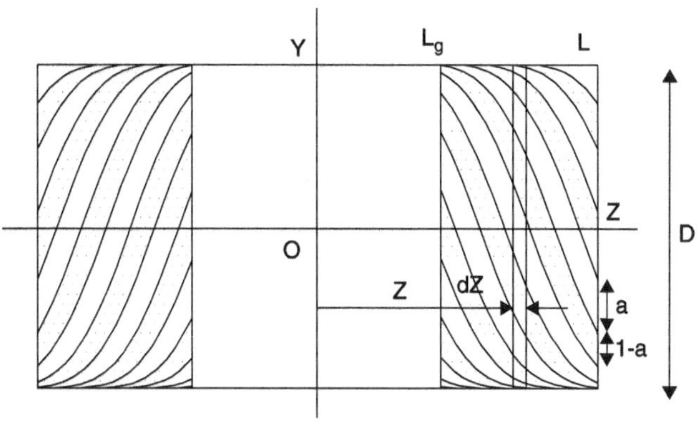

Figure 5.44 Elemental area on partially grooved journal bearing

5.12 Overview

The hydrodynamic herringbone or spiral grooved bearing can achieve low motion errors without the need for components to be manufactured to extreme precision or for the need to have recirculating lubricant supplies. It can also be manufactured from a variety of materials, making it relatively straightforward to integrate directly into rotating equipment. Assuming the bearing can meet load and speed requirements, then groove manufacture is perhaps the greatest challenge to its use. Groove tolerances are not critical and a variety of manufacturing techniques can be used. CNC milling is often used to generate grooves on flat annular thrust bearings, whilst CNC lathes can be used to scribe herringbone grooves into shafts. Both operations require some attention to deburring, but are suitable for producing small batches of components. Bearings manufactured in large volumes tend to use dedicated equipment and processes. For example, [25] describes an electrochemical machining process for generating herringbone and spiral grooves and investigates the consistency of groove geometry over 100 000 cylindrical journals. An alternative process suitable for large batches of bearings is to plate a surface masking the groove pattern [26]. The plating material can have good tribological properties, allowing the use of a wide variety of substrate materials.

5.13 References

1. Steigmeier, E.F. and Auderset, H. (1983), Patent US4526484. 'Ferrofluid thrust and radial bearing assembly'.
2. Whipple, R.T.P. (1949), 'Herringbone pattern thrust bearing', *AERE*, T M, **29**: 110.
3. Muijderman, E.A. (1966), *Spiral Groove Bearings*, London: Macmillan.
4. Elrod, H.D. (1973), 'Some refinements on the theory of the viscous screw pump', *Transactions of the ASME, Series F, Journal of Lubrication Technology*, **95**: 82–3.
5. Booy, M.M. (1966), 'Effect of optic channel ends on screw pump performance', *Transactions of the ASME, Series D, Journal of Basic Engineering*, **88**: 121–31.
6. Hirs, G.G. (1969), 'The load capacity and stability characteristics of hydrodynamic grooved journal bearings', *Transactions of the ALSE*, **12**: 287–96.
7. Chow, C.Y. and Vohr, J.H. (1970), 'Helical groove journal bearing operated in turbulent regime', *Transactions of the ASME, Series F, Journal of Lubrication Technology*, **92**: 346–58.

8. Reinhoudt, J.P. (1972), 'On the stability of rotor bearing systems and on the calculation of sliding bearings', *Philips Research Report*, Suppl. No. 1.
9. Reinhoudt, J.P.A. (1970), 'Grease lubricated helical groove bearing for a satellite flywheel', *Lubr. Engng*. **26**: 95–100.
10. Remmers, G. (1974), 'Grease lubricated helical groove bearings of plastic', *Philips Technical Review*, **34**: 103–5.
11. Pan, C.H. and Sternlicht, B. (1967), 'Thermal distortion of spiral grooved gas lubricated thrust bearings due to self heating', *Transactions of the ASME, Series F, Journal of Lubrication Technology*, **89**: 197–202.
12. Bootsma, J. (1969), 'The effect of viscosity variation with temperature on the performance of spiral groove bearings', *Transactions of the ASLE*, **12**: 287–96.
13. Volman, H.J.W.M. (1975), 'The push-pull spiral groove bearing, a thrust bearing with a self adjusting internal preloading', *Philips Technical Review*, **35**: 11–14.
14. Bootsma, J. (1975), 'Liquid lubricated spiral groove bearings', PhD Thesis, Technische Hogeschool, Delft.
15. Bootsma, J. and Tielmans, L.P.M. (1977), 'Conditions of leakage free operation of herringbone grooved journal bearings', *Transactions of the ASME, Journal of Lubrication Technology*, **99(2)**: 215–22.
16. Leuthold, H., Jennings, D. and Nagarathnam, L. (1995), 'Spindle motors with hydrodynamic bearings and high inertial load', *Proceedings of the 24th Annual IMCSD Symposium*, San Jose, CA, 6–8 June, 257–62.
17. Ku, C.P. (1996), 'Dynamic characteristics of hard disc drive spindle motors – comparison between ball bearings and hydrodynamic bearings', *Transactions of the ASME*, **118**: 402–6.
18. Rahman, M. and Leuthold, H. (1996), 'Computer simulation of a coupled thrust and journal hydrodynamic bearing using a finite element method', *IMCSD Proceedings*, 103–12
19. Malanoski, S.B. and Pan, C.H. (1965), 'The static and dynamic characteristics of the spiral-grooved thrust bearing', *Transactions of the ASME, Journal of Fluids Engineering*, **87(3)**: 547–55.
20. Cheney W. and Kincaid, K. (1999), *Numerical Mathematics and Computing*, 4th edition, Pacific Grove, CA: Brooks/Cole Publishing Company, ITP Inc.
21. Vohr, J.H. and Chow, C.Y. (1965), 'Characteristics of herringbone grooved, gas lubricated journal bearings', *Transactions of the ASME, Journal of Basic Engineering*, **87**: 568–76.
22. Hamrock, B.J. (1994), *Fundamentals of Fluid Film Lubrication*, New York: McGraw-Hill.
23. Wardle, F.P. and Goat, C.A. (2000), 'Dynamic characteristics of oil lubricated herringbone groove bearings', *Institute of Mechanical Engineers Conference*, C576/028.
24. Lund, J.W. (1987), 'Review of the concept of dynamic coefficients for fluid film journal bearings', *Trans. ASME, Journal of Tribology*, **109**: 37–41.
25. Lund, J.W. (1966), 'Self excited stationary whirl orbits of a journal in a sleeve bearing', PhD Thesis, Rensselaer Polytechnic Institute, Troy NY.
26. Wardle, F.P. and Goat, C. (2000), 'Electrochemical machining of herringbone grooved journal bearings', *Proceedings of the 3rd Euspen International Conference*, Eindhoven.

5.14 Notation

a_g – Groove width
a_h – Ridge width
C – Clearance under no load
c, c_1, c_2 – Constants
C_{xx}, C_{yy}, C_{xy}, C_{yx} – damping coefficient
D – Diameter of journal bearing
D_i – Bore diameter of annular thrust bearing
D_o – Outside diameter of annular thrust bearing
F – Static force
F_a – Axial force
F_d – Damping force
F_o – Static force at concentric rotation
F_x, F_y – Component of force in x or y direction
G_1 – Static groove geometry parameter
G_2 – Static groove geometry parameter
G_{1s} – Static seal geometry parameter
G_{2s} – Static seal geometry parameter
g_1 – Dynamic groove parameter
g_2 – Dynamic groove parameter
H – h_g/h_r for a journal bearing under radial load
H_o – h_g/h_r for a journal bearing under no load
h_g – Groove clearance
h_r – Ridge clearance
K – Static stiffness
K_a – Axial Static stiffness
K_{xx} – Angular stiffness about x-axis $= K_a D_p^2/2\pi$
L – Journal bearing length
M – Mass
M_c – critical mass
M_{or}, M_{zr}, M_{og}, M_{zg}, – mass flux component
M'_{or} – Flux, M_{or}/P
M'_{zr} – Flux, M_{zr}/P
M'_{og} – Flux, M_{og}/P
M'_{zg} – Flux, M_{zg}/P
m_g – Mass flow rate
M_{xx} – moment
P_o – Static pressure
p – Dynamic or perturbed pressure
Q – Power loss
R – Radius of journal bearing
R_i – Inner radius of annular thrust bearing
R_m – Transition radius of annular thrust bearing
R_o – Outer radius of annular thrust bearing
T – Total torque
T_g – Torque generated by grooved region of an annular thrust bearing
T_p – Torque generated by plain region of an annular thrust bearing

t – Time
U – Surface velocity
W – Total load
W_o – Total static load
x,y,z – Cartesian coordinates

α – $a_g/(a_g + a_h)$
β – Groove angle
Δ – radial deflection
Δh – Perturbation to bearing gap
Δm – elemental mass flux
ε – groove coordinate
ε_r – Eccentricity ratio
η – Dynamic viscosity
θ – Angular coordinate
μ – Kinematic viscosity
ρ – Lubricant density
ς – Damping constant
τ – Lubricant shear stress
τ_g – Shear stress in lubricant in groove
τ_r – Shear stress in lubricant over ridge
ψ – Mass flux
ψ_g – Mass flux in grooved region
ψ_r – Mass flux in ridge region
ω, ω_1, ω_2, ω_y – Angular speed
ω_c – critical frequency
σ – groove coordinate
Ω – frequency of escatation
ϕ – attitude angle
γ – phase angle
λ – complex number

6

Aerodynamic herringbone and spiral grooved bearings

DOI: 10.1533/9780857092182.385

Abstract: Aerodynamic herringbone and spiral grooved bearings have many similarities to their hydrodynamic equivalents, but in using air as a lubricant they operate with much smaller clearances, typically in the 2 to 10 μm range and are therefore far more demanding in terms of manufacturing precision. They also have a relatively low load carrying capacity, which restricts their use to high speed or light load applications, such as optical scanning, turbo-machinery and instrumentation, where manufacturers have developed their own bearing designs and manufacturing processes. In this chapter, analytical procedures for determining the load capacity and stiffness of annular thrust and cylindrical journal bearing designs are presented.

Key words: self-acting bearings, aerodynamic bearings, herringbone grooved bearings, spiral grooved bearings, chevron grooved bearings, fluid film bearings, high speed bearings, high stability bearings, low motion error.

6.1 Introduction

The analysis and the design principles for aerodynamic herringbone and spiral grooved bearings are similar to those of the hydrodynamic bearings presented in Chapter 5. Here the analyses of the previous chapter are extended to cover the case of a compressible lubricant and the differences in operating characteristics between the two bearing types are highlighted. Again the focus is on annular thrust and cylindrical journal bearing types.

As for the hydrodynamic bearing, there are a variety of aerodynamic bearings currently in use but what distinguishes the herringbone and spiral grooved bearings from these other types are their excellent stability and low motion error characteristics. Using air as a lubricant also gives the aerodynamic bearing the advantages of low friction and clean operation, but on the down side is an inherently low load carrying capacity when compared to the other types of bearing described in this book. Improving the lubricant shear rates in the bearing film by minimising clearances and by increasing surface speed, to some extent mitigates the load capacity limitation but nonetheless they are generally selected for use in medium to high speed applications or in applications where loads are very light.

The reason for devoting separate chapters to hydrodynamic and aerodynamic grooved bearings is to distinguish between some of their principle operating characteristics. With oil as a lubricant, viscosity can be selected to avoid the need for extremely small clearances in hydrodynamic bearings leading to a tolerant bearing design. As oil is incompressible, bearing load capacity and frictional losses both continue to increase with speed. The main disadvantage of high friction in high precision bearing applications is simply to create unwanted thermal distortions. In contrast, the viscosity of air is low and air is compressible. Low viscosity requires aerodynamic bearings to be designed to operate with low clearances to maximise load capacity and stiffness. Although load capacity does increase with speed, the compressibility of air means that it reaches some limiting value, making low load capacity an inherent feature of this type of bearing. The low clearances also mean frictional losses are still significant but now the main effect of an increase in film temperature is to change the bearing clearance. A temperature gradient is set up that usually reduces clearance, in turn reducing the dynamic and thermal stability of the bearing.

The differences between oil and air as a lubricant also influence the choice of groove geometry. In aerodynamic bearings, improvements in load capacity can be achieved using partial groove designs, whereas for hydrodynamic bearings, fully grooved lands maximise dynamic stability and in annular thrust bearings can be designed to give a zero net flow of lubricant.

In this chapter, a means of calculating the load capacity, static stiffness, damping and power loss occurring in aerodynamic grooved bearings is presented or discussed with reference to that for hydrodynamic bearings.

The focus is on flat annular and cylindrical journal bearing geometries. Motion error characteristics are similar to those of the hydrodynamic bearing and are not included here.

6.2 Bearing types

As for hydrodynamic bearings, grooves may be machined into tapered, spherical, cylindrical and flat annular bearing geometries. Also fully grooved or partially grooved bearing designs may be used. For aerodynamic bearings, the latter can have the advantage of increased load capacity and static stiffness at low and moderate speeds and are sometimes preferred in applications where there is a need to maximise the first critical speed or to minimise the speed at which 'lift off' occurs. The lift off speed is the speed at which a full aerodynamic lubricant film is formed and sliding contact and hence wear of the opposing surfaces is eliminated.

Figure 6.1 shows an example of grooved flat annular and cylindrical bearing geometries used on a rotor from an optical scanner required to operate at speeds of up to 40 000 rpm. The shaft diameter is 19 mm and the groove depths are 4 μm used with a 3 to 4 μm bearing gap. Inward pumping spiral grooves together with a seal land are used on the annular thrust bearing, whilst partial herringbone grooves are used on the journal bearings. In this case, the load capacity of both the journal and the thrust bearings was optimised to generate a low lift off speed of approximately 100 rpm when used with a horizontal or vertical axis of rotation.

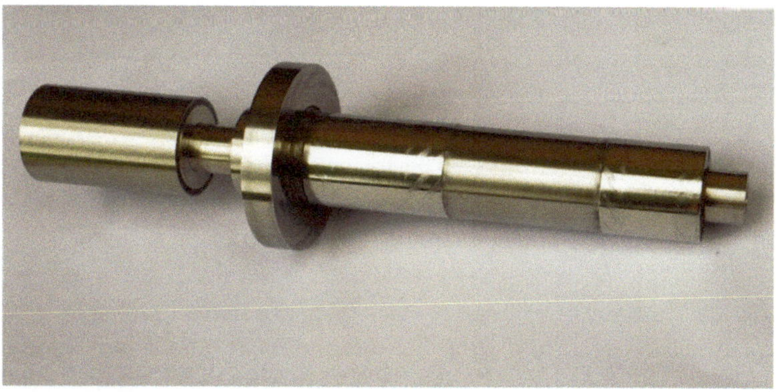

Figure 6.1 Partially grooved bearings on optical scanner

6.3 Past developments

Aerodynamic grooved bearing developments have run in parallel with those of the hydrodynamic grooved bearing, but driven by a different set of applications. The combination of air as a lubricant and a self pressurising bearing has the potential to save weight, reduce the overall size of, and simplify bearing systems. There is no need for a recirculation pump with associated lubricant reservoir and filters, and elaborate sealing arrangements are unnecessary. Furthermore, the low friction characteristics of an air lubricated bearing have the potential to reduce system energy losses. Energy efficiency, space and weight saving are important requirements of the aerospace industry and the grooved aerodynamic bearing was potentially well suited to lightly loaded, high speed turbo-machinery applications, such as fuel pumps and micro-gas turbines. Other areas of application include those where conventional lubricants may break down, such as at high temperatures or in a radioactive environment.

Sternlicht and Elwell [1] recognised the potential benefits of aerodynamic bearings and in 1958 published the results of a theoretical and experimental study of the performance of a plain journal bearing. They found that it had a reasonable load carrying capacity but made a number of recommendations to improve performance, one of which was the use of grooves to improve dynamic stability. Ausman [2,3] also considered the use of plain bearings in high speed turbo-machinery. In 1959, he reviewed the state-of-the-art in aerodynamic bearing analysis and at the same time put forward an approximate theory for predicting the load capacity and static stiffness of plain journal bearings subject to radial load and misalignment [2]. Later, in 1961, he published an improved theory [3], which gave more accurate predictions at higher eccentricities, but again this was limited to plain bearings.

The development of aerodynamic grooved bearings began at about this time and much of the work focused on developing analyses for optimising their design and predicting their performance. In 1965, a comprehensive analysis of the spiral grooved thrust bearing was published by Malanoski and Pan [4], and for the herringbone grooved journal bearing by Vohr and Chow [5]. In both cases, analytical formulae were derived for predicting the static and dynamic characteristics of the bearings albeit for small amplitude deflections and motions and based on the assumption of narrow grooves. The small amplitude limitation was considered by Castelli and Vohr in 1967 [6], who found the large deflection characteristics of grooved journal bearings to be non-linear, increasing in stiffness with amplitude in a manner similar to that of plain bearings. However, it was concluded that the perturbation analysis of [5] was accurate at low deflections and provided a reasonable lower bound solution for bearings operating at higher deflections. The small amplitude journal bearing analysis was verified by Cunningham et al. [7,8] in 1969, who obtained reasonable agreement between the analysis of [5] and experiment in respect of steady state and stability characteristics of six high speed journal bearing designs.

The analyses of [4,5] was extended to a spool bearing arrangement comprising a journal bearing located between two thrust bearings by Smalley in 1969 [9], who in 1972 [10] also developed a more general grooved bearing analysis that could be applied to other geometries such as spherical bearings. The journal bearing analysis of [5] was also used by Hamrock [11] and Fleming [12], to find the optimum journal bearing designs to give maximum load carrying capacity and maximum stability. The papers both contain a series of non-dimensional graphs and tables covering a wide range of bearing designs and operating conditions and as such enable the reader to conveniently select an optimum bearing design for a specific set of operating conditions.

One of the assumptions used in the work of [4,5] is that the bearing has an infinite number of infinitesimally wide grooves. In practice, bearings have a low or moderate number of grooves of finite width. The assumption has been widely debated and generally accepted to be reasonable, although expected to over-predict the load capacity of bearings with low numbers of grooves. This was verified in 1994 by Bonneau and Absi [13], who used finite elements to generate an accurate load–deflection relationship for bearings with 4, 8,12 and 16 grooves and compared them with predictions using [5]. Good agreement was obtained between the latter and finite element analyses for 12 and 16 grooves. Much as expected, the load capacity of bearings with 4 and 8 grooves was somewhat lower.

The use of grooved air bearings in high precision applications such as laser scanners and gyroscopes can be traced back to the 1970s, but there is little in the way of scientific publications on bearing designs for this field of application. Almost all of the analytical development work is related to high speed turbo-machinery. [4] and [5] are particularly relevant to the design of precision aerodynamic bearings, as they provide analytical formulae for predicting the static and dynamic characteristics of grooved annular thrust and grooved journal bearings. Although these require a numerical solution, the simplifying assumptions made in deriving the formulae enable accurate solutions to be obtained with short computation times using well established numerical routines and are the basis of the following sections.

6.4 Static load capacity and stiffness of aerodynamic annular thrust bearings

The following analysis continues on from that of Section 5.5.2 for hydrodynamic annular thrust bearings. For aerodynamic bearings, fluid density is pressure dependent and obeys the isothermal law, $P=\rho RT$. Finding the load capacity and static stiffness of aerodynamic bearings follows the same procedure as for hydrodynamic bearings, by solving Eqs 5.10 to 5.12 but now with $\rho=P/RT$. Considering the outward pumping bearing and rewriting Eq. 5.12:

$$\frac{d\psi}{d\theta}=-\frac{p}{12\eta RT}\left\{G_1 r\frac{dp}{dr}-G_2 r^2 6\eta(\omega_1-\omega_2)\right\} \qquad [6.1]$$

Substituting into Eq. 5.11 and simplifying gives:

$$G_1 r p\frac{dp}{dr}-G_2 6\eta(\omega_1-\omega_2)\, r^2 p=c_1 \qquad [6.2]$$

where c_1 is a constant of integration. For a sealing land, Eq. 6.1 reduces to:

$$h_r^3 rp\frac{dp}{dr}=0$$

which gives a pressure distribution:

$$\frac{p^2-P_a^2}{P_m^2-P_a^2}=\frac{\ln(r/r_o)}{\ln(r_m/r_o)} \qquad [6.3]$$

a pressure gradient:

$$\frac{dp}{dr}=\frac{P_m^2-P_a^2}{2pr\ln(r_m/r_o)} \qquad [6.4]$$

and a mass flux density of:

$$\frac{d\psi}{d\theta}=-\frac{p h_r^3}{12\eta RT}r\frac{dp}{dr} \qquad [6.5]$$

For a grooved land, Eq. 6.2 is not amenable to an exact analytical solution but is readily solved by numerical integration using any of a number of well-known techniques, such as Runge–Kutta [11]. A straightforward approach to bearing design is to determine the pressure distribution across a grooved land as a function of radial mass flux density. The maximum flux density is given by Eq. 6.1 for dp/dr = 0 and $p = P_a$, i.e. the bearing is acting as a viscous pump:

$$\frac{d\psi_m}{d\theta} = \frac{-P_a G_2 r^2 6\eta}{12\eta RT}(\omega_1 - \omega_2) \qquad [6.6]$$

The numerical integration process requires boundary conditions to be defined. Pressure and pressure gradient at the outside diameter or inside diameter of the bearing must be known. Pressure gradient is determined by rearranging Eq. 6.1:

$$\frac{dp}{dr} = \frac{1}{r_o G_1}\left[\frac{-12\eta RT}{P_a}\frac{d\psi_m}{d\theta} + G_2 r_o^2 6\eta(\omega_1 - \omega_2)\right] \qquad [6.7]$$

Equation 6.7 may be integrated numerically from $r = r_o$ to $r = r_m$ to determine P_m for increments in the flux density $d\psi_m/d\theta$, varying from 0 to a maximum given by Eq. 6.6. In the case of a bearing with a sealing land, pressures at the transition diameter are determined as a function of mass flux density for both the grooved land and the seal. Pressure and mass flux must be equal for both lands of the bearing at the transition boundary.

6.5 Design example

An annular thrust bearing with an outside diameter of 40 mm and a bore of 20 mm is required to operate at 20 000 rpm with a ridge clearance of 4 μm. Find the optimum design of inward pumping and outward pumping bearings based on a maximum load carrying capacity criterion. This example is typical of the thrust bearing in an optical scanner and serves to illustrate some of the important differences between aerodynamic and hydrodynamic spiral grooved bearings.

The design procedure follows that of the hydrodynamic bearing considered in Section 5.6. Transition diameter, groove angle, depth and width ratio are the variables that are optimised to give maximum load carrying capacity. Equations 6.6 and 6.7 form the basis of a computer model for determining the pressure distribution in and hence the load supported by a given bearing design and is used to determine load for different values of the four variables. As for the hydrodynamic thrust bearing, the dependence of load on groove design for the aerodynamic bearing is well behaved, so that systematically varying each parameter in turn and determining the optimum value enables a rapid iteration to a final bearing design.

For an inward pumping bearing the optimum design was found to be as in Table 6.1.

For an outward pumping bearing the optimum design was found to be as in Table 6.2.

Table 6.1 Example of inward pumping grooved bearing optimised for load capacity

Inward pumping aerodynamic bearing	
Outside diameter (mm)	40
Bore (mm)	20
Transition diameter (mm)	27
Clearance (μm)	4
Groove depth (μm)	8.7
Groove angle (°)	27.5
Groove width ratio	0.58
Speed (rpm)	20000
Load (N)	57.6

Table 6.2 Example of outward pumping grooved bearing optimised for load capacity

Outward pumping aerodynamic bearing	
Outside diameter (mm)	40
Bore (mm)	20
Transition diameter (mm)	29
Clearance (μm)	4
Groove depth (μm)	9.5
Groove angle (°)	23.5
Groove width ratio	0.6
Speed (rpm)	20000
Load (N)	11.6

Note the large difference in load capacity between the two bearing designs. The inward pumping bearing carries five times the load of the outward pumping bearing design. For the hydrodynamic bearing of Section 5.6, the difference in load capacity between inward and outward pumping bearings was much less, with the former having approximately 50% more capacity than the latter. The outward pumping bearing has two inherent disadvantages compared to the inward pumping bearing: with grooves around its bore, the surface speed over the pump is low and with a sealing land around its outside diameter, the leakage area is high. Coupled with a diverging flow and a low viscosity compressible fluid, the outward pumping aerodynamic bearing simply cannot generate anywhere near the pressure of the inward pumping bearing.

As outward pumping grooves in aerodynamic thrust bearings are so much less effective than inward pumping grooves, a chevron grooved bearing design offers little improvement in load capacity over and above an inward pumping design. Furthermore, as fluid leakage in an aerodynamic bearing is of no significance, the inward pumping design is often favoured for a slight advantage in manufacturing costs.

6.6 Static load capacity and stiffness of aerodynamic herringbone grooved journal bearings

The following analysis continues on from that of Section 5.7 for hydrodynamic bearings. For the aerodynamic bearing, air density is assumed to be related to pressure by $\rho = P/(RT)$. Replacing ρ in Eq. 5.57 and rearranging gives the second-order differential equation for P_1:

$$\frac{L^2C_1\partial^2P_1}{P_a\partial Z^2} + \frac{LC_2\partial P_1}{P_a\partial Z} + \frac{LC_3\partial^2P_1}{P_a\partial Z\partial\psi} + \frac{C_4\partial P_1}{P_a\partial\psi} + \frac{C_5}{P_a}\frac{\partial^2P_1}{\partial\psi^2} + C_6\sin\psi + C_7\cos\psi = 0 \qquad [6.8]$$

The coefficients C_1 to C_T are functions of the concentric pressure gradient, groove geometry, bearing speed and air viscosity and may be shown as:

$$C_1 = \frac{H_o^3 + \alpha(1-\alpha)\sin^2\beta(H_o^3-1)^2}{\alpha + (1-\alpha)H_o^3}\frac{RP_o}{LP_a}(Z) \qquad [6.9]$$

$$C_2 = \frac{\alpha(1-\alpha)\sin\beta\cos\beta(H_o^3-1)(H_o-1)}{\alpha + (1-\alpha)H_o^3}R\lambda_s \qquad [6.10]$$

$$C_3 = \frac{2\alpha(1-\alpha)\sin\beta\cos\beta(H_o^3-1)^2}{\alpha+(1-\alpha)H_o^3}\frac{P_o}{P_a}(Z) \quad [6.11]$$

$$C_4 = \left\{\frac{2[\alpha(1-\alpha)\sin\beta\cos\beta]^2(H_o^3-1)^3(H_o-1)+\alpha(1-\alpha)\sin^2\beta(H_o^3-1)(H_o-1)}{[\alpha+(1-\alpha)H_o^3][H_o^3+\alpha(1-\alpha)\sin^2\beta(H_o^3-1)^2][\alpha+(1-\alpha)H_o^3]}\right\}\lambda_s \quad [6.12]$$

$$-\left[1-\frac{2\omega_3}{\omega_1+\omega_2}\right][1-\alpha+\alpha H_o]\lambda\frac{L}{R}$$

$$C_5 = \frac{H_o^3+\alpha(1-\alpha)\cos^2\beta(H_o^3-1)^2}{\alpha+(1-\alpha)H_o^3}\frac{L}{R}\frac{P_o}{P_a}(Z) \quad [6.13]$$

$$C_6 = \frac{P_o}{P_a}(Z)\,C'_6\left\{\frac{\alpha(1-\alpha)(H_o^3-1)^2\cos^2\beta[(1-\alpha)H_o^2(H_o^2+H_o-1)-\alpha(H_o^2-H_o-1)]}{H_o^3+\alpha(1-\alpha)(H_o^3-1)^2\sin^2\beta}-H_o^2\right\}$$

$$+\frac{P_o}{P_a}(Z)\left[1-\frac{2\omega_3}{\omega_1+\omega_2}\right][1-\alpha+\alpha H_o]\lambda\frac{L}{R} \quad [6.14]$$

and:

$$C'_6 = \frac{-3\alpha(1-\alpha)(H_o-1)^2\sin^2\beta.\lambda_s}{[\alpha+(1+\alpha)H_o^3]^2}$$

$$C_7 = \frac{3(H_o^3-1)(H_o^2-1)R[\alpha(1-\alpha)(H_o-1)\lambda_s\sin\beta\cos\beta]^2}{[\alpha+(1-\alpha)H_o^3]L\,[H_o^3+\alpha(1-\alpha)(H_o^3-1)^2\sin^2\beta]^2} \quad [6.15]$$

$$[H_o^2(H_o^2+1)+\alpha(1-\alpha)(H_o^3-1)^2\sin^2\beta]$$

where:

$$\frac{P_o}{P_a}(Z) = 1+\frac{\alpha(1-\alpha)\sin\beta\cos\beta(H_o^3-1)(H_o-1)}{H_o^3+\alpha(1-\alpha)\sin^2\beta(H_o^3-1)^2}\frac{Z}{L}\lambda_s \quad [6.16]$$

$H_o = h_{go}/h_{ro}$ – the concentric groove to ridge clearance ratio

$$\Lambda = \frac{\eta(\omega_1+\omega_2)}{P_aC^2}R^2 \text{ – the compressibility number} \quad [6.17]$$

$$\Lambda_s = \frac{6\eta(\omega_2-\omega_1)}{P_aC^2}RL$$ – the compressibility number based on relative journal speed [6.18]

The relaxation scheme used to solve Eq. 5.64 for the hydrodynamic grooved journal bearing is also used to solve Eq. 6.8 for the aerodynamic bearing. Boundary conditions also remain the same, i.e.:

$$P_1(\psi,Z) = P_a \text{ at } Z = 0$$

and:

$$\alpha M_{zg1} + (1-\alpha) M_{zr1} = 0 \text{ at } Z = L/2$$

As for the hydrodynamic bearing, integrating pressure over the bearing area in OXZ and OYZ planes gives the radial load components:

$$F_x = 2\int_o^{L/2}\int_o^{2\pi} P_1(Z,\psi)\cos\psi \, dZ \, d\psi$$

$$F_y = 2\int_o^{L/2}\int_o^{2\pi} P_1(Z,\psi)\sin\psi \, dZ \, d\psi$$

The magnitude of the load is:

$$W = (F_x^{\,2} + F_y^{\,2})^{1/2} \qquad [6.19]$$

and the attitude angle is:

$$\phi = \text{Tan}^{-1}(F_y/F_x) \qquad [6.20]$$

6.7 Example of static stiffness and load capacity

The total stiffness of the bearing is simply the ratio of load given by Eq. 6.19 to gap $K_t = W/C$. But as for the hydrodynamic grooved journal bearing, subjecting the aerodynamic bearing to a static applied radial load results in a deflection with components in line and perpendicular to the load vector. Thus the static stiffness is given by:

$$[K] = \begin{bmatrix} K_{xx} & K_{xy} \\ K_{yx} & K_{yy} \end{bmatrix} \qquad [6.21]$$

where K_{xx} and K_{xy} may be expressed in terms of total bearing load, W, eccentricity, e, operating gap, C, and attitude angle ϕ:

$$-K_{xx} = \frac{W\cos\phi}{e} \text{ and } K_{xy} = \frac{W\sin\phi}{e} \qquad [6.22]$$

and the attitude angle is a function of bearing design and speed. For the special case of a bearing operating at zero eccentricity, as is considered here, $K_{xx} = K_{yy}$ and $K_{xy} = -K_{yx}$. The radial load capacity of the bearing is $e_c W$, where e_c is a maximum permissible eccentricity, typically 0.5C for an aerodynamic journal bearing. The analysis described in the preceding sections effectively assumes a linear load–deflection relationship and as for the hydrodynamic bearing will in general, underestimate load capacity. This is supported by the results of [6], where the load–deflection characteristics of aerodynamic herringbone grooved journal bearings were experimentally determined at eccentricities of up to 0.8. At low eccentricities, the load–deflection characteristics were near linear but at higher eccentricities exhibited a load hardening characteristic.

The total static stiffness for 20 mm diameter bearings with the optimum groove design, i.e. groove angle = 30 degrees, width ratio = 0.5 and depth ratio = 1.2, is shown in Figure 6.2. Results for length/diameter ratios of 0.5, 1 and 2 together for grooves rotating (R) or stationary (S) are shown. In all cases, static stiffness increases with speed. It also increases with length/diameter ratio and with the grooves placed on the stationary component.

The attitude angles corresponding to the total static stiffness curves of Figure 6.2 are shown in Figure 6.3 for speeds between 200 rpm and 50 000 rpm. In all cases, the theoretical zero speed, attitude angle is 57.6 degrees and with one exception it decreases in an exponential manner with increasing speed. The exception is the bearing with a short length/diameter ratio of 0.5 and grooves placed on the stationary component. For this case, the attitude angle exhibits a local minimum of 51 degrees at approximately 1000 rpm, increases to 53 degrees at 10 000 rpm and then gradually reduces to 50 degrees at 50 000 rpm. Increasing length/diameter ratio and placing the grooves on the rotating component generally reduces attitude angle at a given speed.

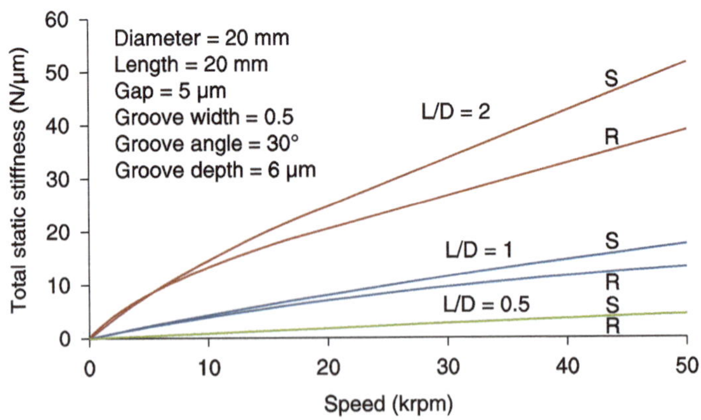

Figure 6.2 Total static stiffness as a function of speed

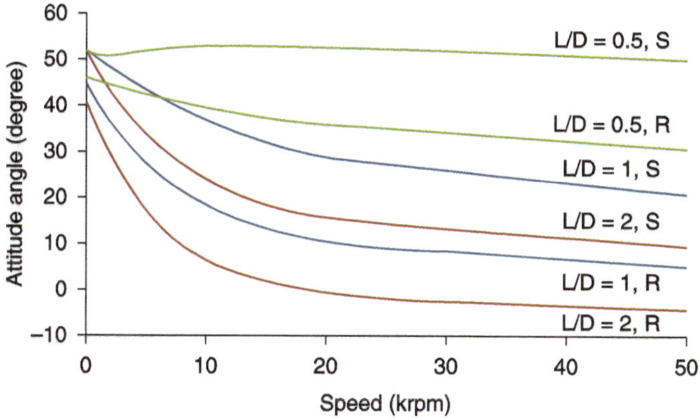

Figure 6.3 Attitude angle as a function of speed: Diameter = 20 mm; Length = 20 mm; Gap = 5 μm; Groove width = 0.5; Groove angle = 30 °; Groove depth = 6 μm

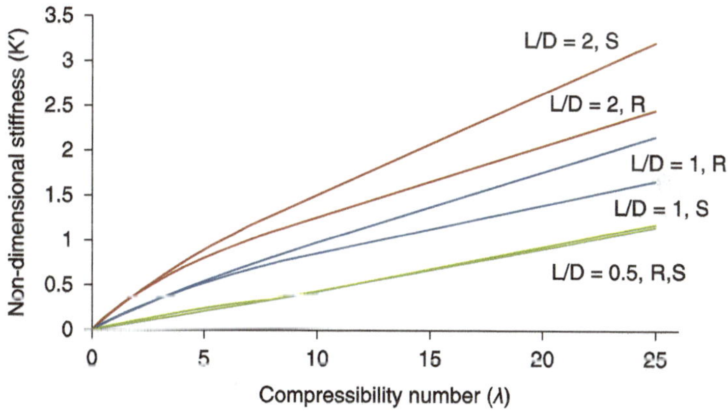

Figure 6.4 Non-dimensional total static stiffness

Total static stiffness and attitude angle may be expressed in non-dimensional forms to enable approximate values of stiffness to be determined for a range of bearing sizes and speeds. In Figure 6.4, the non-dimensional stiffness parameter is:

$$K' = K.C/(LD.P_a) \quad [6.23]$$

and the compressibility number is:

$$\lambda = \frac{6\eta}{P_a}(\omega_1 + \omega_2).\frac{R^2}{C^2} \qquad [6.24]$$

Both Figures 6.4 and 6.5 are for the optimum groove design (angle = 30 degrees, width ratio = 0.5 and depth ratio = 1.2) and the bearing operating at zero eccentricity and provide a first estimate of steady state operating characteristics.

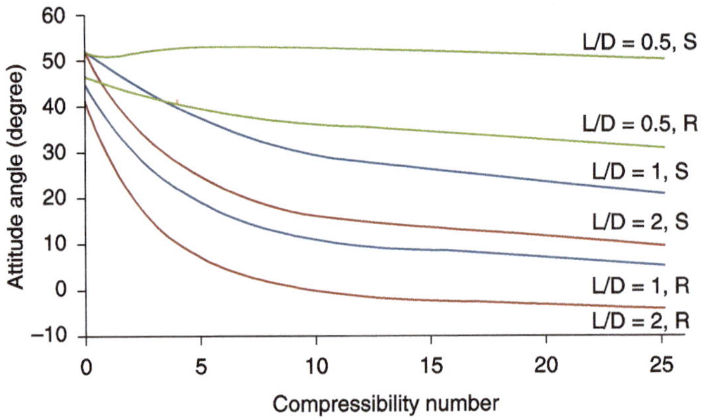

Figure 6.5 Attitude angle as a function of compressibility number

6.8 Damping and dynamic stability

In this section, the dynamic aspects of aerodynamic grooved bearing performance are considered. Their characteristics are similar to the hydrodynamic bearings discussed in Section 5.9, with annular thrust bearings generally exhibiting stable motion and reasonable damping levels, whilst journal bearings are susceptible to half speed and synchronous whirl.

As for hydrodynamic annular thrust bearings, attention is restricted to the damping characteristics of aerodynamic thrust bearings and an expression for damping coefficient is derived for the case of an annular thrust bearing subject to a dynamic, concentrically applied, axial load. The expression is used to determine the damping characteristics of the inward pumping spiral grooved bearing of Table 6.1.

For journal bearings, attention is confined to whirl characteristics. Equation 6.8, describing the dynamic response of an aerodynamic herringbone grooved journal bearing, takes account of whirl motion but neglects squeeze film effects. It can be used to predict amplitudes and frequencies of whirl and to define dynamic coefficients for predicting resonant frequencies in rotor–bearing systems, but the coefficients derived from Eq. 6.8 have zero damping and therefore predict motion amplitudes at resonance to be infinite. Here Eq. 6.8 is used to optimise the dynamic stiffness of the journal bearing used as an example in the previous section.

6.8.1 Aerodynamic thrust bearings

As for the hydrodynamic bearing, Eq. 5.10 for dynamic equilibrium of mass flux and Eq. 5.12 for flux density are the starting points for deriving damping forces in aerodynamic thrust bearings. Damping forces are derived for the case of a small simple harmonic perturbation to the bearing clearance superimposed on the much larger static equilibrium clearance:

$$h_r = h_{ro} + \Delta h.\sin\Omega t \qquad \text{and} \qquad h_g = h_{go} + \Delta h.\sin\Omega t \qquad [6.25]$$

This is assumed to generate a perturbation to the equilibrium pressure distribution:

$$P = P_o + p\,\sin(\Omega t - \phi) \qquad [6.26]$$

The perturbed pressure has in-phase, $p.\sin\Omega t.\cos\phi$ and quadrature, $p.\cos\Omega t.\sin\phi$ components relative to the dynamic motion, where the perturbed pressures are small compared to the static equilibrium pressure P_o.

For aerodynamic bearings air density obeys the isotropic law, $\rho = P/(RT)$, so replacing ρ in Eqs 5.10 and 5.12 and simplifying results in:

$$\frac{\partial}{\partial t}[P(\alpha h_g + (1-\alpha)h_r)] \frac{-1}{12\eta r}\frac{\partial}{\partial r}\left[G_1 r P\frac{\partial P}{\partial r} - 6\eta(\omega_1 - \omega_2)G_2 P r^2\right] = 0 \qquad [6.27]$$

Consider the perturbations to h_g, h_r and P defined in Eqs 6.25 and 6.26 and apply them to the time-dependent term of Eq. 6.27:

$$\frac{\partial}{\partial t}[P(\alpha h_g + (1-\alpha)h_r)] = \frac{\partial}{\partial t}[(P_o + p\,\sin(\Omega t - \phi))\{\alpha(h_g + \Delta h\,\sin\Omega t)$$
$$+(1-\alpha)(h_r + \Delta h\,\sin\Omega t)\}]$$

Ignoring second-order terms and differentiating with respect to time gives:

$$\frac{\partial}{\partial t}[P(\alpha h_g + (1-\alpha)h_r)] = P_o\Omega\Delta h\cos\Omega t + (\alpha h_g + (1-\alpha)h_r)\Omega p\cos(\Omega t - \phi) \qquad [6.28]$$

Let g_1 be a perturbation to G_1 and g_2 be a perturbation to G_2, then consider perturbations to the remaining two terms in Eq. 6.27:

$$G_1 r P\frac{\partial P}{\partial r} = r(G_1 + g_1)(P_o + p\,\sin(\Omega t - \phi))\left(\frac{\partial P_o}{\partial r} + \frac{\partial p}{\partial r}\sin(\Omega t - \phi)\right)$$

Ignoring second-order terms and simplifying:

$$G_1 rP\frac{\partial P}{\partial r}=r\left[G_1P_o\frac{\partial P_o}{\partial r}+G_1P_o\frac{\partial p}{\partial r}\sin(\Omega t-\phi)+G_1\frac{\partial P_o}{\partial r}p\sin(\Omega t-\phi)+g_1P_o\frac{\partial P_o}{\partial r}\right] \quad [6.29]$$

Similarly:

$$6\eta(\omega_1-\omega_2)G_2Pr^2=6\eta\,(\omega_1-\omega_2)[(G_2+g_2)(P_o+p\,\sin(\Omega t-\phi)]$$

Again, ignoring second-order terms:

$$6\eta(\omega_1-\omega_2)G_2Pr^2=6\eta\,(\omega_1-\omega_2)[P_oG_2+g_2P_o+G_2p\,\sin(\Omega t-\phi)] \quad [6.30]$$

Equations 6.28 to 6.30 may be substituted into Eq. 6.27 to give the equation describing dynamic equilibrium. The static terms in Eqs 6.29 and 6.30 equate to zero and may be ignored:

$$P_o\Omega\Delta h\cos\Omega t+(\alpha h_g+(1-\alpha)h_r)\Omega p\cos(\Omega t-\phi)=\frac{\partial}{\partial r}\left[r\left\{G_1P_o\frac{\partial p}{\partial r}\sin(\Omega t-\phi)\right.\right.$$
$$\left.+G_1\frac{\partial P_o}{\partial r}\sin(\Omega t-\phi)+g_1P_o\frac{\partial P_o}{\partial r}\right\} \quad [6.31]$$
$$\left.-6\eta(\omega_1-\omega_2)r^2(G_2p\,\sin(\Omega t-\phi)+g_2P_o)\right]\Big/(12\eta r)$$

g_1 and g_2 are given by Eqs 5.75 and 5.76 (Chapter 5), respectively and contain the term $\sin\Omega t$. Thus a solution to Eq. 6.31 is obtained by separating terms containing $\cos\Omega t$ and $\sin\Omega t$. Putting $p_q=p\sin\phi$ and $p_r=p\cos\phi$, then extracting the terms containing $\cos\Omega t$ gives:

$$P_o\Omega\Delta h+\Omega\left(\alpha h_g+(1-\alpha)h_r\right)p_r=\frac{1}{12\eta r}$$
$$\frac{\partial}{\partial r}\left[r\left(G_1P_o\frac{\partial p_q}{\partial r}+G_1\frac{\partial P_o}{\partial r}p_q\right)-6\eta(\omega_1-\omega_2)r^2G_2p_q\right]$$

Differentiating with respect to r and rearranging, noting that g'_1 and g'_2 are independent of r:

$$\frac{G_1P_o}{12\eta}\cdot\frac{\partial^2 p_q}{\partial r^2}+\left[\frac{G_1}{12\eta}\frac{\partial P_o}{\partial r}+\frac{G_1P_o}{12\eta r}-(\omega_1-\omega_2)\frac{rG_2}{2}\right]\cdot\frac{\partial p_q}{\partial r}$$
$$+\left[\frac{G_1}{12\eta r}\frac{\partial P_o}{\partial r}-(\omega_1-\omega_2)\frac{G_2}{4}\right]p_q= \quad [6.32]$$
$$P_o\Omega\Delta h+\Omega[\alpha h_g+(1-\alpha)h_r]p_r$$

Similarly, extracting the terms containing $\sin\Omega t$ gives:

$$\Omega(\alpha h_g + (1-\alpha)h_r)\,p_q = \frac{1}{12\eta r}\frac{\partial}{\partial r}\left\{ r\left[\frac{G_1 P_o}{\partial r}\frac{\partial p_r}{\partial r} + G_1 P_o \frac{\partial P_o}{\partial r} p_r + g_1' P_o \frac{\partial P_o}{\partial r}\right] - 6\eta(\omega_1 - \omega_2) r^2 [G_2 p_r + g_2' P_o] \right\}$$

Differentiating with respect to r and rearranging:

$$\frac{G_1 P_o}{12\eta}\cdot\frac{\partial^2 p_r}{\partial r^2} + \left[\frac{G_1}{12\eta}\frac{\partial P_o}{\partial r} + \frac{G_1 P_o}{12\eta r} - (\omega_1 - \omega_2)\frac{r}{2}G_2\right]$$

$$\frac{\partial p_r}{\partial r} + \left[\frac{G_1}{12\eta r}\frac{\partial P_o}{\partial r} - (\omega_1 - \omega_2)\frac{G_2}{4}\right] p_r = \qquad [6.33]$$

$$\Omega[\alpha h_g + (1-\alpha)h_r]p_q$$

Equations 6.32 and 6.33 take the form:

$$C_1 \frac{\partial^2 p_r}{\partial r^2} + C_2 \frac{\partial p_r}{\partial r} + C_3 p_r = C_4 p_q$$

$$C_1 \frac{\partial^2 p_q}{\partial r^2} + C_2 \frac{\partial p_q}{\partial r} + C_3 p_q = C_4 p_r + C_5$$

Eliminating p_r results in a fourth-order linear differential equation for p_q:

$$C_1^2 \frac{\partial^4 p_q}{\partial r^4} + 2C_1 C_2 \frac{\partial^3 p_q}{\partial r^3} + 2(C_1 C_2 + C_2^2)\frac{\partial^2 p_q}{\partial r^2} + 2C_2 C_3 \frac{\partial p_q}{\partial r} + (C_3^2 - C_4^2)p_q = C_3 C_5 \qquad [6.34]$$

The coefficients are at most functions of groove geometry, speed, lubricant viscosity, radius and the steady state pressure distribution. Equation 6.34 may be integrated numerically using a relaxation technique to obtain the out-of-phase perturbation pressure distribution p_q across the grooved bearing land. To obtain a solution, it is necessary to define boundary conditions at the bore and outside diameter of the land. As the land must be part of an inward pumping, outward pumping or chevron grooved bearing for which one of the boundaries is a transition diameter, the static and perturbed pressures for both lands at this diameter must be equal. Furthermore, since the pressures are equal, Eq. 5.89 giving the perturbed pressure at the transition diameter for a hydrodynamic bearing also applies to the aerodynamic bearing. The other boundary is at atmospheric pressure, for which the perturbed pressure must be zero. Thus the boundary conditions are $p = p_m$ at $r = r_m$ and $p = 0$ at $r = r_i$ or $r = r_o$ where p_m is determined from Eq. 5.89.

Damping force is found by numerically integrating the perturbed pressure over the bearing area:

$$F_d = 2\pi \int_{ri}^{ro} r p_q . dr \quad [6.35]$$

and the damping coefficient is $F_d/\Omega\Delta h$, where Ω is the frequency of oscillation and Δh the amplitude.

6.8.2 Damping characteristics of aerodynamic thrust bearings

Damping coefficient is determined for the inward pumping thrust bearing of Section 6.5 (Table 6.1). The effect of ridge clearance and rotational speed on damping coefficient is shown in Figures 6.9 and 6.10. Damping increases linearly with speed of rotation and reduces exponentially with clearance. The speed dependent characteristic of the aerodynamic bearing contrasts with that of the hydrodynamic bearing considered in Section 5.9.2, for which damping reduced almost linearly with increasing speed. The difference is that for the compressible fluid characteristic of an aerodynamic bearing, the increase in pressure produced by increasing speed also increases fluid density.

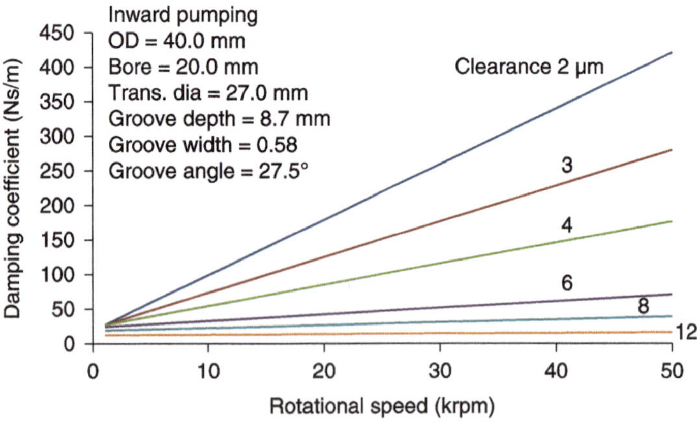

Figure 6.6 Effect of speed on damping coefficient for inward pumping aerodynamic thrust bearing

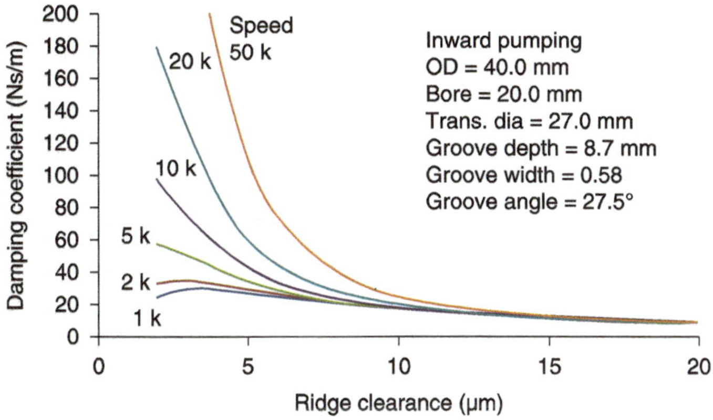

Figure 6.7 **Effect of clearance on damping coefficient for an inward pumping aerodynamic bearing**

6.8.3 Optimum dynamic stiffness of journal bearings

Groove design may be optimised with respect to a number of criteria. Hamrock [11] optimised groove design for maximum radial load capacity and considered bearings with length/diameter ratios from 0.25 to 2, fully and partially grooved, operating over a wide range of speeds. The results are presented in tabular and graphical form simplifying bearing selection. Similarly, Fleming [12] optimised groove design for maximum stability, considering a similar range of conditions to [11] and presenting the results in a similar form. Here attention is confined to fully grooved bearing designs with length/diameter ratios of between 0.5 and 2 operating under no load at speeds of up to one million DN, where DN = bore in mm × speed in rpm. The optimisation criterion is minimum dynamic stiffness, more specifically, maximising the total dynamic stiffness occurring at the half speed whirl condition. This criterion is most relevant to precision bearing applications, where low motion errors, particularly sub-synchronous errors, are important.

Figures 6.8 to 6.10 give an example of the effect of the groove parameters angle, width and depth on the minimum dynamic stiffness of a fully grooved aerodynamic journal bearing having a length to diameter ratio of 1 operating over a practical range of speeds between 5000 and 50 000 rpm. It can be seen that dynamic stiffness is strongly dependent on speed and each of the three groove parameters exhibits an optimum value at which the stiffness is a maximum. The optimum values are groove angle = 30 degrees, groove to groove plus ridge width ratio = 0.5 and groove depth/gap ratio = 1.2. These do not vary significantly over the range of speeds considered and compare with 25 degrees, 0.5 and 1.2 for the hydrodynamic journal bearing of Section 5.8 operating under very different conditions. Furthermore, a deviation from the optimum parameter value of up to 10% produces a relatively small change in dynamic stiffness, generally less than 5%. Extending the results to length/diameter ratios of 0.5 and 2 shows only a small variation in optimum parameter values. Groove depth/gap ratio increases to 1.3 and groove angle to 32 degrees for a length/diameter ratio of 2. Similarly, placing the grooves on the stationary journal bore instead of the rotating shaft does not significantly change the optimum groove parameter values for the range of conditions considered.

The results can be scaled to other bearing diameters and clearances using Eq. 6.8, knowing the compressibility number, λ. Figures 6.8 to 6.11 correspond to $2.5 < \lambda < 25$. Keeping the ratio of bearing gap/radius constant gives the same λ and the same dynamic stiffness for the same groove design and length/diameter ratio.

The whirl response of the optimised grooved bearing is compared with that of the equivalent plain bearing in Figure 6.11. The latter has a theoretically zero dynamic stiffness at a whirl speed ratio of 0.5 for the three rotational speeds considered. In fact, the plain bearing has a zero

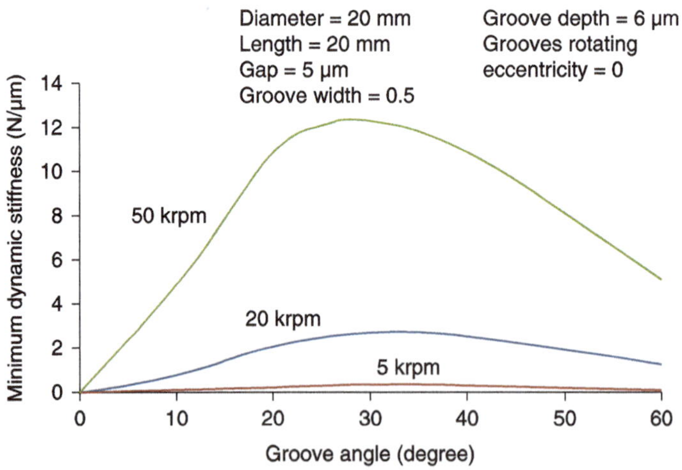

Figure 6.8 **Effect of groove angle on minimum dynamic stiffness of an aerodynamic journal bearing**

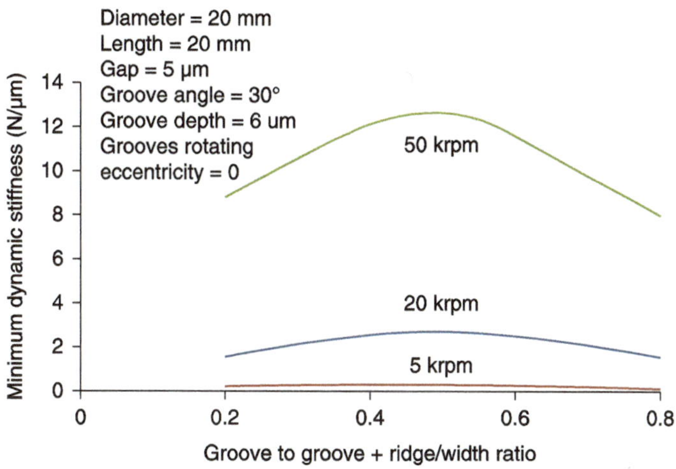

Figure 6.9 Effect of groove width on minimum dynamic stiffness of an aerodynamic journal bearing

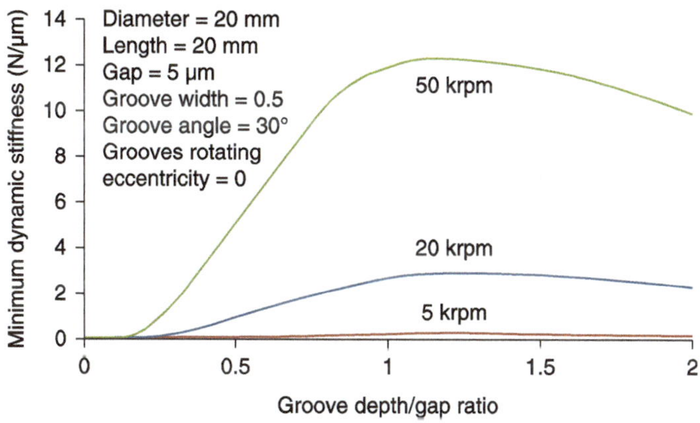

Figure 6.10 Effect of groove depth on minimum dynamic stiffness of an aerodynamic journal bearing

half speed whirl stiffness at all rotational speeds, a characteristic completely unsuited to high precision applications. Nonetheless, plain aerodynamic journal bearings are used in optical scanning applications but with some means of stabilisation [14], for example, uses a straight axial groove along the bore of the bearing journal to break up the axial symmetry of the pressure field and create an internal radial load, which controls the position of the rotor's centre of rotation.

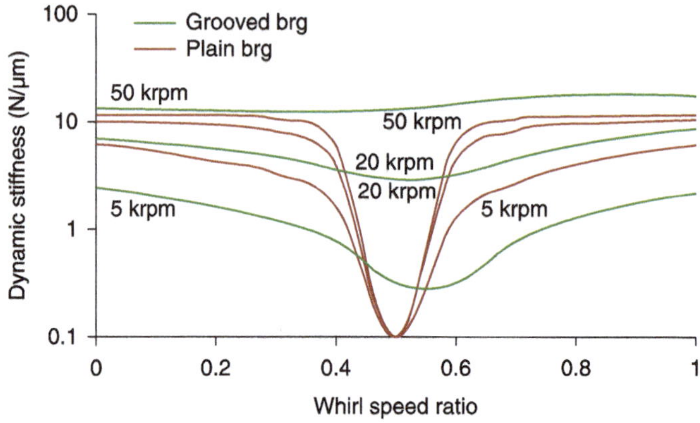

Figure 6.11 Whirl response of optimised groove journal bearing: Diameter = 20 mm; Length = 20 mm; Gap = 5 µm; Grooves Rotating, Groove width = 0.5; Groove angle = 30°; Groove depth = 6 µm

In contrast, the grooved bearing retains a finite dynamic stiffness at all whirl speed ratios and all practical rotational speeds. The minimum dynamic stiffness is speed dependent and can occur at whirl speed ratios other than 0.5. For the three rotational speeds considered, the minimum dynamic stiffnesses are as in Table 6.3.

The dynamic response at 50 000 rpm takes a different form to the responses for the lower speeds of 5000 and 20 000 rpm (Figure 6.11). These show a minimum dynamic stiffness much lower in value than the static stiffness corresponding to zero whirl speed. The minimum dynamic stiffness also occurs at or close to a speed ratio of 0.5. For the higher rotational speeds, the minimum dynamic stiffness of 12.3 N/µm compares to a static stiffness of 13.3 N/µm and occurs at a speed ratio of 0.4. For

Table 6.3 Minimum dynamic stiffness of grooved bearing

Rotational speed (rpm)	Dynamic stiffness (N/um)	Speed ratio
50 000	12.3	0.4
20 000	2.9	0.5
5000	0.32	0.55

higher speed ratios, the dynamic stiffness climbs to a maximum value of 17.5 N/μm at a speed ratio of 0.85, far exceeding the static stiffness.

Another significant feature of the grooved bearing is that at high speeds its static stiffness exceeds that of the equivalent plain bearing, whereas at low speeds it is significantly less. For example, from Figure 6.11, the static stiffness of the grooved bearing is 13.3 N/μm at 50 000 rpm compared to 11.2 N/μm for the plain bearing, whereas at 10 000 rpm the grooved bearing has a static stiffness of 4.3 N/μm compared to 8.63 N/μm for the plain bearing.

6.8.4 Dynamic coefficients and stability of journal bearings

Dynamic coefficients, enabling the whirl response to be determined for a rotor-bearing system of which a grooved journal bearing is part, may be derived from Eq. 6.8 by the method explained in Section 5.9.3.2 for hydrodynamic grooved bearings. The stability criterion derived in Section 5.9.3.4 for the hydrodynamic bearing also applies to aerodynamic bearings. It should be noted that the coefficients derived apply to concentric rotation and do not take account of squeeze film damping.

6.9 Thermal characteristics

The mechanism of heat generation is the same for both hydrodynamic and aerodynamic types and the equations for torque and power loss of hydrodynamic bearings derived in Section 5.11 also apply to aerodynamic bearings.

Although air has a low viscosity, aerodynamic bearings are generally used at high speeds and require low clearances and relatively large bearing areas to generate load carrying capacity, in practice power loss and heat generation is still significant. However, the mechanism by which it affects bearing performance is somewhat different to that of the hydrodynamic bearing. Heat loss from the bearing is again primarily by conduction through the housing and shaft as air flow through the bearing is negligible and the specific heat of air is low. Conduction rates through the housing are almost always different from those through the shaft and an increase in the bearing operating temperature is accompanied by a temperature gradient across the bearing. When conduction rates through the housing are higher than those through the shaft the temperature gradient acts to reduce the bearing's operating clearance. This may well have a benefit in increasing bearing stiffness but it also produces further increases in shear rates, power loss and operating temperature. As design clearances are just a few microns or so, the loss in clearance can be critical and at a sufficiently high speed thermal instability can occur, eventually leading to bearing seizure. Thus for high speed journal bearings, it is important that the designed clearance of the bearing is chosen to allow for this effect.

In this section, formulae for bearing torque and power loss are presented for annular thrust and cylindrical journal bearing types. Used with finite element analyses, they enable the overall temperature distribution and thermal distortion of the bearing system and hence the effect on bearing clearance to be determined.

6.9.1 Torque and power loss produced by annular thrust bearings

The analysis is the same as for hydrodynamic bearings, only now the viscosity of oil is replaced by the viscosity of air. The relevant formulae for torque and power loss are quoted from Section 5.11.2. For a bearing of bore diameter D_i and outside diameter D_o:

$$T = \pi\eta\omega[\alpha/h_g + (1-\alpha)/h_r]\,[D_o^{\,4} - D_i^{\,4}]/32 \qquad [5.132]$$

and the power loss of the bearing is:

$$Q = \pi\eta\omega^2[\alpha/h_g + (1-\alpha)/h_r]\,[D_o^{\,4} - D_m^{\,4}]/32 + \pi\eta\omega^2\,[D_m^{\,4} - D_i^{\,4}]/32h_r \qquad [5.134]$$

where D_m is the groove–seal transition diameter. Similar expressions may be written for an outward pumping bearing and a chevron grooved bearing where, unusually, different depths or widths may be used on inboard and outboard grooves.

As for hydrodynamic bearings, a symmetrical, double-sided bearing, where the no load clearance on each side of the bearing is h_r, the power loss is 2Q. The power loss remains at 2Q under the action of an externally applied axial load, but a greater proportion of the loss is generated on the loaded side of the bearing. The resulting uneven temperature distribution can in some instances significantly 'bow' the thrust rotor, affecting bearing stiffness and load carrying capacity.

6.9.2 Torque and power loss produced by journal bearings

Again the torque and power loss are quoted from the hydrodynamic bearing analysis of Section 5.11.3:

$$T = \pi\omega\eta D^3 \{(L - L_g)[\alpha/2h_g + (1 - \alpha)/2h_r] + L_g D^3/2h_r\} \quad [5.136]$$

and the power loss is:

$$Q = \pi\omega^2\eta D^3 \{(L - L_g)[\alpha/2h_g + (1 - \alpha)/2h_r] + L_g/2h_r\} \quad [5.137]$$

The application of a radial load to the bearing will create a variation in bearing clearance with angular position, but the total torque and power loss of the bearing remains unchanged.

6.10 Overview

In lightly loaded or high speed applications, aerodynamic herringbone and spiral grooved bearings have the advantage of not requiring any liquid lubricant. Thus there is no need for any auxiliary support equipment and sealing is not critical, enabling them to reduce the overall weight and size of the equipment they are part of. Also they do not generate asynchronous motion error and in having an axi-symmetric pressure distribution under 'no load' they provide good averaging of the effect of bearing surface geometry errors on synchronous motion error.

A potential disadvantage of this type of bearing is the manufacturing precision required to make them operate reliably and to generate designed levels of performance. Bearing clearances are of the order of 10 μm or less and must be controlled to a small fraction of this if they are to generate the stiffness and load capacity required of them. Similarly, the flatness of thrust bearing and roundness of journal bearing surfaces must be controlled, preferably to well below 1 um peak-valley. Furthermore, the design of the bearing arrangement and the manufacture of associated components must ensure the bearings can be set up with negligible misalignment.

With low bearing clearances, groove depths are shallow, demanding high precision if put in by a mechanically based process. They are more easily generated by chemical or electrochemical processes. Electro-discharge processes can also readily achieve the required accuracy, but tend to be slower or more expensive.

For bearings encountering a large number of start-stop cycles, the tribological properties of their surfaces are critically important, material type and surface finish must ensure extremely low wear rates and any wear debris to be negligible in size compared to bearing clearances. The materials must be corrosion resistant and have a high surface hardness and/or low friction characteristics. Tungsten carbide, martensitic stainless steel and coatings such as diamond-like carbon and Tufram are examples of materials that have been used successfully.

6.11 References

1. Sternlicht, B. and Elwell, R.C. (1958), 'Theoretical and experimental analysis of hydrodynamic gas-lubricated journal bearings', *Transactions of the ASME*, **80**: 865–78.
2. Ausman, J.S. (1959), 'Theory and design of self acting gas lubricated journal bearings including misalignment effects', *1st International Symposium on Gas Lubricated Bearings*, ACR-49, Office of Naval Research, 161–92.
3. Ausman, J.S. (1961), 'An improved analytical solution for self acting gas lubricated journal bearings of finite length', *Transactions of the ASME, Journal of Basic Engineering*, **83**: 188–93.
4. Malanoski, S.B. and Pan, C.H. (1965), 'The static and dynamic characteristics of the spiral-grooved thrust bearing', *Transactions of the ASME, Journal of Basic Engineering*.
5. Vohr, J.H. and Chow, C.Y. (1965), 'Characteristics of herringbone grooved', *Transactions of the ASME, Journal of Basic Engineering*, **87**: 568–76.
6. Castelli, V. and Vohr, J.H. (1967), 'Performance characteristics of herringbone grooved journal bearings operating at high eccentricity ratios and with misalignment', MTI Report, MTI-67TR15.
7. Cunningham, R.E., Fleming, D.P. and Anderson, W.J. (1969), 'Experiments on the steady state characteristics of herringbone grooved air lubricated journal bearings', NASA Technical Note, NASA TN D-5386.
8. Cunningham, R.E., Fleming, D.P. and Anderson, W.J. (1960), 'Experimental stability studies of the herringbone grooved gas lubricated journal bearing', *Transactions of the ASME, Journal of Lubrication Technology*, **91**: 52–9.
9. Smalley, A.J. (1969), 'Static and dynamic characteristics of the gas-lubricated spiral-grooved spool bearing for motion in the axial direction', *Transactions of the ASME*, **91**: 104–12.
10. Smalley, A.J. (1972), 'The narrow groove theory of spiral grooved gas bearings: development and application of a generalised formulation for numerical solution', *Transactions of the ASME, Journal of Lubrication Technology*, **94**: 86–92.
11. Hamrock, B.J. and Fleming D.P. (1971), 'Optimisation of self-acting herringbone grooved journal bearings for maximum radial load capacity', *Proceedings of the 3rd Gas Bearing Symposium*, University of Southampton, March.

12. Fleming, D.P. and Hamrock, B.J. (1974), 'Optimisation of self-acting herringbone grooved journal bearings for maximum stability', *Proceedings of the 6th International Gas Bearing Symposium*, University of Southampton, March.
13. Bonneau, D. and Absi, A. (1994), 'Analysis of aerodynamic journal bearings with small number of herringbone grooves by finite element method', *Transactions of the ASME, Journal of Tribology*, **116**: 698–704.
14. Tempest, M.C. and Dimoft, F. (1997), Patent No. US593230A, 14 January.

6.12 Notation

a_g – Groove width
a_h – Ridge width
C – Clearance under no load
c – Constant
D – Diameter of journal bearing
D_i – Bore diameter of annular thrust bearing
D_o – Outside diameter of annular thrust bearing
F – Static force
F_d – Damping force
F_x,F_y – Component of force in x or y direction
G_1 – Static groove parameter
G_2 – Static groove parameter
g_1 – Dynamic groove parameter
g_2 – Dynamic groove parameter
H – h_g/h_r for a journal bearing under radial load
H_o – h_g/h_r for a journal bearing under no load
h_g – Groove clearance
h_r – Ridge clearance
L – Journal bearing length
m_g – Mass flow rate
P – pressure
P_o – Static pressure
p – Dynamic or perturbed pressure
Q – Power loss
R – Radius of journal bearing
R_i – Inner radius of annular thrust bearing
R_m – Transition radius of annular thrust bearing
R_o – Outer radius of annular thrust bearing
T – Total torque
T – Torque generated by grooved region of an annular thrust bearing
T_p – Torque generated by plain region of an annular thrust bearing
U – Surface velocity
W – Total load
W_o – Total static load
x,y,z – Cartesian coordinates

α – $a_g/(a_g + a_h)$
β – Groove angle
ε – Eccentricity
η – Dynamic viscosity
θ – Angular coordinate
μ – Kinematic viscosity
ρ – Lubricant density
τ – Lubricant shear stress
τ_g – Shear stress in lubricant in groove
τ_r – Shear stress in lubricant over ridge
ψ – Mass flux
ψ_g – Mass flux in grooved region
ψ_r – Mass flux in ridge region
ω – Angular speed
ω_1 – Angular speed of grooved surface
ω_2 – Angular speed of non-grooved surface
ω_3 – Whirl speed

Index

FSC
www.fsc.org
MIX
Papier | Fördert gute Waldnutzung
FSC® C083411

Kolibri 360 GmbH
Ferdinand-Jühlke-Straße 7
99095 Erfurt, Deutschland
produktsicherheit@kolibri360.de